Mechanical and Aerospace Engineering

Mechanical and Aerospace Engineering

Edited by **Roman Fritz**

Willford Press
New York

Published by Willford Press,
118-35 Queens Blvd., Suite 400,
Forest Hills, NY 11375, USA
www.willfordpress.com

Mechanical and Aerospace Engineering
Edited by Roman Fritz

International Standard Book Number: 978-1-68285-201-9 (Hardback)

Printed in China.

Contents

Preface

This book encompasses the recent studies in the field of mechanical and aerospace engineering. The chapters that are included in this book on theories of fluid mechanics, aerodynamics, control engineering design, temperature, etc. are bound to provide detailed insights to the readers. Some of the diverse topics covered in this book address the varied branches that fall under this category. With state-of-the-art inputs by acclaimed experts of this field, this book targets students and professionals.

Various studies have approached the subject by analyzing it with a single perspective, but the present book provides diverse methodologies and techniques to address this field. This book contains theories and applications needed for understanding the subject from different perspectives. The aim is to keep the readers informed about the progress in the field; therefore, the contributions were carefully examined to compile novel researches by specialists from across the globe.

Indeed, the job of the editor is the most crucial and challenging in compiling all chapters into a single book. In the end, I would extend my sincere thanks to the chapter authors for their profound work. I am also thankful for the support provided by my family and colleagues during the compilation of this book.

Editor

1

OPTO-MECHANICAL DESIGN AND ANALYSIS OF MOUNTED OFF-AXIS MIRRORS FOR SPACE APPLICATIONS

Kaushik M. M[1], M. M. M. Patnaik[2], Venkateswaran R[3]

[1] *M.Tech. Student (Machine Design), Department Of Mechanical Engineering, KSIT, Bangalore-62*
[2] *Associate Professor, Department Of Mechanical Engineering, KSIT, Bangalore-62*
[3] *Head, OEND, Scientist/Engineer-SG, LEOS-Indian Space Research Organisation, Bengalore-58*
Author Correspondence: kaushikmm@hotmail.com

Abstract

Off-Axis mirrors are used in space borne telescopes to minimize the aberrations for improved optical performance. This telescope consists of concave hyperbolic primary and convex hyperbolic secondary mirrors which help in capturing the light from far away sources and obtain the images on to the detector. There is a need to design and analyze the opto-mechanical system to mount the primary mirror of diameter 146 mm and secondary mirror of diameter 49 mm used in this project. The opto-mechanical system needs to withstand the environmental effects during launching of the rocket as well as during space operating conditions. Mainly the system encounters the following types of environmental conditions: inertia loads, vibrations and temperature excursions. The system needs to be designed and analyzed to withstand the above conditions for failure as well as for distortion free wave front error on the optical surfaces at all operating conditions. The objectives of the project are as follows: 1) Prevention of mirror from failure for any operating conditions i.e. during launching of satellite in a rocket and satellite operation in space and 2) Optimization of geometrical parameters of the opto-mechanical components for minimum weight, which helps in improving agility and life of the satellite. In order to achieve the above objectives, two different kinds of mounting configuration have been considered, namely leaf spring ring mount and bipod ring mount. The opto-mechanical system is modeled using Solid Works (CAD Software). Two types of finite element analysis i.e. 1) modal analysis under clamped/free-free condition and 2) linear static analysis under thermal and inertia loads have been carried out using MSC Patran(FEA pre-processor & post-processor software) and MSC Nastran(FEA processor software) to study the stresses induced due to different environmental conditions. Several iterations have been carried out to arrive at the optimum design. Analytical calculations are done and compared with the FEA results in order to ensure validity of the results.

Keywords: Off-Axis Mirrors, Opto-Mechanical system, Mounting Configurations, Optimization for Weight, Finite Element Analysis

1. Introduction

Opto-mechanical design of optical instruments is a tightly integrated process involving many technical disciplines. It begins with a statement of need for a particular hardware item and a definition of goals and firm requirements for the system's configuration, physical characteristics and performance in a given environment where it is intended to function. The design effort proceeds through a logical sequence of major steps and concludes only when the instrument is awarded a pedigree establishing its ability to meet its specifications and to be produced in the required quantity.

The mirrors used in this project have applications in U.V telescopes. Satellite based U.V telescopes are developed for various science and astronomical studies. One of them is the Solar U.V telescope, which is used to view the sun disc in the U.V region and to study the solar dynamics. The Solar U.V telescope consists of an opto-mechanical system to house the primary and secondary mirror. Optical systems are widely used in scientific, defense, and commercial applications. By their nature, optical systems often calls for stringent alignment and geometrical requirements that must be met despite the presence of static, dynamic, and thermal loads. These requirements often couple the optical, structural, and thermal domains and demand a coupled approach to optimize the system's performance.

As an example, an optical instrument inside a spacecraft must meet a number of requirements:

- Be robust enough to withstand the static and dynamic loading encountered during launch and ground handling.
- Function after exposure to extreme temperatures and vacuum conditions.
- Meet optical requirements in the presence of any dynamic, static, or thermal operating loads by minimizing distortions and relative displacements
- Be designed such that its requirements for prelaunch adjustment, calibration, and testing can be met in the context of the larger system
- Be lightweight enough to meet mass requirements

To achieve these objectives, an instrument's design process undergoes through various phases that involve both analysis and testing. Once the results are found to be satisfactory, a prototype will be made and experimental testing will be carried out to assess the performance of the system under simulated conditions. Succeeding which, the actual model will be made and the telescope will be ready to be integrated with the spacecraft for the final testing in integrated mode and then for the launching.

This paper presents brief details of the opto-mechanical design concepts which could be employed for mounting the off-axis mirrors for space applications .Some of the finite element analysis exercises and their results are also presented along with the analytical calculations.

2. Design Constraints

- Since the diameter of the primary mirror is 146 mm, a gap of 57 mm from the bottom part of the mirror to the base of the mount is maintained based on the preliminary design. The base to mirror centre distance amounts to 130mm. The base to mirror centre distance (130 mm) must be maintained in the secondary mirror mount so that the light rays falls appropriately on the reflective surfaces to obtain the images.

- The mass of the mirror mount is dependent upon the total load carrying capacity and the number of payloads being carried by the rocket. Along with this telescope, many payloads are being carried by the rocket and the weight of this telescope is governed by the weight of the other payloads. It has been decide that the total mass of the telescope (Primary mirror mount + Secondary mirror mount) should be less than 6.5 Kg.

3. Opto-Mechanical Design

3.1. Primary Mirror Opto-Mechanical Design

A. Leaf Spring Ring Mount: The components of primary mirror opto-mechanical system shown in figure 1. It can be employed as a design concept to mount the primary mirror having a diameter of 146 mm. Leaf spring ring mounting configuration is considered as a first option.

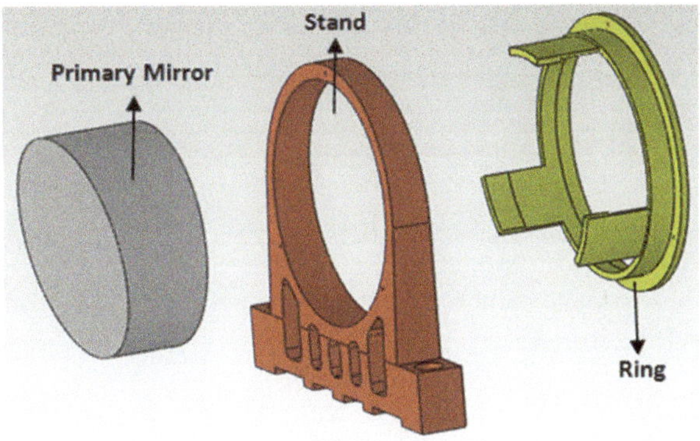

Component	Mass (grams)
Mirror	2354.89
Ring	623.08
Stand	1695.43
	Total mass = 4673.39

Figure 1: Exploded view of the primary mirror opto-mechanical system

The major contribution of the overall mass of the assembly is the mass of the stand and there is scope for reducing the mass of the mounting structure. However, the mass of the mirror is fixed and it cannot be altered once the mirror design is finalized. The opto-mechanical system must be optimized for weight. It is always desirable that the mass shall be as minimum as possible without sacrificing the strength and stiffness. The initial mass before optimization is found to be 5.345 kgs. The final mass after optimization is found to be 4.673 kgs. We find that a percentage reduction of about 12.6 % is possible. To obtain the final design, several iterations are carried out. Figure 2 and 3 shows the preliminary and final design of the primary mirror opto-mechanical system respectively.

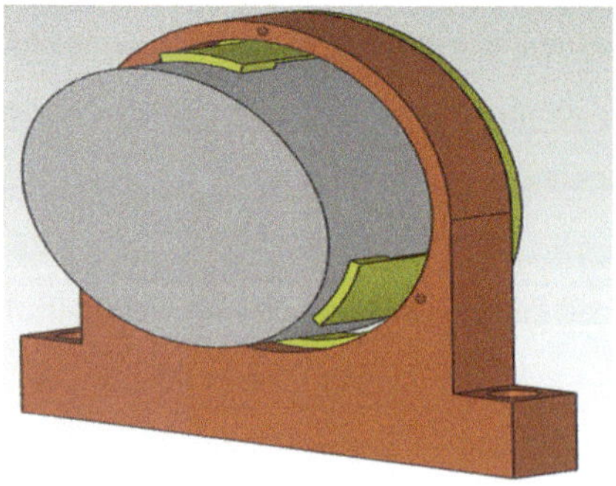

Figure 2: Preliminary design of the primary mirror opto-mechanical system

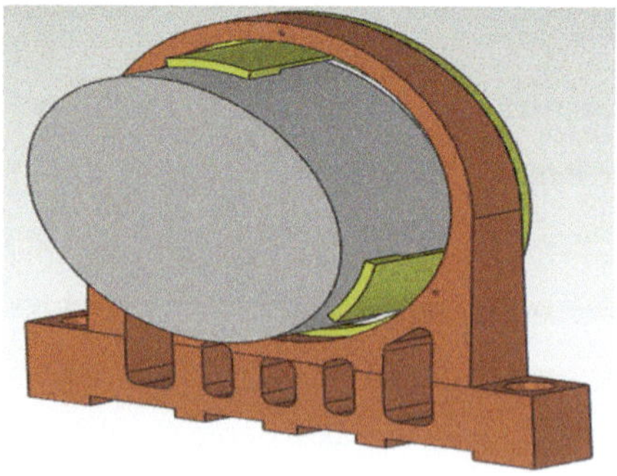

Figure 3: Final design of the primary mirror opto-mechanical system

B. Bipod –Mirror Mount: The leaf spring ring mount is replaced with a bipod type ring mount having the same contact area with the mirror. Bipod type ring mounting configuration is considered as the second option. The performance of both the type of mounts is compared. Figure 4 shows the bipod ring for the primary mirror.

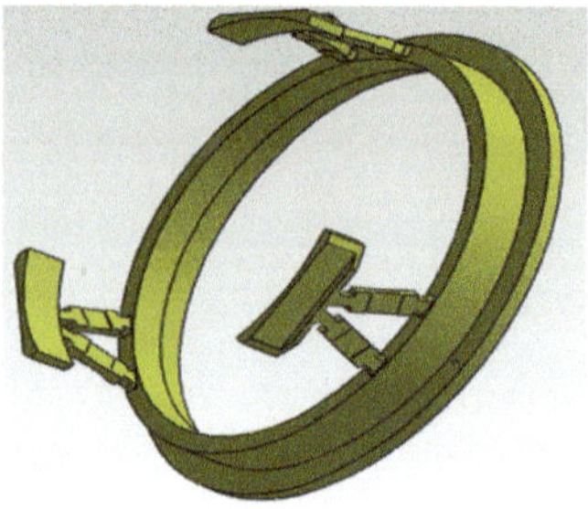

Figure 4: Bipod Ring for primary mirror

The mass of the stand is slightly modified to accommodate the bipod ring. Figure 5 shows the bipod ring mount assembly for primary mirror.

Component	Mass (grams)
Mirror	2354.89
Ring	665.70
Stand	1668.07
	Total mass = 4688.66

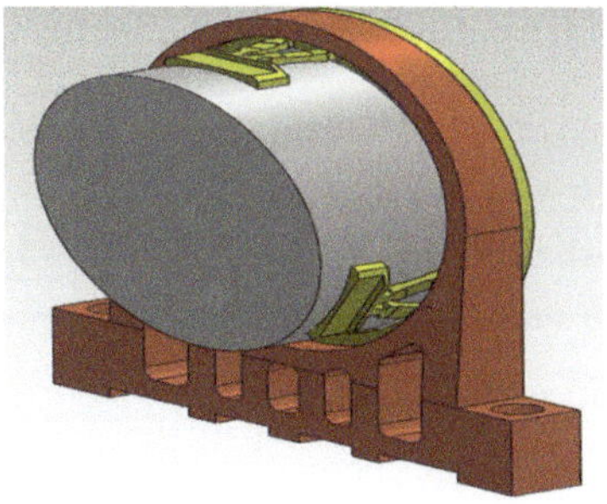

Figure 5: Bipod ring mount assembly for primary mirror

3.2. Secondary Mirror Opto-Mechanical Design

The components of secondary mirror opto-mechanical system, shown in figure 6, could be employed as a design concept to mount the secondary mirror having a diameter of 49 mm. Leaf spring ring mounting configuration is considered. The opto-mechanical system is optimized for weight by designing a mount having mass as minimum as possible. Figure 7 shows the secondary mirror opto-mechanical assembly. For secondary mirror, analysis is carried out only for leaf spring ring mount since the weight of secondary mirror is less compared to primary mirror.

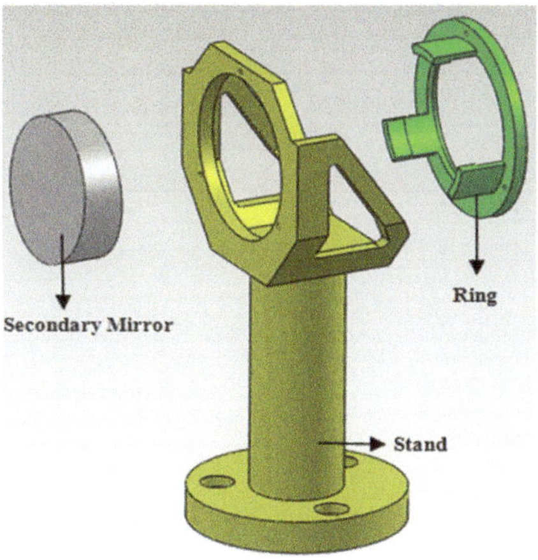

Figure 6: Exploded view of secondary mirror opto-mechanical

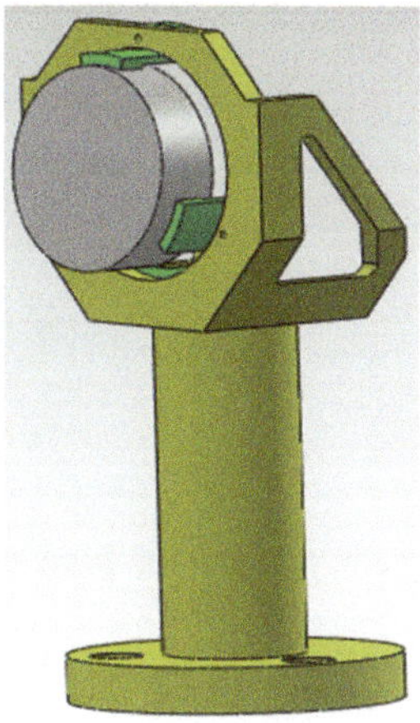

Figure 7: Secondary mirror opto-mechanical assembly

Component	Mass (grams)
Mirror	65.08
Ring	53.59
Stand	382.82
	Total mass = 501.49

4. Finite Element Analysis of the Opto-Mechanical System

4.1. FEA Procedure Employed

The CAD model is imported into MSC Patran (FEA Pre-Processor) where the components are meshed using CQUAD 4 elements by extracting the surfaces. The CQUAD 4 elements are then transformed into CHEXA 8 elements using the sweep/transform- extrude, arc, loft, etc tools provided. Since accuracy is of the highest concern, hexahedral mesh is the most preferable one. Rigid body elements can be used to connect different elements if required. Material properties, loads /boundary conditions are applied and the bulk data file (bdf file) is generated. The bulk data file is imported into MSC Nastran (Processor) and MSC Access database (xdb file) is generated. The MSC Access database (xdb file) is imported back to MSc patran for post-processing purposes. The following material properties of the components are used in the finite element analysis and is given in table 1

Table 1: Material properties of components

Component	Mirror	Ring	Stand	Glue	Screws
Material	Zerodur	Invar	Titanium	EC2216 B/A	S.S
Young's Modulus(E) (Gpa)	90.3	148	108	0.689	193
Poisson's Ratio(υ)	0.243	0.29	0.31	0.43	0.3
Density ρ (gm/cm^3)	2.53	8	4.5	1.25	8
CTE (x 10^{-6}/°C)	0.05	1.3	8.6	102	16
Yield Strength (Mpa)	5-10	300	800	17.2 at 24°C	900

4.2. FE Model Assumptions

Following assumptions are made while carrying out the finite element analysis.

1. Mirror is assumed to be linearly elastic even though it is brittle, because any brittle material posses elastic property when the deformations are in nano scale.

2. Glue layer is not modeled in the FE Models, as the glue layers are very thin, very compliant and low modulus (E < 1000 psi) rubber like and nearly incompressible ($\nu > 0.49$). Each of the above features causes some difficulty in modeling the adhesive layer. The sudden change in element size required to describe very thin layers can cause geometrical modeling problems. Due to computational limitations, the optics and mount cannot be modeled with such a fine resolution.

3. Screws are replaced with rigid body elements.

4.3. Acceptance Criteria

The following are the stress levels within which the individual components of the assembly can be subjected to while applying all types of loading conditions.

- Stress in mirror < 5 Mpa
- Stress in ring < 300 Mpa
- Stress in stand < 800 Mpa
- Stress in glue < 5 Mpa
- Natural Frequency of assembly > 200 Hz

Environmental noise due to the launching of the rocket is considered while calculating the natural frequency. The frequency range of the noise is between 10-100 Hz. Hence, the component must be designed in such a way that its natural frequency is more than 100 Hz. The opto-mechanical mount is designed to obtain a natural frequency of above 200 Hz since the analysis does not consider the material non-homogeneity, voids, cracks, etc. The validity of the results is checked by testing the components after its assembly. Since the experimental results will be closer to the FEA results, a margin of more than 100 Hz is considered as the design target. (i.e.> 200 Hz)

4.4. Analysis of Primary Mirror Mounts

4.4.1. Leaf Spring Ring Mount

4.4.1.1. Modal Analysis

Boundary condition: The base is fixed with 6 degrees of freedom. Figure 8 shows the finite element model of primary mirror with leaf spring ring mount. Figure 9 shows the Modal analysis result. Natural frequency obtained is 242.02 Hz

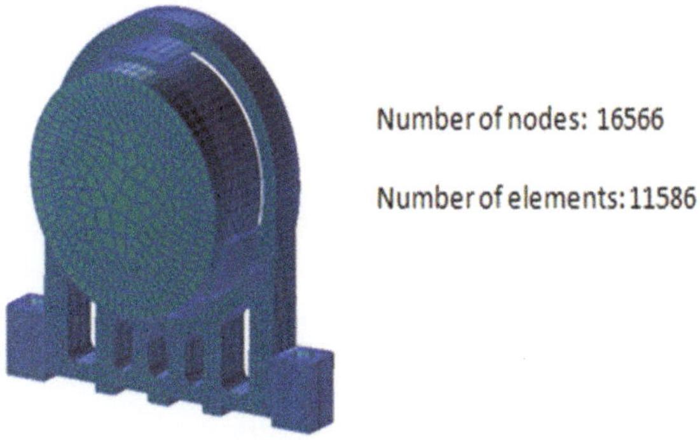

Figure 8: Finite element model of primary mirror with leaf spring ring mount

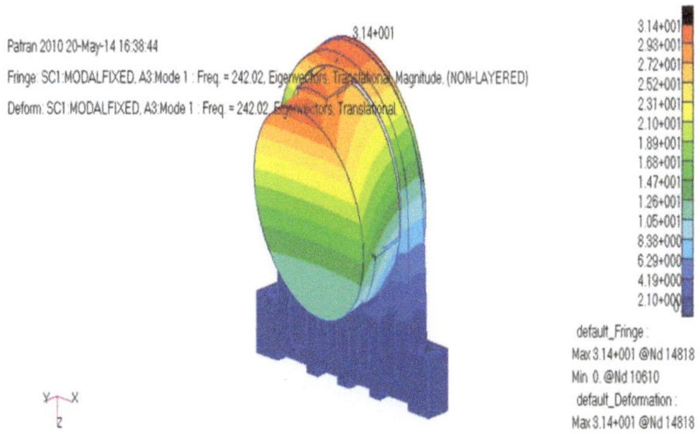

Figure 9: Modal analysis result of primary mirror with leaf spring ring mount

Several iterations are carried out to optimize the opto-mechanical system for weight. The variation of natural frequency with reduction in mass is presented in the table 2.

Table 2: Variation of natural frequency with reduction in mass

Iteration	Mass (Grams)	Natural Frequency (Hz)	Generalized Stiffness Values
1	5345.22	295.84	3.45E+06
2	4978.10	274.48	2.97E+06
3	4837.87	271.92	2.92E+06
4	4723.18	254.47	2.56E+06
5	4673.39	242.02	2.31E+06

From table 2, we can conclude that the results comply with the fact that natural frequency is directly proportional to the stiffness of the opto-mechanical system. Further reduction in mass of the stand will lead to reduction of natural frequency. After five iterations, the design is considered optimum keeping in view that the natural frequency target to be above 200Hz.

4.4.1.2. Linear Static Analysis

Boundary condition for quasi-static loads: The random vibration/dynamic load that arises during the launch is converted to an equivalent Quasi-Static load using Miles' equation [1]. It gives an approximate method to find whether the component would withstand the dynamic loads. This eliminates in carrying out the dynamic analysis. However, for accurate prediction of stress distribution, dynamic analysis may be conducted.

From Miles' equation we have,

$$G_{RMS} = \sqrt{\frac{\pi\, f_n\, Q\, [ASD_{input}]}{2}} \qquad (1)$$

Where,

G_{RMS} = Root Mean Square Acceleration in G's

f_n= Natural frequency = 242 Hz

$Q = \frac{1}{2\zeta}$= Transmissibility (or amplification factor) at f_n

where ζ is the critical damping ratio.

A critical damping ratio of 0.05 is considered since the transmissibility / amplification is high at this ratio.

$[ASD_{input}]$ = Input Acceleration Spectral Density at f_n in units of $\frac{g2}{Hz}$ = 0.1

$$G_{RMS} = \sqrt{\frac{\pi * 242 * 0.1}{2 * 2 * 0.05}} = 19.5$$

The mirror should be designed to withstand a minimum random vibration level of 3 x 19.5 = 58.49 times gravity. The primary mirror mount is designed to withstand an inertia load of 70 g.

Boundary condition for thermal loads: Taking 20°C as the ambient temperature. The system is subjected to a temperature variation of +30°C and -30°C. (ΔT=30°C). Figure 10 shows some of the FEA results of the primary mirror (with leaf spring ring mount) opto-mechanical components. Figure 10 (a) shows stress in the mirror due to 70g inertia load in X direction. Figure 10 (b) shows stress in the ring due to 70g inertia load in X direction. Figure 10 (c) shows stress in the stand due to 70g inertia load in X direction. Figure 10 (d) shows stress in the mirror due to plus 30°C thermal load. The stresses in mirror, leaf spring ring mount and stand with 70g inertia load imposed in X, Y and Z directions and due to variation in temperature are presented in section 4.4.1.3, summary of results.

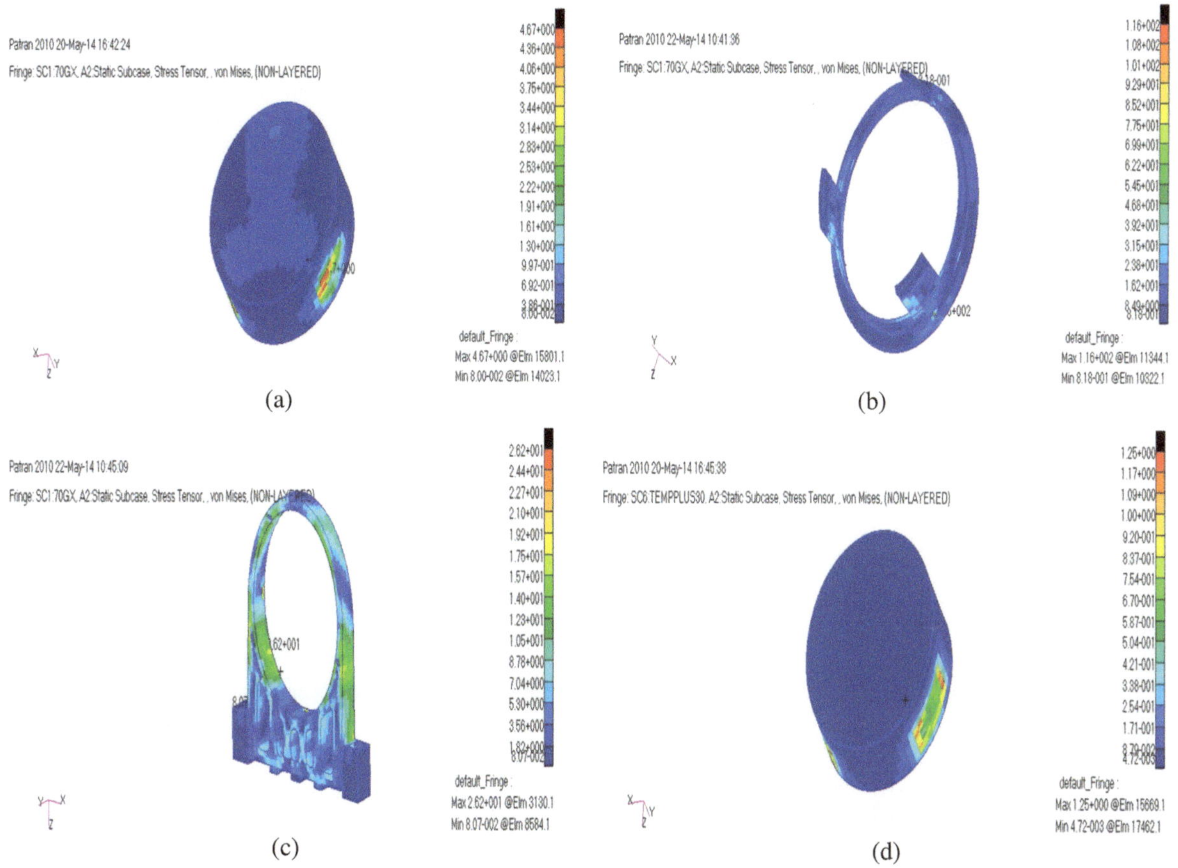

Figure 10: FEA results of the primary mirror (with leaf spring ring mount) opto-mechcanical components

4.4.1.3. Summary of results

Table 3 and 4 shows the summary of results for the components of primary mirror with leaf spring ring mount subjected to quasi-static and thermal loads respectively. The stresses in the individual components are given.

Table 3: Quasi-Static Analysis results for components of primary mirror with leaf spring ring mount

Sl. No.	Component	Von Mises Stress(Mpa) in component for 70g load			Allowable stress(Mpa)
		X	Y	Z	
1	Mirror	4.67	3.89	4.82	5
2	Ring	116	98.5	120	300
3	Stand	26.2	28	72.6	800

Table 4: Thermal stress analysis for components of primary mirror with leaf spring ring mount

Sl. No.	Component	Von Mises Stress(Mpa) in component due to thermal loads		Allowable stress(Mpa)
		+30°C	-30°C	
1	Mirror	1.3	1.3	5
2	Ring	33.2	33.2	300
3	Stand	84.9	84.9	800

4.4.2. Bipod Ring Mount

4.4.2.1. Modal Analysis

Boundary condition: The base is fixed with 6 degrees of freedom. Figure 11 shows the finite element model of primary mirror with bipod ring mount. Figure 12 shows the Modal analysis result. Natural Frequency obtained is 230.64 Hz

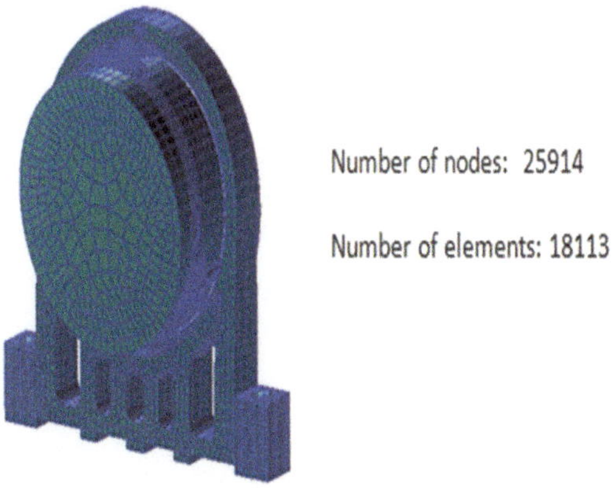

Figure 11: Finite element model of primary mirror with bipod ring mount

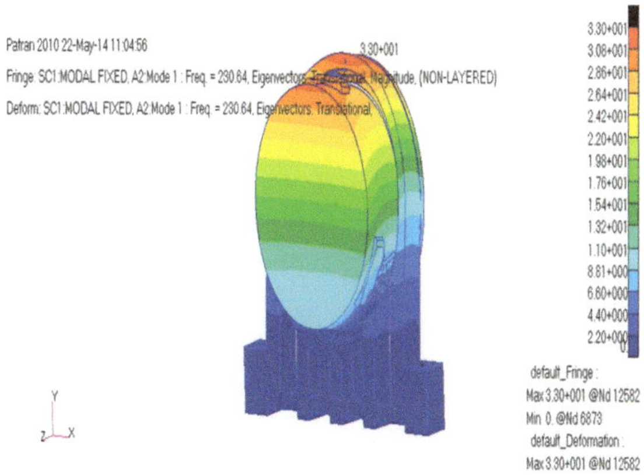

Figure 12: Modal analysis result of primary mirror with bipod ring mount

4.4.2.2. Linear Static Analysis

Boundary condition for quasi-static loads: Applying Miles' equation (equation 1) for the primary mirror with bipod ring mount configuration, we get

$$G_{RMS} = \sqrt{\frac{\pi * 230.64 * 0.1}{2 * 2 * 0.05}} = 19.03$$

The mirror should be designed to withstand a minimum random vibration level of 3 x 19.03= 57.10 times gravity.

The Primary mirror mount is designed to withstand an inertial load of 70 g.

Boundary condition for thermal loads: Taking 20°C as the ambient temp. The system is subjected to a temperature variation of +30°C and -30°C. (ΔT=30°C). Figure 13 shows some of the FEA results of the primary mirror opto-mechanical components.

Figure 13 (a) shows stress in the mirror due to 70g inertia load in Z direction. Figure 13 (b) shows stress in the bipod ring due to 70g inertia load in Z direction. Figure 13 (c) shows stress in the mirror due to minus 30°C thermal load. Figure 13 (d) shows stress in the stand due to plus 30°C thermal load. The stresses in mirror, bipod ring mount and stand with 70g inertia load imposed in X, Y and Z directions and due to variation in temperature are presented in the section 4.4.2.3, summary of results.

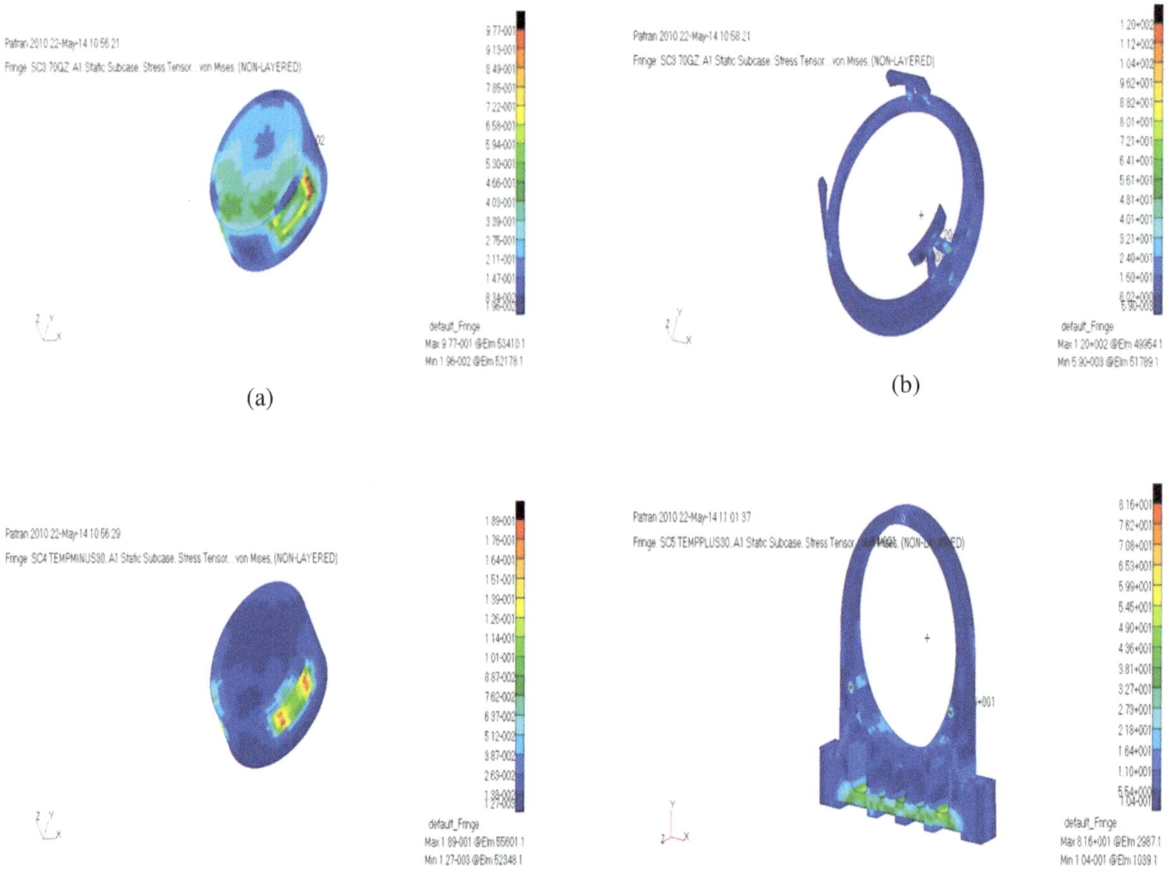

(a) (b) (c) (d)

Figure 13: FEA results of the primary mirror (with bipod ring mount) opto-mechcanical components

4.4.2.3. Summary of results

Table 5 and 6 shows the summary of results for components of primary mirror with bipod ring mount subjected to quasi-static and thermal loads respectively. The stresses for all the individual components are given.

Table 5: Quasi-Static Analysis results for components of primary mirror with bipod ring mount

Sl. No.	Component	Von Mises Stress(Mpa) in component for 70g load			Allowable stress(Mpa)
		X	Y	Z	
1	Mirror	1.60	1.51	0.98	5
2	Ring	272	262	120	300

3	Stand	20.7	29.5	73.2	800

Table 6: Thermal stress analysis for components of primary mirror with bipod ring mount

Sl. No.	Component	Von Mises Stress(Mpa) in component due to thermal loads		Allowable stress(Mpa)
		+30°C	-30°C	
1	Mirror	0.2	0.2	5
2	Ring	32.6	32.6	300
3	Stand	81.6	81.6	800

A comparison of stresses induced in the mirror for leaf spring and bipod ring mount are shown in the form of graphs (figure 14 and 15) when they are subjected to inertia and thermal loads respectively.

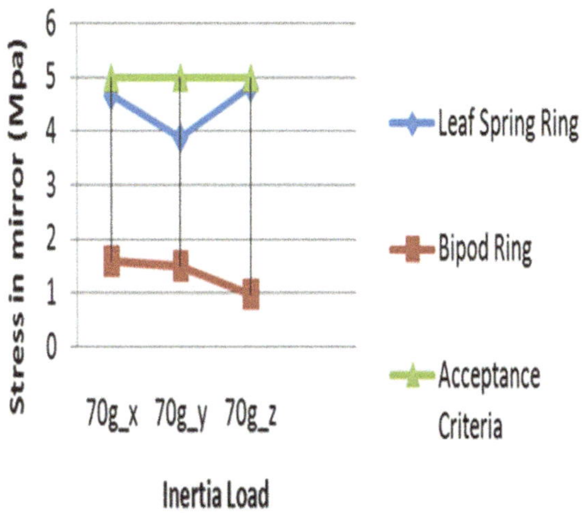

Figure 14: Stress in mirror for inertia loads with leaf spring and bipod ring mounts

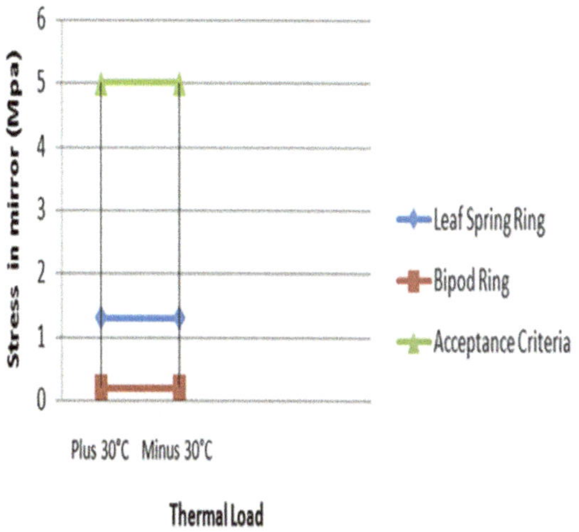

Figure 15: Stress in mirror for thermal load with leaf spring and bipod ring mounts

4.5. Analysis of Secondary Mirror Mount

4.5.1. Modal Analysis

Boundary condition: The base is fixed with 6 degrees of freedom. Figure 16 shows the finite element model of secondary mirror with leaf spring ring mount. Figure 17 shows the Modal analysis result. Natural Frequency obtained is 499.11 Hz

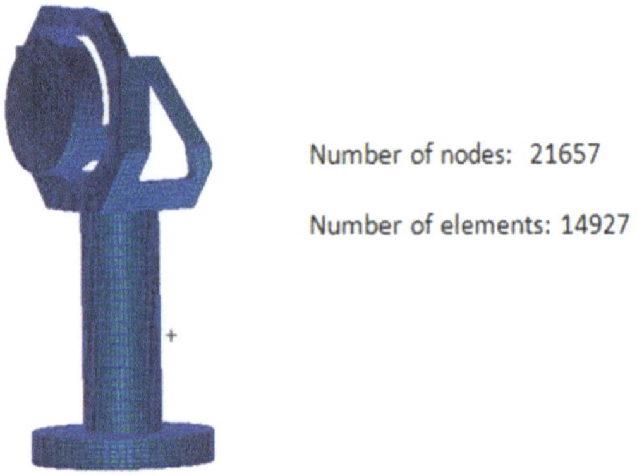

Figure 16: Finite element model of secondary mirror opto-mechanical system

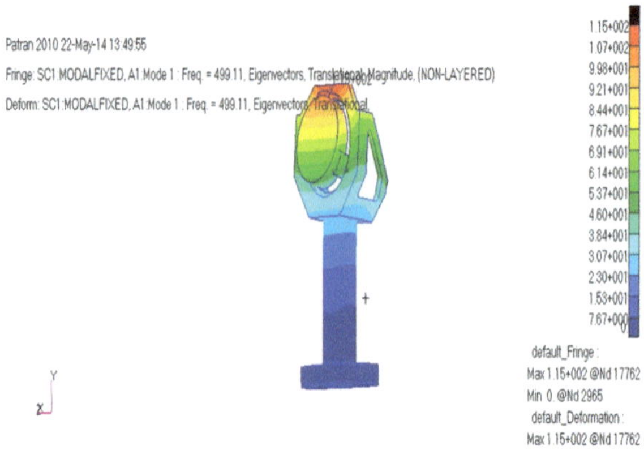

Figure 17: Modal analysis result of secondary mirror opto-mechanical system

4.5.2. Linear Static Analysis

Boundary condition for quasi-static loads: Applying Miles' equation (equation 1) for the secondary mirror opto-mechanical system, we get

$$G_{RMS} = \sqrt{\frac{\pi * 499.11 * 0.1}{2 * 2 * 0.05}} = 28$$

The mirror should be designed to withstand a minimum random vibration level of 3 * 28 = 84 times gravity. The secondary mirror mount is designed to withstand an inertial load of 90 g.

Boundary condition for thermal loads: Taking 20°C as the ambient temp. The system is subjected to a temperature variation of +30°C and -30°C. (ΔT=30°C). Figure 18 shows some of the FEA results of the secondary mirror opto-mechanical components.

Figure 18 (a) shows stress in the mirror due to 90g inertia load in Y direction. Figure 18 (b) shows stress in the ring due to 90g inertia load in Z direction. Figure 18 (c) shows stress in the stand due to 90g inertia load in X direction. Figure 18 (d) shows stress in the mirror due to minus 30°C thermal load. The stresses in mirror, leaf spring ring mount and stand with 90g inertia load imposed in X, Y and Z directions and due to variation in temperature are presented in section 4.5.3, summary of results.

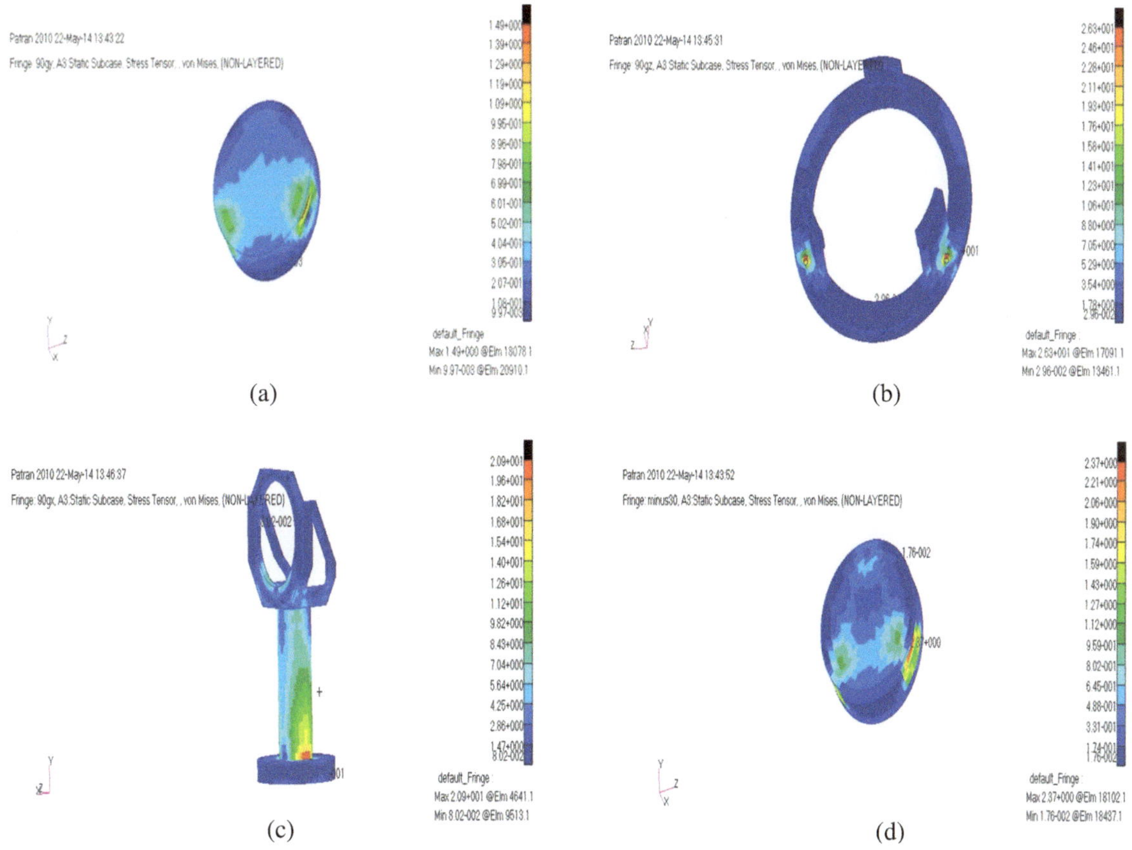

Figure 18: FEA results of the secondary mirror opto-mechcanical components

4.5.3. Summary of results

Tables 7 and 8 show the summary of results for components of secondary mirror subjected to quasi-static and thermal loads respectively. The stresses for all the individual components are given.

Table 7: Quasi-Static Analysis results for components of secondary mirror opto-mechanical system

Sl. No.	Component	Von Mises Stress(Mpa) in component for 90g load			Allowable stress(Mpa)
		X	Y	Z	

1	Mirror	1.71	1.49	1.02	5
2	Ring	12.4	18.7	26.3	300
3	Stand	20.9	5.93	20	800

Table 8: Thermal stress analysis results for components of secondary mirror opto-mechanical system

Sl. No.	Component	Von Mises Stress(Mpa) in component due to thermal loads		Allowable stress(Mpa)
		+30°C	-30°C	
1	Mirror	2.37	2.37	5
2	Ring	31	31	300
3	Stand	87.2	87.2	800

5. Analytical Calculations

5.1. Glue Area

Guidelines for determining the appropriate adhesive area for bonding prisms to mechanical mounts have appeared in the literature (Yoder, 1988, 2002) [2]. In general, the minimum area of the bond, Q_{MIN}, is determined by equation 2.

$$Q_{MIN} = W\frac{a_G f_s}{J} \tag{2}$$

Where,

W *is* the weight of the optic

a_G is the worst-case expected acceleration factor

f_s is the safety factor

J is the shear or tensile strength of the adhesive joint (usually approximately equal).

A safety factor of at least 2 and possibly as large as four to allow for some unplanned, non optimum condition, such as inadequate cleaning during processing, is to be taken.

For primary mirror, we have

$$Q_{MIN} = 2.3\frac{9.81 * 70 * 2}{17.2} = 183.65 \text{ mm}^2$$

Since the load is shared by 3 blades, a minimum bond area of 183.65/3 = 62.25 mm^2 per blade is required to withstand the inertia loads. The bond area chosen for primary mirror mount is 611.56 mm^2.

For secondary mirror, we have

$$Q_{MIN} = 0.065\frac{9.81*90*2}{17.2} =6.67 \text{ mm}^2$$

For the secondary mirror a minimum bond area of 2.22 mm^2 per blade is required to withstand the inertia loads. The bond area chosen for secondary mirror mount is 76.97 mm^2.

5.2. Stresses in Cemented and Bonded Optics Due to Temperature Changes

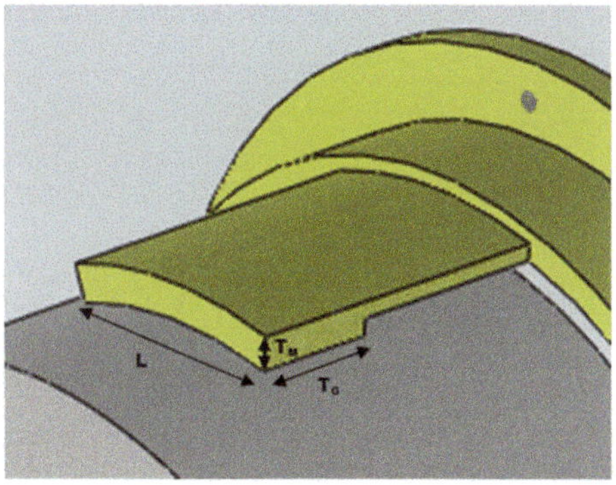

Fig. 19: Estimation of shear stress in adhesive

Vukobratovich (2002, 2003) [3] presented an analytical method, based on work by Chen and Nelson (1979), for estimating the shear stress developed in a bonded joint as a result of differential dimensional changes at temperatures other than that at assembly. Figure 19 shows the dimensions of the leaf spring which is used for estimating the shear stress in the adhesive. The pertinent equations are as follows:

$$S_s = \frac{(\alpha_M - \alpha_G)\, \Delta T\, S_e \, tanh(\beta L)}{\beta \, t_e} \tag{3}$$

$$S_e = \frac{E_e}{2(1+ \nu_e)} \tag{4}$$

$$\beta = \left[\left(\frac{S_e}{t_e}\right)\left(\frac{1}{E_M\, t_M} + \frac{1}{E_G\, t_G}\right)\right]^{1/2} \quad (5)$$

Where,

S_s is the shear stress in the joint

α_M and α_G is the CTEs of the metal and glass, respectively

ΔT is the temperature change from assembly temperature

S_e is the shear modulus of the adhesive

tanh is the hyperbolic tangent function

L *is* the largest dimension (length, width, or diameter) of the bond

t_e is the thickness of the bond

E_M and E_G is Young's modulus for the metal and glass, respectively

v_e is the Poisson's ratio for the adhesive

t_M and t_G is the thicknesses of the metal and glass components, respectively.

For primary mirror, we have

$$E_e = 689.5\ Mpa\ ;\ \ v_e = 0.43;\ \alpha_M = \ 1.26 * 10^{-6}\ /°C\ ;\ \alpha_G = 5 * 10^{-8}\ /°C\ ;\ L = 38.22mm\ ;\ t_M = 5\ mm$$

$$t_G = 16\ mm\ ;\ t_e = 0.1mm\ ;\ S_e = \frac{689.5}{2(1 +\ 0.43)} = 241.08\ Mpa$$

$$\beta = \left[\left(\frac{241.08}{0.1}\right)\left(\frac{1}{147000*5} + \frac{1}{90600*16}\right)\right]^{1/2} = 0.070\ /mm\ \ ;\beta L = 0.070 * 38.22 = 2.6754$$

$$S_s = \frac{(1.26 * 10^{-6} -\ 5 * 10^{-8})(30)(241.08)\ tanh(2.6754)}{0.070 * 0.1} = 1.23\ Mpa$$

This stress causes no damage to the mirror.

For secondary mirror, we have

L=24.50 mm ; $t_M = 2.94\ mm\ ;\ \ t_G = 6\ mm\ ; t_e = 0.1mm\ ; S_s = 0.86\ Mpa$

This stress causes no damage to the mirror.

5.3. Thermal Stresses Developed Due to Temperature Excursions

Tensile or compressive stress may be caused by restricting thermal expansion. Thermal stress can be calculated as

$$\sigma = E\varepsilon \qquad (6)$$

$$= E\alpha\Delta T$$

Where

σ = stress due to temperature expansion (MPa)

E =Young's Modulus (Mpa)

ε = strain

α = temperature expansion coefficient (/°C)

ΔT = temperature difference (°C)

Differential displacement (ΔL) = $\Delta\alpha$ x ΔT x (Mirror Diameter /2)

$$= (1.26 - 0.05) \text{ x } 30 \text{ x } (146/2) \text{ x } 10^{-6} = 2.6495 \text{ x } 10^{-3}$$

$$\varepsilon = \Delta L / L \qquad (7)$$

$$= (2.6495 \text{ x } 10^{-3}) / (146) = 1.815 \text{ x } 10^{-5}$$

$$\sigma = E\varepsilon = 90600 \text{ x } 1.815 \text{ x } 10^{-5} = 1.64 \text{ Mpa}$$

The analytical and FEA results are compared and shown in the table 9 below.

Table 9: Comparison of Analytical and FEM Results due to Thermal Load

Component	Stress by Analytical Calculation	Stress from FEA Result
Mirror	1.6 Mpa	1.3 Mpa

5.4. Stress Induced in Mirror due to Inertia Loads

Stresses are induced in the mirror when the mirror is subjected to inertia loads during launching of the rocket. This section gives an analytical way to calculate the stress.

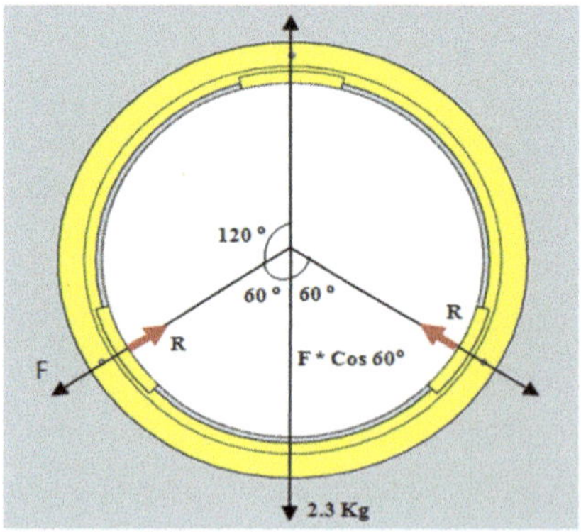

Figure 20: Stress in mirror due to inertia loads

Consider the free-body diagram shown in figure 20. The weight of the mirror is 2.3 Kg and it acts vertically downwards. F is the force acting on the blade due to weight of the mirror.

Force = Mass * Acceleration

Resolving the force (F), we get F * cos 60° = m*g

= 2.3 * 9.81=22.563 N

For 70 g load, F * cos 60° = 22.563*70

F= 3158.82 N

Since the load is shared by the two bottom blades,

$$\frac{F}{2} = \frac{3158.82}{2} = 1579.41\ N$$

R is the reaction force acting on the mirror.

$$Stress = \frac{Reaction\ Force}{Contact\ area\ of\ blade} = \frac{1579.41}{611.56} = 2.58\ Mpa\ \approx 2.6\ Mpa$$

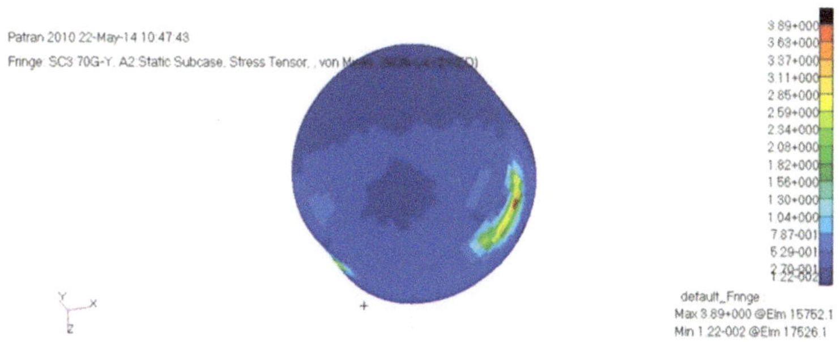
Figure 21: Primary mirror subjected to 70g inertia load in –Y direction

The analytical and FEA results are compared and shown in the table 10 below

Table 10: Comparison of Analytical and FEM Results due to Inertia Load

Component	Stress by Analytical Calculation	Stress from FEA Result
Mirror	2.6 Mpa	3.89 Mpa

In the analytical approach for calculation of stress in the mirror due to inertia loads, it is assumed that the mirror is of symmetric mass. However, the mirror used in the finite element model (figure 21) is of asymmetric mass. This explains the variation of analytical results with FEA results.

6. Conclusion

1. The opto-mechanical design implemented for primary and secondary mirrors in this project has yielded successful results. The overall weight of the assembly (primary mirror mount + secondary mirror mount) has met the mass requirements (<6.5 kg) by suitable optimization.
2. The opto-mechanical system should withstand the vibrations encountered during launching of the rocket. The natural frequency of the system (>200 Hz) should be away from the operating range of frequency (10-100 Hz). It should also survive the effects due to random vibration as the system is designed for higher inertia loads using Miles' equation [1].
3. The linear static analysis result of the opto-mechanical assembly has met the required targets. Mirror being a brittle material, the main concern in this project is whether the mirror would withstand the changes in environmental conditions. The stresses induced in the mirror due to thermal and inertial loads are well within the acceptance criteria (< 5 Mpa). The stresses induced in the mounts are also within the yield stress of the material which indicates the design of the assembly is safe.
4. Two types of ring mounts are compared. Namely leaf spring ring mount and bipod ring mount. From the finite element analysis results it is observed that the stresses induced in the mirror due to inertia and

thermal loads reduce when bipod ring mount is used. Bipod ring mount has given better results when compared to leaf spring ring mounts.

5. The materials/material properties selected for this work is ideal. Invar can be used as ring mount as it is compatible with Zerodur (Mirror). Titanium can be used as it has high strength/stiffness ratio. The glue material (EC2216) employed in the contact area is also compatible as the shear stress induced in the glue due to temperature excursions is low (< 5 Mpa), which keeps the mirror safe and it has better shear strength than most of the adhesives.
6. Analytical calculations for thermal stresses developed due to temperature excursions and stress induced in the mirror due to inertia loads comply with the FEA results. The validity of the FEA results is established. Hence, it can be concluded that the mounting configuration employed in this work can be used for mounting small off-axis mirrors for the optical systems used for the space applications.

References

[1] John W. Miles, November 1954, "Structural Fatigue under Random Loading", Journal of the Aeronautical Sciences, pg. 753.

[2] Paul. R. Yoder Jr., 2006, "Opto-Mechanical Systems Design" Third Edition, SPIE Press.

[3] Vukobratovich, D., 2003, "Introduction to optomechanical design", *SPIE Short Course SC014*.

[4] Anees Ahmad, 1997, "Handbook of Opto-mechanical Engineering" CRC Press.

[5] Katie Schwertz, 2010, "Useful Estimations and Rules Of Thumb for Optomechanics", University of Arizona.

[6] Warren C. Young & Richard G. Budynas, "Roark's Formulas for Stress and Strain" 7th Edition McGraw-Hill Publications.

[7] Nithin S Gokhale et al, 2008, "Practical Finite Element Analysis" Finite to infinite publications.

[8] Ryan Simmons, "Basics of Miles' Equation", http://femci.gsfc.nasa.gov/random/MilesEqn.html.

[9] M. Attaba et al, 2002, "Finite Element Stress and Vibration Analyses for a Space Telescope '', University of Surrey.

[10] Francisc M. Tapos et al., 2005, "High bandwidth fast steering mirror", Proceedings of SPIE Vol. 5877.

A Brief Author Biography

1st Author – Kaushik M. M, M-Tech (Machine Design), Department Of Mechanical Engineering, K.S. Institute of Technology, Bengaluru-560062, Karnataka, India. E-mail: kaushikmm@hotmail.com

2nd Author – M. M. M. Patnaik, Associate Professor, Department Of Mechanical Engineering, K.S Institute of Technology, Bengaluru-560062, Karnataka, India. E-mail: mmmpatnaik@gmail.com

3rd Author – Venkateswaran R, Head, OEND, Scientist/Engineer-SG, LEOS-ISRO, I Phase, Peenya Industrial Area, Bengaluru-560058, Karnataka, India. E-mail: rv_leos@leos.gov.in

Control of Cavity Acoustics by Surface Waviness In Landing Configurations

Laurent Dala
CSIR, DPSS/Aeronautics Systems, Pretoria 0001, South Africa
Mechanical & Aeronautical Engr. Dept., University of Pretoria, Hatfield 0083, Pretoria, South Africa
E-mail: LDala1@csir.co.za

Abstract

The aviation industry's interest towards cavity control methods rose accordingly due to the acoustic attenuation potential of such methods. However cost effective control devices which are also efficient over a wide range aircraft operations are yet to be developed.

This paper investigates the use of a sinusoidal surface modification application upstream of a cavity as a passive acoustics control device in approach conditions. Optimum sinusoidal amplitude and frequency were previously determined by the means of a two-dimensional computational fluid dynamics analysis for a cavity with a length to depth ratio of 4. A complete three-dimensional CFD analysis of this configuration as well as a base-line case without control device was carried out in this study. The overall sound pressure level was reduced with the surface modification at the majority of the points investigated.

Keywords: Applied Aeronautics, Aeroacoutics, Landing Configurations, Cavity Noise.

1. Introduction

Research has been made on flow over cavities for over a century, however after the Second World War the interest toward cavity aerodynamic and acoustics has been rising due to possible application to aircrafts. The oscillation within and surrounding cavities became an issue due to the generalisation of bomb bays and wheel wells, for instance. Along with a reduction of aircraft performance by drag increase, the presence of open cavity can result in the generation of adverse acoustic effects. Generation of high acoustic peaks results in high applied pressure which could induce moments depending on the location of the cavity as well as compromise aircraft structure at the cavity surrounding.

Various flow control solutions have been elaborated over the years, from simple passive solutions generally based on geometry modifications to expensive active method requiring addition to operate. However, a cost effective solution, which is efficient over a wide range of operations, is yet to be discovered. Most of the research previously made related to weapon and bomb bays. However, the need of the modern aircraft industry to reduce the acoustic generation of their designs due to the rising concerns regarding noise pollution surrounding airports resulted in a recent augmentation of subsonic flows over cavities studies.

This study was carried out as a continuation of the two-dimensional computational fluid dynamics (CFD) analysis on subsonic flow over cavities and attenuation of self-sustaining oscillation occurring within them as described in [1, 2]. The attenuation method investigated consists of the application of a sinusoidal surface modification ahead of the cavity use as a control device.

This research aims to improve the study made on the previously determined optimum control device by carrying out a full three-dimensional analysis.

2. CAVITY FLOW FIELDS

Noise pollution became a major concern over the years. Organisations such as the Advisory Council for Aeronautics Research in Europe (ACARE), defined a range of objectives for environmental improvements by 2020 as detailed in [3], including a 10 dB reduction per fixed wing aircraft operation. Topology of cavity flow fields, especially in subsonic conditions, are reviewed in this section in order to point out the main flow characteristics, as well as the oscillation prediction methods.

2.1. Time-Averaged Flow

Despite the fact that flows over and surrounding cavities are unsteady by nature, information such as flow patterns and pressure distribution can be drawn out of time-averaged flow. As described in [4], the types of flows over rectangular planform cavities can be divided into two main categories with a range of transitional ones between them. The length to depth ratio *l/h* is the cavity characteristic which mainly determines the flow type within it, which is

2.1.1. Open and Closed Flow Types

Open flow occurs within cavities with low *l/h*. As the flow enters the cavity the shear layer is separated from the leading edge, bridges the cavity and impinges near or at the top of the rear wall generating a stagnation point, and consequently a vortex is generated within the cavity, as shown in Figure 1.

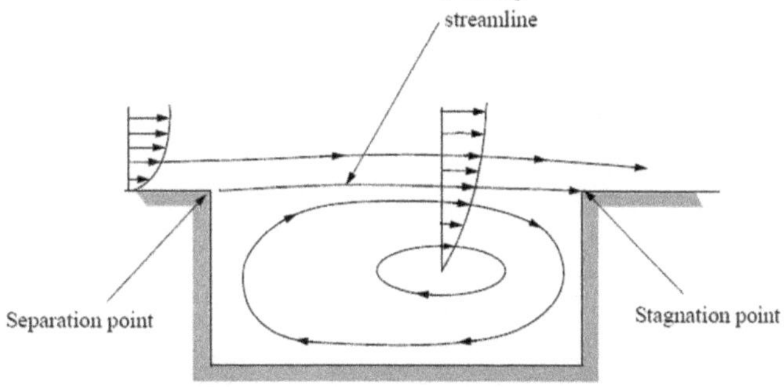

Figure 1: Subsonic open flow over rectangular cavity [5]

Furthermore, it is important to note that in the case of really low l/h multiple superposed vortices can be generated.

Closed flow occurs within cavity with large l/h. As the flow enters the cavity the shear flow separates at the front wall and then impinges on the cavity floor. Then at the proximity of the rear wall the flow separates and hits the rear wall top region creating a stagnation point. This results in the formation of two vortices, next to the front and rear walls of the cavity, as presented in Figure 2.

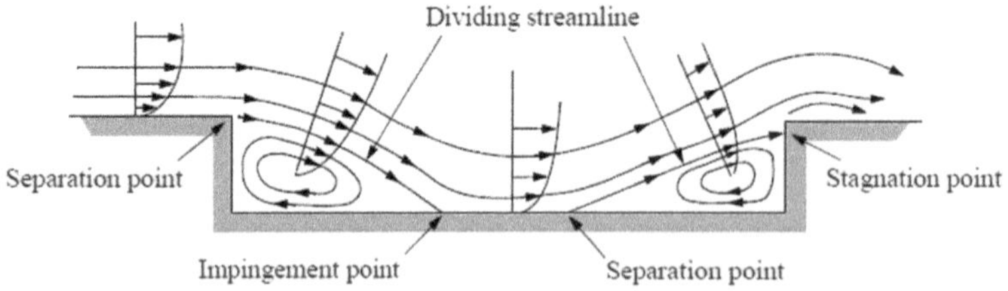

Figure 2: Subsonic closed flow over rectangular cavity [5]

2.1.2. Determination of Flow Types

Throughout the years categorisation of flow regimes [6-8], at subsonic speeds, for different l/h cavity ratios has been achieved by analysing the floor pressure distribution curves. An example of this is the work described in [9] and is a main source of information in cavity flow type characterisation. Figure 3 shows the floor pressure graphs for subsonic speeds for an increasing l/h as given in ESDU 02008 defined in [10] by using data collected in [9].

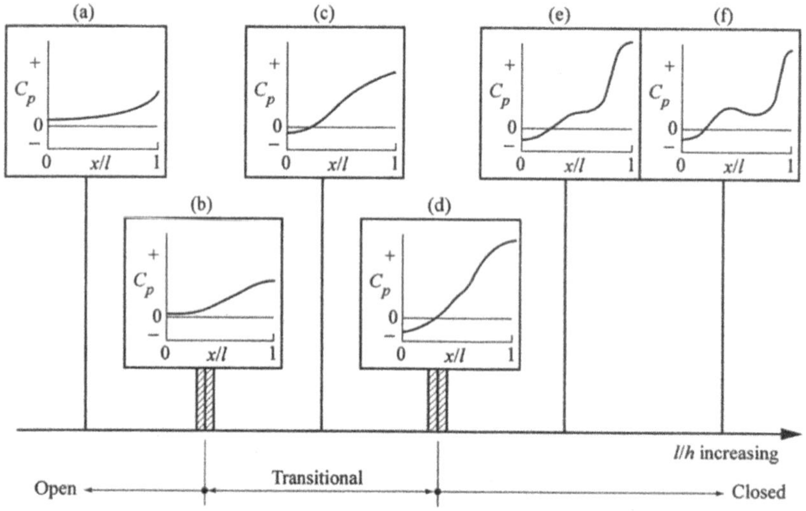

Figure 3: Cavity floor pressure distribution for cavity flow field at types at subsonic speeds [9]

Pressure distributions in Figure 3 present the following characteristic according to the topology of the Cavity flow field in Subsonics.

(a) Open flow

- The pressure is roughly linear for a substantial part of the floor
- Closer to the rear wall the pressure distribution has a concave-up shape

(b) Open/Transitional flow boundary (shown hatched to suggest uncertainty)

- Pressure over the floor remained almost the same to that of open flow apart from the final portion which has changed to a concave-down shape.

(c) Transitional flow

- At this stage of the flow the front part of the flow exhibits a negative pressure with a slight increase in pressure at the rear part. The overall shape of the graph is unchanged.

(d) Transitional/Closed flow boundary (shown hatched to suggest uncertainty)

- The pressure coefficient graph has a higher gradient than before changing from a negative value, at the front of the floor, to a positive value at the rear of the wall. It is also possible to note the start of the creation of an inflection, at mid cavity floor.

(e) and (f) Closed flow

- Further increase in l/h causes the inflection occurring at mid-length to reach a zero gradient.
- This flat part of the graph changes to a minimum, close to the front of the cavity floor, followed by a change to maximum just before the rear of the floor.
- The maximum floor pressure is unchanged at the rear part of the cavity floor.

2.1.3. Three-dimensional Effects

In [6], the notion of three-dimensional flow away from the cavity borders was first introduced. This research on flows over cavity with large width to length w/l ratio using oil based medium on the cavity floor and lateral pressure distribution measurements led to discover cellular structured flows within the cavity. At a given w/l, the flow went from a mainly two-dimensional pattern to a three-dimensional one as the depth was gradually increased. The variation of pressure along the cavity width generated a wave-like oil pattern of the cavity floor. The pressure distribution along the cavity central line is therefore demonstrated not to be always representative of the pressure distribution within high w/l cavities.

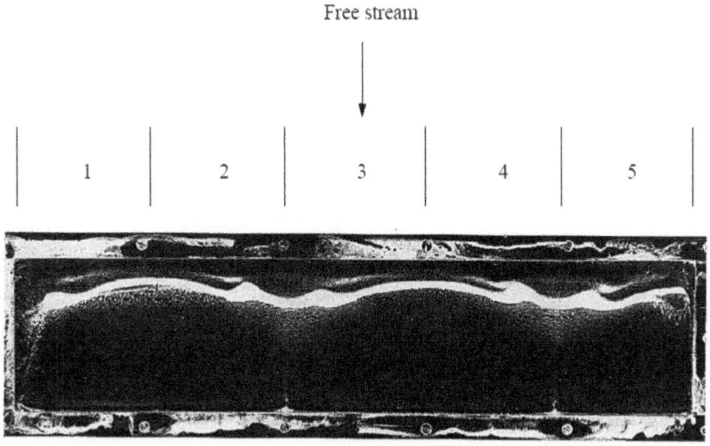

Figure 5: Example surface flow on cavity floor [6]

Effects of the decrease of length to depth l/h ratio on cellular structure formation were investigated. Wave structure was first observed at l/h =2.2, with further decrease in l/h steady wave pattern were only found when the wave length was allowing a finite number of cells. Intermediary stages were found to be unsteady. The following equation developed in [6] allowed determining the cell span s in function of the cavity dimensions:

$$^{s}/_{l} = 0.6\left(1 + ^{l}/_{h}\right) \tag{1}$$

The side walls have an important influence on three dimensional flow characteristics within cavity. Logically, flow over cavities with high *w/h* only experience effects of wall at their proximity. By opposition, flow over cavities with low *w/h* will be dominated by them. The schematic diagrams in Figure 6 of flow structures for small width open and closed rectangular cavities at subsonic speeds described in [10] illustrate the possible three-dimensional nature of the main flow.

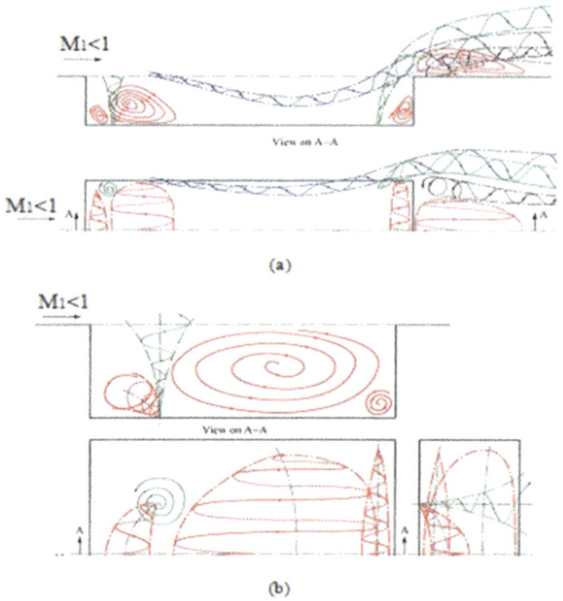

Figure 6: Main three-dimensional flow structures for rectangular closed (a) and open (b) cavities (adapted from [10])

Those structures have been extrapolated from various theoretical prediction results as well as flow visualisations as described in [11-14].

The flow patterns suggest the following main flow structures within the closed and open cavities:

- Closed cavity shown in Figure 6 (a)
  There are two main regions of vortical flow aft of the front wall and forward of the rear wall as shown in Figure 2. The impingement line is curved due to the side wall effects. A very weak lateral vortex forms between the front wall and the forward separation line, making a counter rotating flow region. On the cavity base just aft of the front wall and close to the side walls, there are two weak separation nodes, which are associated with a pair of counter rotating vertical vortices. There are two other separation nodes on the cavity floor, close to the side walls and at the extremities of the separation line, which are indicated by a pair of counter rotating vertical vortices. While the shear layer cascades over the front wall, the flow is entrained over the side walls. Due to the recompression action of shear layer rising to pass over the rear wall, some of the entrained flow is ejected from the side walls.

- Open cavity described in Figure 6 (b)

There are tandem rotating vortices, in the central spanwise region. There is also a small, weak corner vortex, which has a sense of rotation opposite to that of the strong rear vortex. There are two separation nodes on the cavity floor just aft of the front wall, which indicate a pair of counter rotating vertical vortices. For deep open cavities, one large captive vortex usually dominates the flow and the two recirculation areas associated with corner vortices are either small or negligible. Therefore, the pressure on the centreline is quasi-uniform, and then increases near the downstream wall of the cavity as shown in Figure 3. However, for open-transitional flow regimes, the upstream recirculation zone has a non-negligible size. In these cases, the pressure distribution shows a lower pressure near the upstream wall when compared to the open cavity flow type.

2.2. Unsteady Flow field

Unsteady cavity flow field consists of a combination of both random and periodic pressure fluctuations. Pressure fluctuations are by nature random for cavities with closed flow. On the other hand periodic fluctuations, responsible for cavity tones, are prevalent for open flow. Nevertheless such fluctuations are also common with transitional flows and can even occur with closed flow. In [15], it was determined and classified as resonant, cavity configuration and subsonic flow likely to produce significant tones. The figures below illustrate typical spectra for cavity with closed and open flows as well as a section of the resonant range classification presented in [15] corresponding to the cavity dimensions of interest for this study.

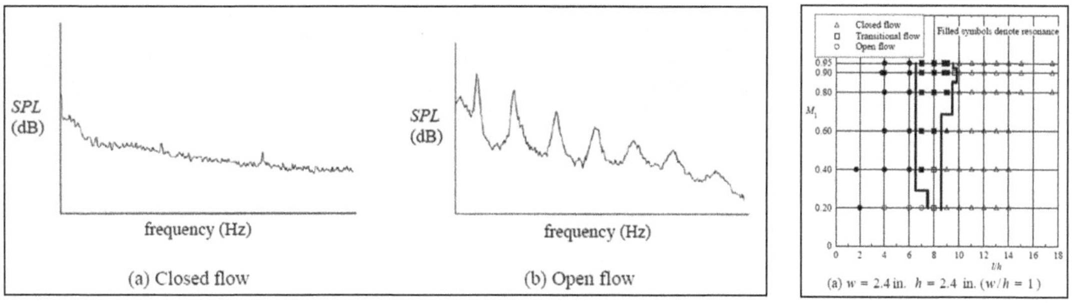

Figure 7: Cavities with closed and open flows spectra examples and resonance range with cavities at various subsonic speeds [15]

2.2.1. Self-Sustaining Oscillations and Types of Oscillations

The terms of self-sustaining oscillation is self-explanatory, however the principles behind it are rather complex. The impingement of the shear layer on the rear wall at the proximity of its edge generates an upstream propagation of the disturbances called feedback mechanism. These returning pressure perturbations generate vorticity fluctuations near the shear layer separation at the front wall, amplifying the instabilities within the shear layer which then impinge on the rear wall, closing the self-sustaining cycle.

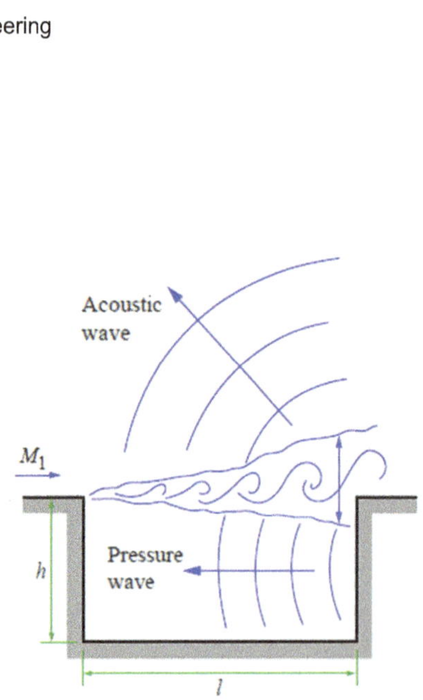

Figure 8: Self-sustaining oscillations representation [16]

In [16], the following three classes of cavity oscillations: fluid-dynamic, fluid-resonant and fluid-elastic were defined.

- Fluid-dynamic oscillation, represented in the figure below, is an oscillation generated only by the shear flow instability.

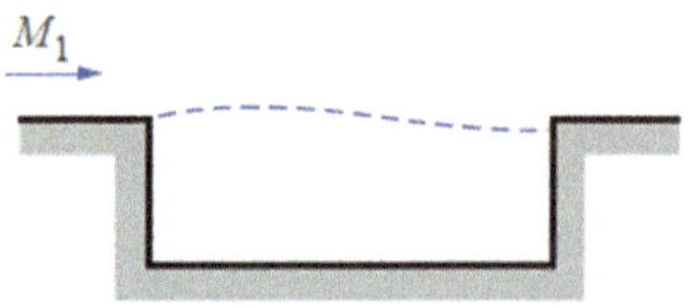

Figure 9: Fluid-dynamic oscillations [17]

According to [18], fluid-dynamic oscillations will only occur if the ratio between the cavity length to acoustic wavelength is very small. The primary mechanism for this type of oscillation is the amplification of unstable disturbances in the shear layer. The additional feedback condition is also important in the generation of oscillations. The feedback is enhanced by the upstream propagation of disturbances, due to the cavity trailing edge. Pressure perturbations originating from the trailing edge of the cavity produce fluctuating vortices close to the shear layer, which in turn enhance the disturbances in the shear layer. This process is repeated over and over again.

- Cavity oscillations are classified as fluid-resonant if the oscillations are affected by resonant wave effect also known as standing waves. In this case the shear layer oscillation is couple with either one or more acoustic resonant modes. These modes are length, depth or width related as represented in the following picture.

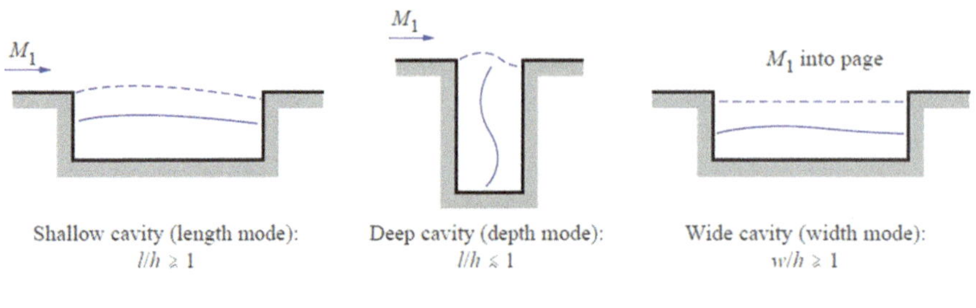

Figure 10: Fluid-resonant oscillations [17]

Fluid-resonant oscillation can only occur if one of the cavity dimensions is of the same order or greater than acoustical wavelength; the standing waves generated then occur in the longest of the cavity dimensions. Finally it should be noted that fluid-resonant oscillation generally occurs at high speed.

- Cavity oscillations are classified as fluid-elastic if the oscillations within the shear layer are coupled with the motion a part or the entirety of the cavity walls as represented in the figure below.

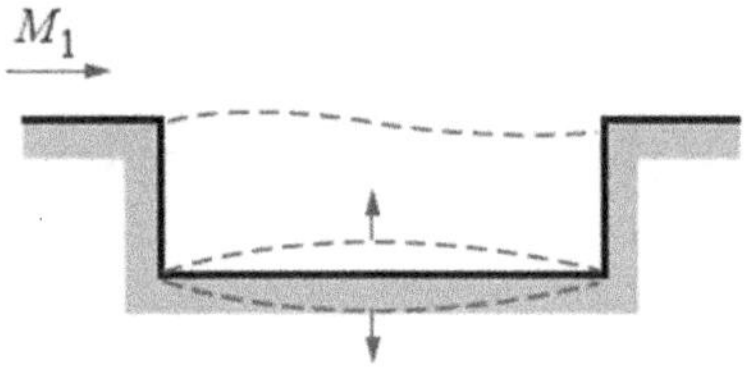

Figure 11: Fluid-elastic oscillations [17]

- Normal Acoustic Mode Oscillations consist of an excitation of the cavity normal acoustic mode by the shear layer instabilities. Such oscillation can occur at any speed however it generally takes place with deep cavities at very low speed. Note that a feedback mechanism is not required to generate normal mode resonance; on the other hand it can amplify the resonance if present.

2.2.2. Prediction of Oscillations

In [19], an equation was derived enabling the prediction of mode frequencies. Two assumptions were made to derive the equation; the first one is that the frequency of the acoustic pressure wave response is equal to the vortex shedding frequency.

$$St = f_m \frac{l}{U_\infty} = \frac{m-\xi}{\left(\frac{1}{k_v}+M_\infty\right)} \tag{2}$$

Where St is the Strouhal number, m is the tonal mode number and f_m is the frequency of the tonal mode. Both k_v and ξ were determined empirically from wind tunnel testing as detailed in [19]. k_v is the ratio of the average downstream vortex convection speed to that of the free stream and finally ξ is an empirical parameter used to take in account the phase difference acoustic wave arrival upstream and the vortex impingement on the rear wall.
ξ was determined to be dependent on the length to depth ratio l/h however it was shown in [19] that ξ=0.25 a satisfactory value to assume independently to both l/h and width to depth ratio w/h. He also deducted the value of 0.57 for k_v from tests of $l/h=4$ cavities.
Those two values have been widely accepted however it is possible that they might vary with increase of the boundary layer thickness ahead of the cavity δ. As detailed in [20], seven cavities with l/h=4.5 and w/h=1 were tested at Mach 0.6 with seven different scaling factors to vary the cavity depth to boundary layer thickness ratio δ/l. The results obtained were in agreement with the k_v value put forward in [19]. However the results given in [20] indicated that the value of ξ increased with increase of δ.
When compared with test results, the predictions obtained by the equation defined in [19] with ξ = 0.25 and k_v = 0.57, were found to be in coherent up to approximately Mach 1.5. The relationship between the speed of sound a and the acoustic pressure wave c described in [21] is given by:

$$c = a\left[1 + \tfrac{1}{2}r(\gamma-1)M^2\right]^{\frac{1}{2}} \tag{3}$$

Where r is the temperature recovery factor and γ is the specific heat ratio. The assumption taken in [19] that $a= c$ consisted as assuming that r being equal to 0. But as detailed in [21], r was determined to be equal to 0.9 and 0.95, respectively at Mach 0.8 and 3. Eq. (2) was rearranged for supersonic flows assuming r = 1, as shown in [21].

$$St = \frac{m-\xi}{\frac{1}{k_v} + \frac{M}{\sqrt{1+\frac{1}{2}(\gamma-1)M^2}}} \tag{4}$$

3. MODEL DESCRIPTION AND SIMULATION METHODOLOGY

3.1. Cavity Geometry and Monitoring Points

The CAD Package used to create the geometry and domains studied is UGS Solid Edge V20 and Ansys ICEM CFD. The domain consists of a rectangular shaped cavity included within a larger domain. The dimensions of the square cavity were taken from [22]. The width w and length l equal to 50.8mm (2inch) and with a depth h equal 12.7mm (0.5inch), resulting with a length to depth ratio l/h of 4 and a width to length w/l ratio of 1.

For the purpose of this study, two cases were examined. The first case which is designated as the "flat" case investigated and was used as a baseline is the square cavity described in [22].

The second case which is designated as the "wavy" case investigated the cavity acoustic control solution consisting of a sinusoidal surface modification upstream of the cavity over a length of 254 mm. Only one length of the wavy surface was investigated in this research which was based on the test case described in [22].

In [1, 2], different surface modifications were simulated by varying the amplitude and frequency in function of the length to depth ratio l/h. Hughes in [2] showed that the amplitude and the frequency of the wavy surface does not depend on the length to depth ratio. It was then shown that the case with amplitude and frequency respectively equal to 3.25mm and 4.92×10^{-3} cycles per mm was observed as achieving the best reduction in cavity resonance at Mach 0.3.

The surface modification was applied on the entire width of the domain. However it should be noted that this optimum design has the cavity leading edge higher than the trailing edge by 3.25mm, generating a step on the side of the cavity as shown in Figure below

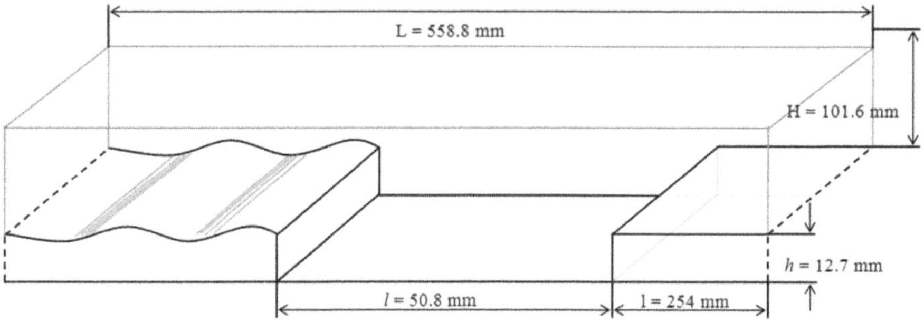

Figure 12: Cavity dimensions for l/h = 4

Pressure fluctuations with time were acquired via a total of 70 monitoring points. As detailed in [1], 20 monitoring points were used for the 2D cases; every tenth of the cavity length one monitoring point was placed in the alignment of the cavity edges and one on the cavity floor. This discretisation of control points inside of the cavity was driven by the sound pressure level measurements. A further study will require control points on the forward wall in order to characterise the cavity floor pressure.

In this 3D analysis similar sets of points were placed, one on the cavity central line, one half way between the central line and the cavity side wall and the third one on the cavity side wall. Finally a set of 10 monitoring point was place at 12.7 mm from the cavity side wall at the level of the cavity edges. The following Figure illustrates the location along the cavity length and width as well as the logic behind the numbering of the monitoring points.

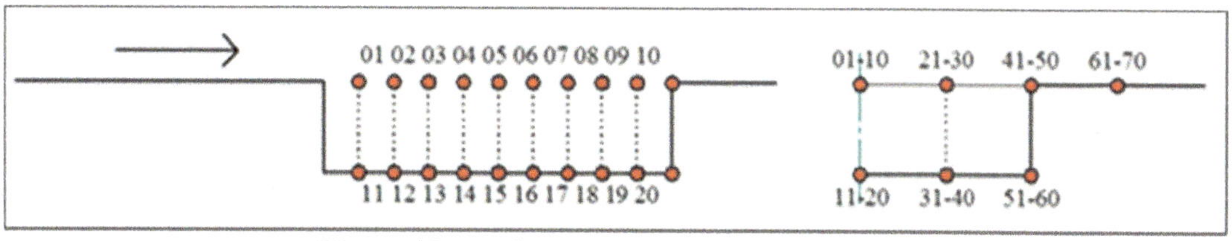

Figure 13: Monitoring points location and numbering

For the first set, the points bridging the cavity were numbered from 1 to 10 in the direction of the flow along with the cavity floor ones numbered from 11 to 20 in the direction of the flow. The other sets were numbered similarly, with points 21 to 40, 41 to 60 and 61 to 70 corresponding respectively to the set half way between central line and side wall set and cavity side set.

3.2. Computational Mesh

Figure 14 shows the computational domain and the grid structure for the Flat Case. The flat case mesh replicates the 2D parameters detailed in [1]. The 3rd dimension mesh has been obtained using the consecutive extrusions by normal element. For the first extrusion the entirety of the 2D surface mesh was selected then 20 layers with 1mm spacing between each other were extruded in the positive Z direction. Then all the cells of the 2D surface mesh except the cavity ones are selected and 40 layers with 1mm spacing between each other were extruded in the negative Z direction.

The following figures and tables are respectively a scaled sketch of the cavity section and flat case sub-divided into blocks with labelled edges and table of the meshing parameters associated to the different edges.

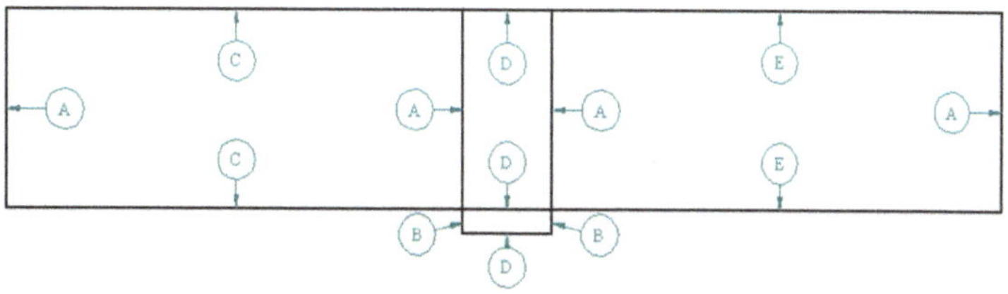

Figure 14: Side View of the Meshing Domain Sub-divided Into Four Blocks with Labelled Edges

Edges	**A**	**B**	**C**	**D**	**E**
Nodes	101	50	254	254	88
Mesh Law	Geometric 1	BiGeometric	Geometric 2	Geometric 1	BiGeometric
Spacing 1	0.05	0.05	0.05	0.05	0.05

Ratio	1.2	1.2	1.2	1.2	1.2
Spacing 2	0.05	0.05	0.05	0.05	0.05
Ratio 2	1.2	1.2	1.2	1.2	1.2
Max Space	1	1	1	1	1

Table 1: Meshing Parameters

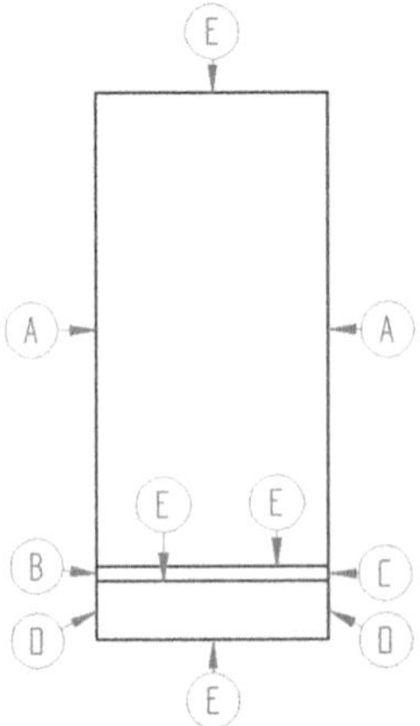

Figure 15: Front View of the Meshing Domain Sub-divided Into Three Blocks with Labelled Edges

Edges	**A**	**B**	**C**	**D**	**E**
Nodes	101	26	26	50	88
Mesh Law	Geometric 1	Geometric 2	Geometric 1	BiGeometric	BiGeometric
Spacing 1	0.05	0.05	0.05	0.05	0.05
Ratio	1.2	1.2	1.2	1.2	1.2
Spacing 2	0.05	0.05	0.05	0.05	0.05
Ratio 2	1.2	1.2	1.2	1.2	1.2
Max Space	1	1	1	1	1

Table 2: Meshing Parameters

This represents a total of 60 cells for the width of the model, a total of 75 cells were initially considered to match the general mesh density being used. However this reduction of 20% of the total number of cells within the mesh is justified by the proportional reduction of 44 hours to complete each CFD run as well as a reduction of 26Go of hard drive space required to store data produces by each run. The following Figure presents the detailed mesh for the Flat plate case.

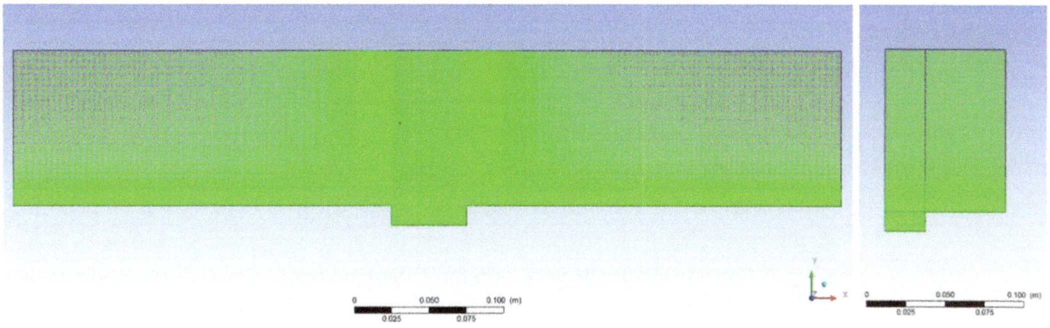

Figure 16: Side and rear view of the flat case mesh

The Computational mesh for the wavy case was generated to keep the same grid resolution defined for the flat case. Figure 17 and Figure 18 show the Mesh topology for the wavy case.

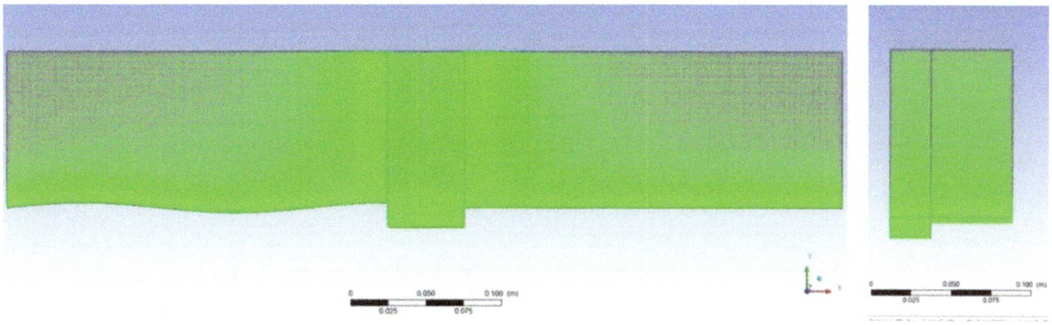

Figure 17: Side and rear view of the wavy case mesh

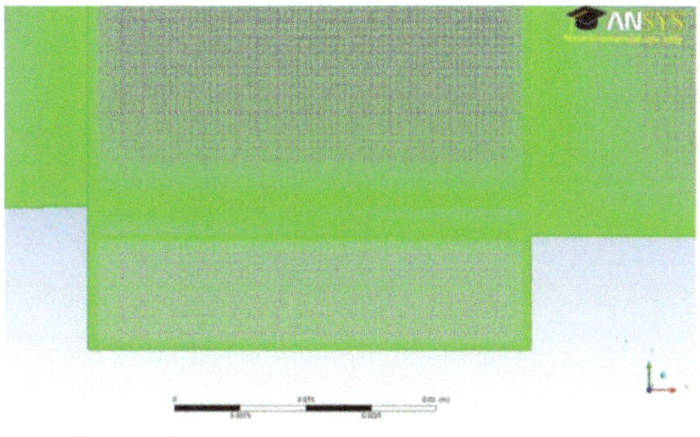

Figure 18: Wavy case central section mesh

The time to run the solver being a crucial parameter of this study, the solver control was set in a way to optimize the time required without neglecting the accuracy. The convergence criterion was set as 1.E-4 RMS with a minimum and maximum coefficient loops respectively set as 2 and 5. This limits the approximately hundred first time steps to do too much iteration and forces the other one to do two iterations per time step.

The following table compiles the solver control parameters incremented for pre-processing, common to both flat and Wavy Case.

	Tab	Setting	Value
Solver Control	**Basic Settings**	Advance Scheme > Option	Cartesian
		Transient Scheme > Option	Second Order Backward Euler
		Convergence Control > Min Coeff. Loops	2
		Convergence Control > Max Coeff. Loops	5
		Convergence Criteria > Residual Type	RMS
		Convergence Criteria > Residual Target	E-4

Table 3: Solver Control Parameters

3.3. Turbulence modeling

Anysy CFX Pre v 11 was used for the computations. Due to the topology of cavity flows Unsteady Reynolds-Averages Navier-Stokes (RANS) modeling and Shear Stress transport (SST) modeling approaches were used. This turbulence model was chosen in order to represent appropriately the incoming boundary layer and, particularly, to resolve the subsequent shear layer and its instabilities, which is directly coupled with the oscillating pressure pattern in the cavity.

Furthermore, time dependency is a crucial factor of this study due the unsteadiness of the flow field consisting of random and periodic pressure fluctuation occurring within the cavity. A transient simulation was therefore used, enabling the determination and exportation of the almost instantaneous pressure variations with time. Each time step was set as 5x10-5 seconds with initial time equal to zero and a the maximum time duration was set as 2048 time steps, cumulating a simulation time of 10.24 milliseconds.

3.4. Data processing

The pressure fluctuations as a function of time, P(t), were recorded from the monitor points and were post processed as to achieve the power spectral density (PSD) and subsequently the sound pressure level as a function of frequency. The PSD describes how the power of a signal or time series, in our case the time dependent pressure, is distributed with frequency.

The PSD spectrum is determined by means of a Fast Fourier Transform (FFT) utilizing windowing as to smoothen the PSD estimate, in our case the Hanning window. The FFT of a signal P(t) is defined as

$$\mathrm{FFT}(t,f) = \int_{-\infty}^{\infty} P(\tau)w(\tau - t)e^{-j2\pi f\tau}d\tau \qquad (5)$$

where *w(t)* is the Hanning window function, applied to obtain a clear definition of the fundamental frequencies.
A relation exists between the SPL and the PSD which is defined

$$\mathrm{SPL} = 10\log\left(\frac{\mathrm{PSD}}{P_{ref}^2}\right) \qquad (6)$$

where P_{ref} is the reference pressure for threshold of human hearing and has a value of 2×10^{-5} Pa.
Time averaging the pressure fluctuations one would obtain the mean pressure

$$\overline{P} = \frac{1}{N-N_0}\sum_{k=N_0}^{N} P_k(t) \qquad (7)$$

where N is the total number of samples taken and N_0 is the first sample number at which the time averaging starts. The root mean square pressure is therefore calculated via the equation:

$$P_{rms} = \frac{1}{N-N_0}\sum_{k=N_0}^{N}(P_k(t) - \overline{P})^2 \quad (8)$$

Another important parameter which can be calculated is the overall sound pressure level (OASPL). This is defined by

$$OASPL = 20\log\left(\frac{P_{rms}}{P_{ref}}\right) \quad (9)$$

4. RESULTS AND ANALYSIS

All the computations were performed to simulate the approach conditions of an aircraft. Therefore a freestream Mach number of 0.3 (99.44 $m.s^{-1}$) and a temperature of the air equal to 25 °C (298.15K) were used as initial conditions.

A total of 70 monitoring sub-divided into four sets along the cavity width were used for each case, however this study focused on two points at which the maximum SPL occurs, one located within the shear layer, the other one located on the cavity floor. As shown in [1], the monitoring points obtaining the maximum SPL values were the points 4 and 15. In the first set of monitoring points for the 3D cases, the maximum SPL occurs at point 5 and 15, though the SPL occurring at point 16 is comparable to the one occurring at point 15. This study therefore focused on points 4, 5 and 15 for the validation with the 2D cases then only on points 5 and 15 for 3D cases comparisons.

4.1. Validation of results

The following sections will in order aim to validate the 3D results obtained, analyse the evolution of the SPL along the width of the cavity and finally determine the effectiveness of the control method.

4.1.1. Flat Plate Case

Comparison of the two and three dimensional SPL spectrograms revealed a good conservation of the dominant tones for points 4 and 5. At monitoring point 4, the two tones respectively corresponding to the 1st and 2nd Rossiter modes were reduced by 14dB and slightly shifted by only 10Hz in the 3D case. Such reduction in SPL is to be expected when comparing 2D with 3D case due to dissipation in the 3rd dimension. Tonal peaks at monitoring point 5 on the other hand were moderately increased in 3D, by 5dB for the dominant peak corresponding to the 1st Rossiter mode. The amplitude of the second dominant peak is similar in the 2D and 3D results. Finally monitoring point 15 spectrograms have similar features in the two cases. However a systematic dumping is perceptible in 3D, this is due to the dissipation acting on the cavity floor. Even so the 2nd dominant peak is still observable.

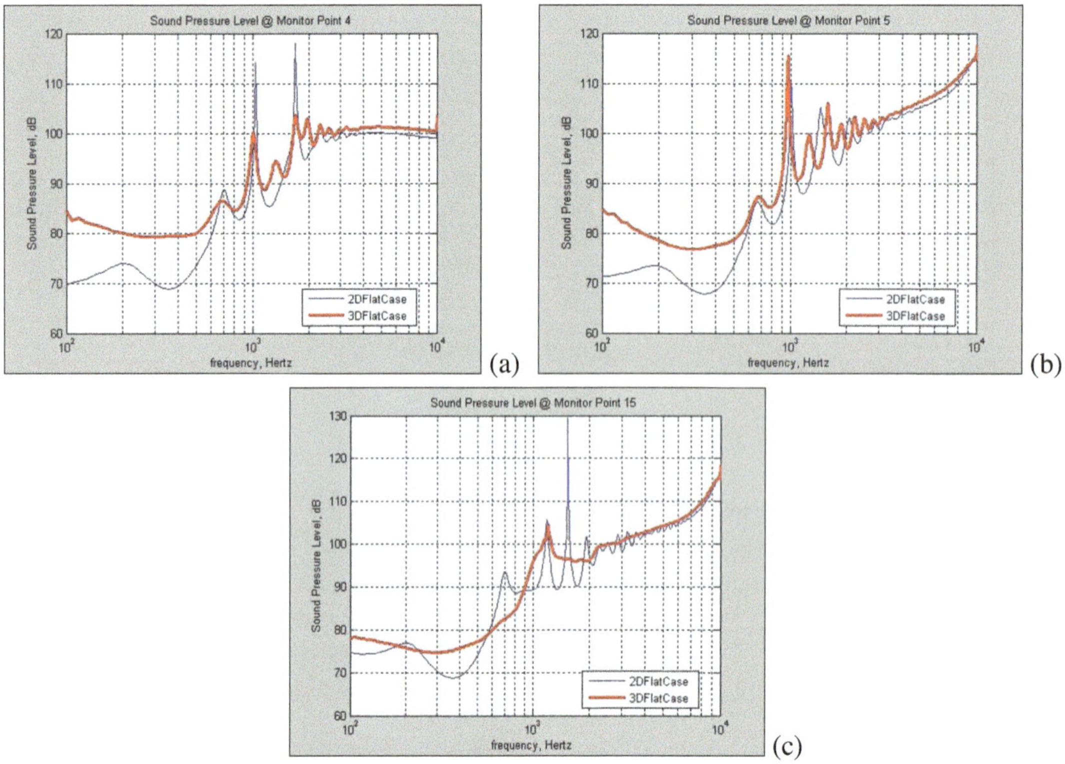

Figure 19: 2D/3D Comparison, Flat plate case SPL spectrograms at point 4(a), 5(b) and 15 (c)

4.1.2. Wavy Plate Case

Once compared, two and three dimensional SPL spectrograms of points 4, 5 and 15 showed similar repartition to some extent of SPL over frequencies. The lack of fluctuation in SPL for monitoring point 4 could be explained by being relatively distant from the shear layer, this is due to inclined shear layer generated by the difference in height of 3.25 mm between the front and rear wall. The increase in SPL fluctuation as the points are located downstream would tend to confirm this theory. The reason why this phenomenon is more pronounce in the 3D case is that the shear layer height varies along the cavity width and was noticed to be higher around the central line and at the side of the cavity.

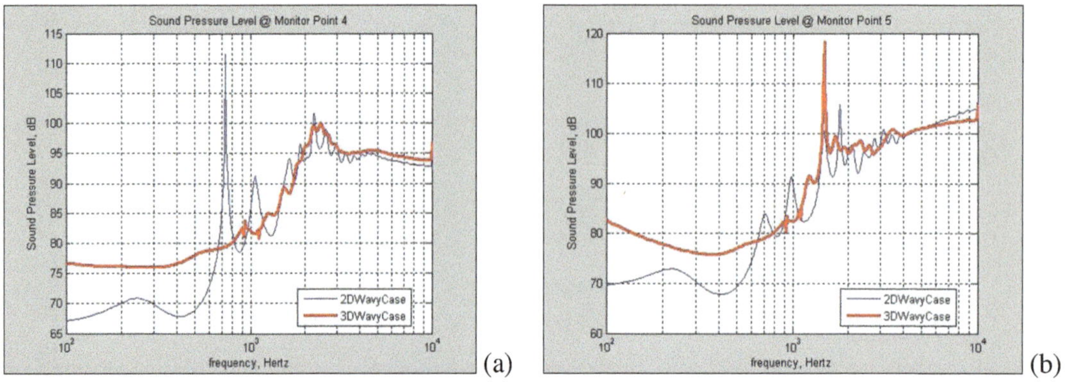

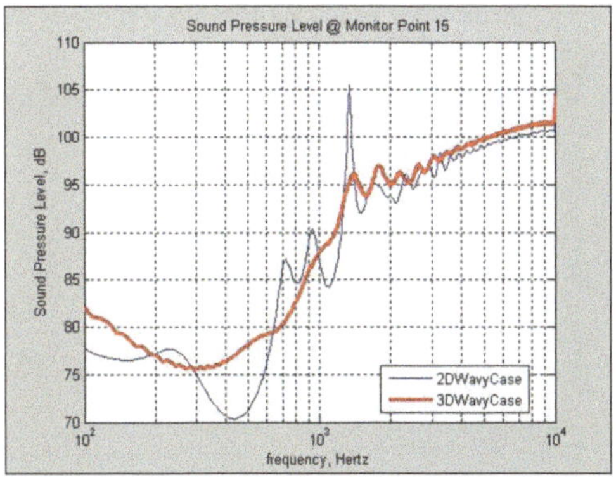

(c)

Figure 20: 2D/3D Comparison, Wavy plate case SPL spectrogram at point 4 (a), 5 (b) and 15 (c)

The spectrogram of point 5 which displays a tonal peak increased to 118.3dB at 1484.4Hz. The dominant peak amplitude therefore differs by 12dB between the two cases. As for monitoring point 15, the differences between the 2D and 3D correlate with the differences obtained above. The dominating tone occurring at a frequency 1347Hz is still noticeable but is reduced from 105.5 to 96.2 dB.

4.2. 3D effects and Sound Pressure Level (SPL)

This section analyses the relationship between the position along the cavity width and the SPL spectrogram extracted. To do so the central line monitoring points 5 and 15 was compared with corresponding points located along the width.

4.2.1. Monitoring point 5 Comparison

Comparison SPL spectrograms of points 5 and 25, located half-way between central line and side wall, revealed similar characteristics. In both cases the dominant tonal peak is however attenuated, by 13 and 15 dB respectively for the flat and the wavy case.

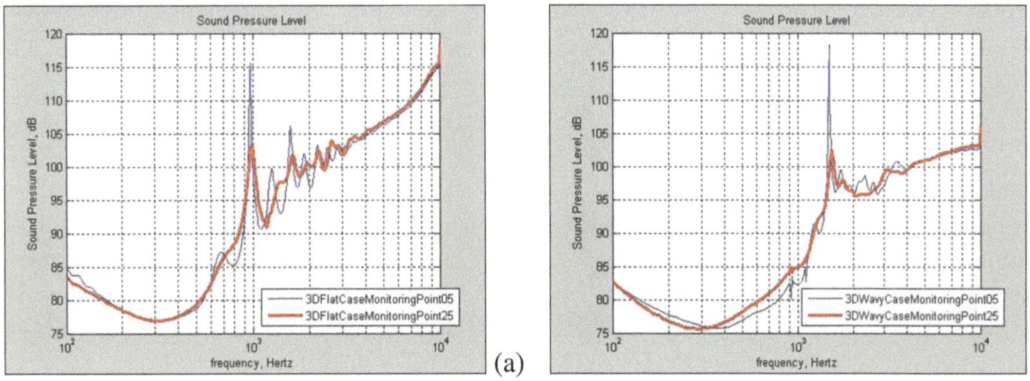

(a) (b)

Figure 21: Comparison of points 05/25, SPL spectrograms for the wavy plate (a) and flat plate (b)

Relationship between spectrogram of monitoring points 5 and 45, located on the top of the cavity side wall, are less evident to determinate than for the other comparisons previously made. In the flat case the original dominant tone occurring at point 5 is attenuated at point 45, however three other tones and two of them have similar amplitude as the point 5 dominant one, within 2dB. In the wavy case the peak present at point 5. However the amplitude of points 45 dominant peak reaches the impressively high value of 141.8dB.

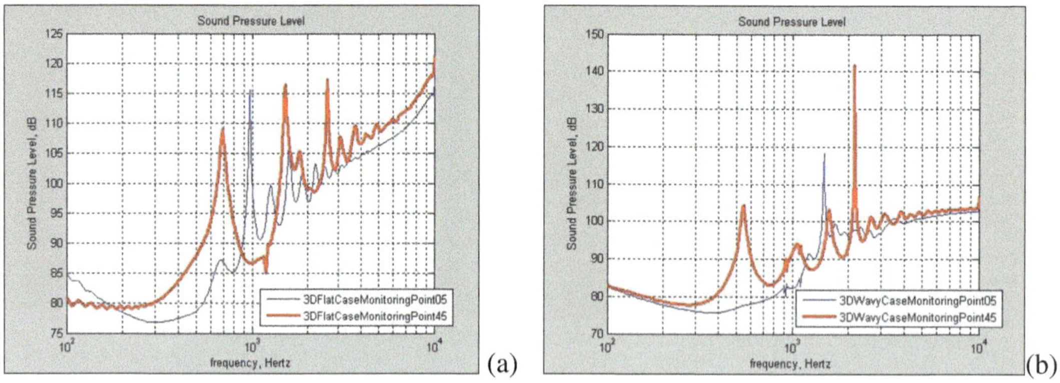

Figure 22: Comparison of points 05/45, SPL spectrograms for the wavy plate (a) and flat plate (b)

Comparison of SPL spectrograms for points 5 and 65 showed that at the resonant tone vanished point 65, even so the overall SPL (OASPL) seems to be increased. In the wavy case two tonal peaks are generated, probably due to the impingement of the shear layer at the proximity.

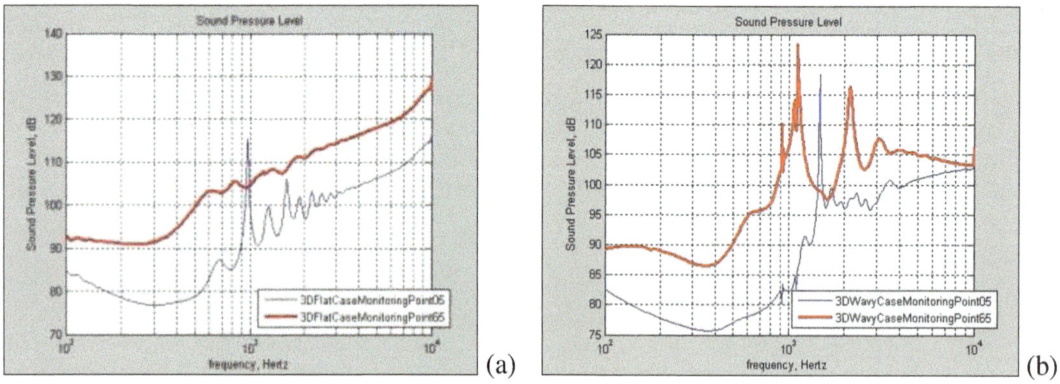

Figure 23: Comparison of points 5/65, SPL spectrograms for the wavy plate (a) and flat plate (b)

4.2.2. Monitoring point 15 Comparison

By opposition to monitoring 5, the comparisons of monitoring point 15 are pretty straight forward. The dissipation on the cavity floor results in almost constant SPL spectrogram characteristics along the width for each case as the following figure demonstrates. The only noticeable difference in a tonal peak increase at the point 55 located against the cavity side wall. The increases are by 2 and 5dB respectively for the flat and wavy case reaching a 106.7 dB, which is a negligible increase compare to the level reached at point 45.

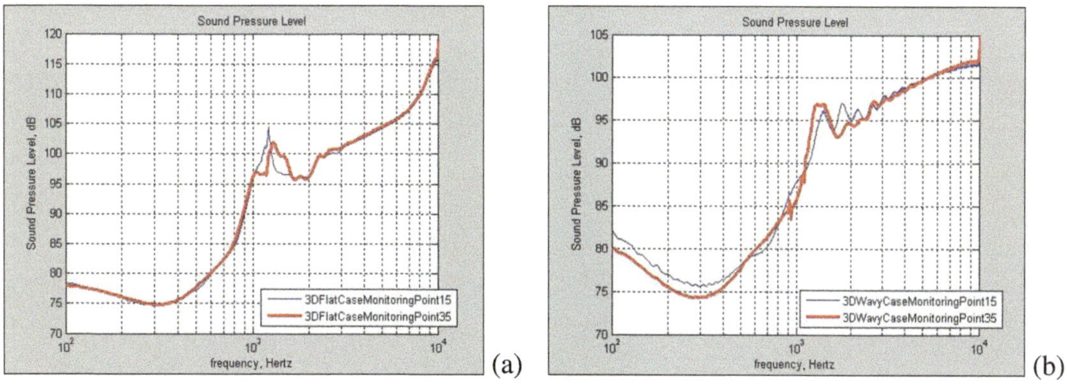

Figure 24: Comparison of points 15/35, SPL spectrograms for the wavy plate (a) and flat plate (b)

4.2.3. Analysis on modification effectiveness

The evaluation of the effectiveness of the control method used in this project is done by comparing the SPL spectrograms for the two points generating the maximum SPL values of the two cases studied. At point 5, the dominant tone which corresponds to the 1st Rossiter mode in the flat case has been shifted by 500Hz and reaches 118.3dB, representing a 3dB increase. However the overall SPL seems to have been decreased, decrease is also noticeable at point 15.

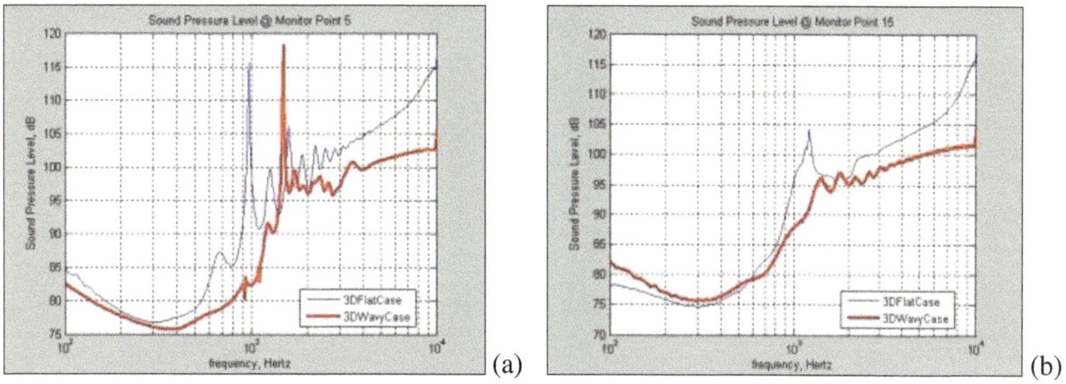

Figure 25: Comparison of flat and wavy 3D case, SPL spectrograms at point 5 and 15

As a conclusion, the optimum surface modification previously developed by the use of a two dimensional analysis which provided a reduction in tonal peak by up to 24dB does not achieve such results in three dimensions. In fact the three-dimensional analysis shows an increase of a couple decibels in peak tones. On the other hand an decrease in OASPL extracted for the majority of the monitoring points.

4.3. Flow visualisation

Figures 24a-h show the flow structures generated by both cases. The reference case (flat case) is located on the left hand side and the wavy case is located on the right hand side.

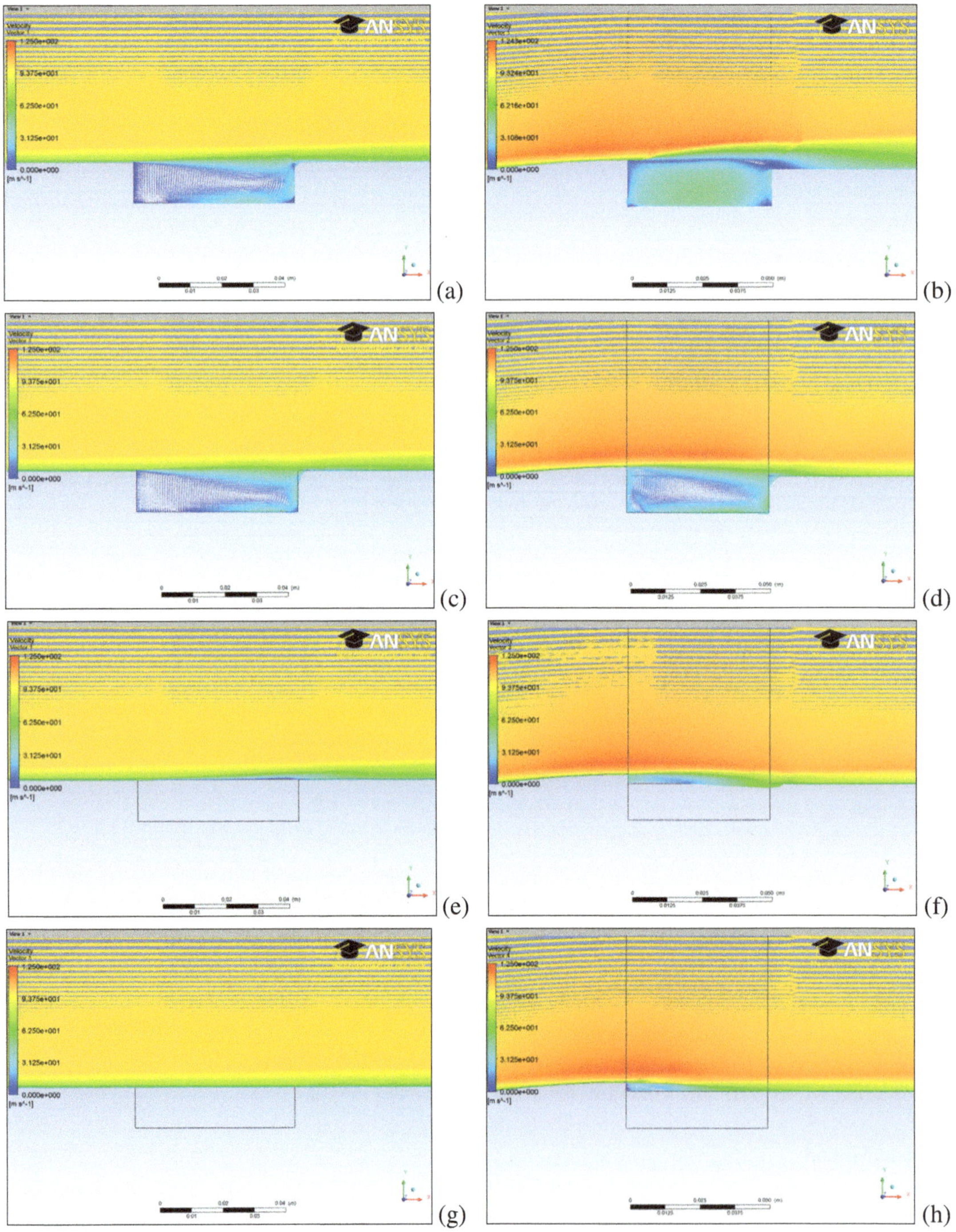

Figures 26 (a-h): Velocity Vectors at 10ms for both Flat Case (left column) and Wavy Case (right column) located respectively from top to bottom at the cavity central line, half-way between central line and side wall, the side wall and at 12.7 mm to the side-wall

The large rear vortex exhibited by Figures 26a, c and d proves the presence that both cases experience an open flow. One of the major differences between the two cases is the fact that the applied surface waviness accelerated the flow ahead of the cavity by up to 24.8m.s^{-1}. Additionally the control device achieved to move up the rear wall the impingement of the shear layer at the level of the central line as shown in Figure 26a and in Figure 26b. Such a displacement in the location should theoretically reduce the feedback mechanism, However this might only be due to the 3.25mm difference in height between the front and rear wall.

Figures 26e and 26g display the flow at respectively at the level span wise of the cavity sidewall and at 12.7mm away from the cavity in the baseline case. In both cases a light thickening of the boundary layer downstream the cavity is noticeable.

Finally Figure 26f and Figure 26h show the flow at respectively at the level span wise of the cavity sidewall and at 12.7mm away from the cavity in the baseline case. This figure exhibit the presence of a recirculation located just at the corner of the 3.25mm step generated by the surface modification. In addition to this a low velocity flow tangential to the free stream flow can be observed at the proximity of the step, as shown in Figure 27. This flow is exiting the cavity and could be increasing vortices at the proximity of the cavity.

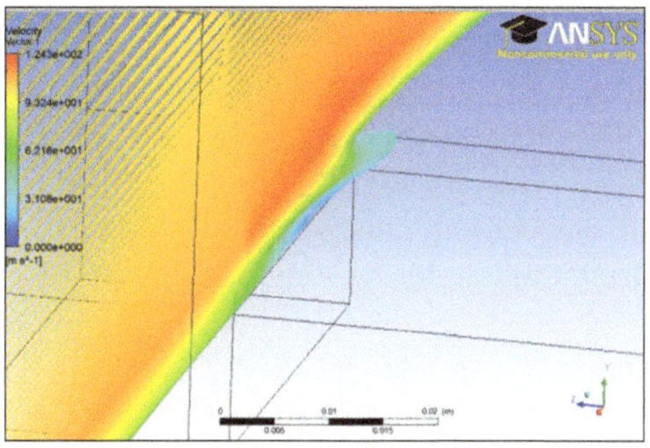

Figure 27: Tangential Flow

CONCLUSIONS

A complete three dimensional analysis of both the baseline and the modified configuration were carried out. Pressure fluctuations were extracted from a total of 70 points for each case and the resulting SPL spectrograms were validated by the previous 2D study. Additionally the flow regime was demonstrated with success by the use of CFD flow visualisation.

On the other hand the two-dimensionally optimised sinusoidal surface modification did not achieve a similar attenuation level in three dimensions. The overall sound pressure level seems to be reducing with the surface modification at the majority of the points investigated. However the dominant peak was increased by 3dB and reached 118.3 dB at the centre of the cavity. Additionally the strongest peak observed occurred on the sidewall of the cavity and has a high value of 141.8dB.

The 3.25 mm step appears to be the aspect of the design of the surface modification generating the most adverse effects. The presence of this step prevents the shear flow from bridging the cavity completely resulting on flow going out of the cavity at the proximity of the front wall and increasing the vortices at the proximity of the side wall. Therefore the two locations which should be focused on in further research are the side wall and the ending of the surface waviness.

To fully understand the wavy surface impact on the cavity flow control, the following further work is required:

- Improving the accuracy at the proximity of the cavity side walls and around the central line with the use of an inflation factor for the extrusion of the 3D meshes
- Increasing the mesh density and study the influence on accuracy of the results
- Investigating the application of the surface modification on only the width of the cavity
- Improving issues generated by the step by either:
    - Incorporating a 3.25mm step to the flat case
    - Investigating an application of the surface modification shifted 3.25mm downward
- Wind Tunnel testing of an optimum 3D model

ACKNOWLEDGMENTS

I would like to thank Simon J-F.C. Denis for his valuable contributions in the acoustic processing of the CFD data.

REFERENCES

1. Hughes G., Mamo, T. and Dala L., Use of Active Surface Waviness for Control of Cavity Acoustics in Subsonic Flows, 15th AIAA/CEAS Aeroacoustics Conference (30th AIAA Aeroacoustics Conference), Miami, Florida, AIAA-2009-3202, 2009.
2. Hughes G., A Computational Fluid Dynamic Study of Self-Sustaining oscillations Present in Subsonic Open Cavity Flow and their Subsequent Control, MEng Aerospace Systems Engineering, University of the west of England, May 2008.
3. ACARE Strategic Research Agenda Vol. 1 & 2 [online agenda], URL: http://www.acare4europe.org/html/documentation.asp [accessed 12th December 2009].
4. Charwat, A.F., Roos, J.N., Dewey, C.F. and Hitz, J.A., An Investigation of Separated Flows - Part I: The Pressure Field, Journal of Aerospace Sciences, Vol. 28, No. 6, 1961, pp.457 – 470.
5. Henderson, J., Investigation of Cavity Flow Aerodynamics Using Computational Fluid Dynamics, PhD Dissertation, University of Glasgow, 2001.
6. Maull, D.J., and East, L.F., Three-Dimensional Flow in Cavities, Journal of Fluid Mechanics, Vol. 10, No. 4, 1963, pp. 620 - 632.
7. Czech, M., The Acoustics and Aerodynamics of Turbulent Flow Over Yawed Rectangular Cavities, PhD Dissertation, University of Surrey, 2000.
8. Zhang, J., Morishita, E., Okunuki, T. and Itoh, H., Experimental and Computational Investigation of Supersonic Cavity Flows, AIAA/NAL-NASDA-ISAS 10th International Space Planes and Hypersonic Systems And Technologies Conference, Kyoto, Japan, AIAA-2001-1755, 2001.
9. Plentovich, E.B., Stallings, R.L. and Tracy, M.B., Experimental Cavity Pressure Measurements at Subsonic and Transonic Speeds, NASA Technical Paper. 3358, 1993.
10. ESDU, Aerodynamics and Aero-Acoustics of Rectangular Planform Cavities, Part I: Time Averaged Flow, ESDU data sheet 02008, 2004.
11. Baysal, O., Srinivasan, S. and Stallings, R.L., Unsteady Viscous Calculations of Supersonic Flows Past Deep and Shallow Three-Dimensional Cavities, AIAA-88-0101, 1988.
12. Rizzeta, D.P., Numerical Simulation of Supersonic flows over Three-Dimensional Cavity, AIAA Journal, Vol. 26, No. 7, 1988, pp.799-807.

13. Srinivasan, S., Baysal, O. and Plentovich, E.B., Navier-Stokes Calculations of Transonic Flows Past Open and Transitional Cavities, Advances and Applications in Computational Fluid Dynamics Symposium, O. Baysal, ed., FED.Vol. 66, ASME, 1988, pp. 169–179.
14. Suhs, N.E., Computations of Three-Dimensional Cavity Flow at Subsonic and Supersonic Mach Numbers, AIAA 19th Fluid Dynamics, Plasma Dynamics and Lasers Conference, Honolulu, Hawaii, AIAA-87-1208, 1987.
15. Tracy, M.B. and Plentovitch, E.B., Characterisation of Cavity Flow Fields Using Pressure Data Obtained in the Langley 0.3-meter Transonic Cryogenic Tunnel, NASA Technical Memorandum 4436, 1993
16. Naudascher, E., From Flow Stability to Flow-Induced Excitation, Journal of the Hydraulics Division, Vol. 93, No. 4, 1967, pp. 15-40.
17. ESDU, Aerodynamic and Aero-Acoustics of Rectangular Planform Cavities, Part II Unsteady Flow and Aero-Acoustics, ESDU data sheet 04023, 2005.
18. Rockwell, D. and Naudascher, E., Review - Self-Sustaining Oscillations of Flow Past Cavities, Journal of Fluids Engineering, Vol. 100, No.2, 1978, pp.152-165.
19. Rossiter, J.E., Wind-tunnel experiments on the flow over rectangular cavities at subsonic and transonic speeds, Aeronautical Research Council Reports and Memoranda No. 3438, 1964.
20. Dix, R.E. and Bauer, R.C., Experimental and Theoretical Study of Cavity Acoustics, AEDC-TR-99-4 (AD-A384010), 2000.
21. Heller, H.H., Holmes, D.G. and Covert, E.E., Flow-Induced Pressure Oscillations in Shallow Cavities, Journal of Sound and Vibration, Vol. 18, No.4, 1970, pp.545-553.
22. Debiasi, M. and Samimy, M., Logic-Based Active Control of Subsonic Cavity Flow Resonance, AIAA Journal, Vol. 42, No. 9, 2004, pp.1900-1909.

Spacecraft Electric Propulsion – A review

Vatsal Sheth
Undergraduate student
Mechanical Engineering Dept.
Dwarkadas J. Sanghvi College of Engineering, Mumbai
vatsal.sheth@djsce.edu.in

Abstract

A short review of the status of electric propulsion (EP) is presented to serve as an introduction to the more specialized technical papers. The principles of operation and the several types of thrusters that are either operational or in advanced development are discussed. The stimulus for development of electrically driven space propulsion systems is nothing less fundamental than Newton's laws of dynamics. Since a rocket propelled spacecraft in free flight derives its only acceleration from discharge of propellant mass, its equation of motion follows directly from conservation of the total momentum of the spacecraft and its exhaust stream.

Keywords: Electric propulsion, thrusters, rocket, propellant, spacecraft

1. Introduction:

Electric propulsion is a technology aimed at achieving thrust with high exhaust velocities, which results in a reduction in the amount of propellant required for a given space mission or application compared to other conventional propulsion methods. Reduced propellant mass can significantly decrease the launch mass of a spacecraft or satellite, leading to lower costs from the use of smaller launch vehicles to deliver a desired mass into a given orbit or to a deep-space target. In general, electric propulsion (EP) encompasses any propulsion technology in which electricity is used to increase the propellant exhaust velocity. Electric thrusters typically use much less propellant than chemical rockets because they have a higher exhaust speed (operate at a higher specific impulse) than chemical rockets. Due to limited electric power the thrust is much weaker compared to chemical rockets, but electric propulsion can provide a small thrust for a long time. Electric propulsion can achieve high speeds over long periods and thus can work better than chemical rockets for some deep space missions.

History:

The idea of electric propulsion for spacecraft dates back to 1911, introduced in a publication by Konstantin Tsiolkovsky. Earlier, Robert Goddard had noted such a possibility in his personal notebook. The first in-space demonstration of electric propulsion was an ion engine carried on board the SERT-1 (Space Electric Rocket Test) spacecraft, launched on 20 July 1964. The Ion Auxiliary Propulsion System (IAPS) project from 1974 to 1983 developed an 8-centimeter mercury Ion Propulsion System (IPS) for satellite station keeping. The NASA Solar Technology Application Readiness (NSTAR) project developed a 30-centimeter IPS that was used as the main propulsion on the Deep Space 1 (DS1) spacecraft from 1998 to 2001. DS1 was the first use of electric propulsion for spacecraft main propulsion. The NSTAR thruster on DS1 propelled the spacecraft 263,179,600 kilometres (163,532,236 miles) at speeds up to 4,500 meters per second (10,066 mph). Over the entire mission, the NSTAR thruster demonstrated 200 starts and 16,246 hours of operation.

2. Components:

The ion propulsion system (IPS) consists of five main parts: the power source, power processing unit (PPU), propellant management system (PMS), the control computer, and the ion thruster. The IPS power source can be any source of electrical power, but solar and nuclear are the primary options. A solar electric propulsion system (SEP) uses sunlight and solar cells for power generation. A nuclear electric propulsion system (NEP) uses a nuclear heat source coupled to an electric generator. The PPU converts the electrical power generated by the power source into the power required for each component of the ion thruster. It generates the voltages required by the ion optics and discharge chamber and the high currents required for the hollow cathodes. The PMS controls the propellant flow from the propellant tank to the thruster and hollow cathodes. Modern PMS units have evolved to a level of sophisticated design that no longer requires moving parts. The control computer controls and monitors system performance. The ion thruster then processes the propellant and power to perform work. Modern ion thrusters are capable of propelling a spacecraft up to 90,000 meters per second (over 200,000 miles per hour (mph). To put that into perspective, the space shuttle is capable of a top speed of around 18,000 mph. The tradeoff for this high top speed is low thrust (or low acceleration). Thrust is the force that the thruster applies to the spacecraft. Modern ion thrusters can deliver up to 0.5 Newtons (0.1 pounds) of thrust, which is equivalent to the force you would feel by holding nine U.S. quarters in your hand. To compensate for low thrust, the ion thruster must be operated for a long time for the spacecraft to reach its top speed. Ion thrusters use inert gas for propellant, eliminating the risk of explosions associated with chemical propulsion. The usual propellant is xenon, but other gases such as krypton and argon may be used.

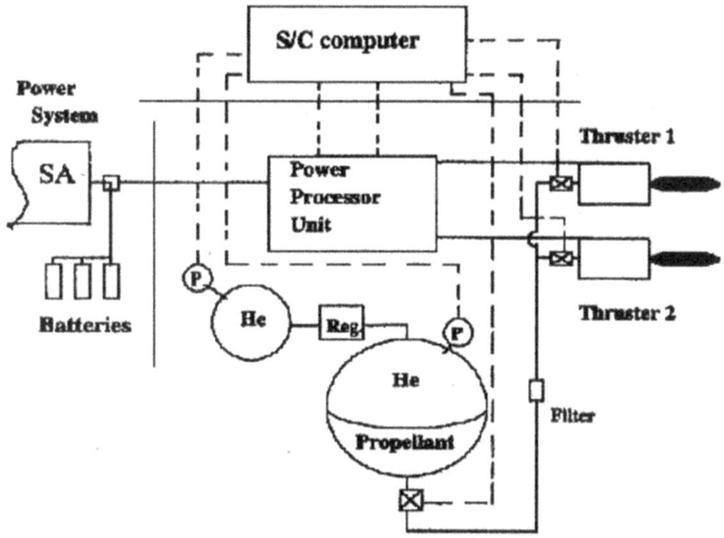

Figure 1: Schematic of a typical EP system

3. Conceptual Subdivision:

Historically, conceptually, and pragmatically, this field has tended to subdivide in the following categories:

1. **Ion/plasma drives:** This type of rocket-like reaction engine uses electric energy to obtain thrust from propellant carried with the vehicle. They don't necessarily have a rocket nozzle.

i) *Electrothermal propulsion*, wherein the propellant is heated by some electrical process, then expanded through a suitable nozzle

ii) *Electrostatic propulsion*, wherein the propellant is accelerated by direct application of electrostatic forces to ionized particles

iii) *Electromagnetic propulsion*, wherein the propellant is accelerated under the combined action of electric and magnetic fields

2. Non-ion drives:

i) *Electrodynamic tethers(EDTs)*, are long conducting wires, such as one deployed from a tether satellite, which can operate on electromagnetic principles as generators, by converting their kinetic energy to electrical energy, or as motors, converting electrical energy to kinetic energy.

Over their periods of development, each of these approaches has spawned its own array of technical specialties and subspecialties, its own balance sheet of advantages and limitations, and its own cadres of proponents and detractors, but in serious assessment, each has validly qualified for particular niches of application, many of which do not seriously overlap. Throughout the history of EP development, the original subdivision of the field into electrothermal, electrostatic, and electromagnetic systems has remained useful, and this subdivision will be respected through the balance of this article. It should be recognized, however, that in virtually all practical systems, two or even all three of these processes function in some concert to accelerate, channel, and expand the propellant flow, and in many cases it is the efficacy of this cooperation that determines the utility of any given device.

Table 1: Ion/plasma drives

	Electrothermal	Electrostatic	Electromagnetic
Acceleration force	Pressure, ∇p Electrically heat propellant and use nozzle expansion	Electrostatic, $\vec{F}_e$ Static Electric field alone accelerates charged particles	Lorentz, $\vec{F}_m$ Magnetic and Electric fields accelerate charged particles
Specific Impulse(Isp)	300-1,500	2000-100,000+	1,000-10,000
$\frac{\text{Thrust}}{\text{Weight}}$	$<10^{-3}$	$<10^{-4}$	$<10^{-4}$-10^{-6}

Electrothermal Propulsion:

The electrothermal category groups the devices where electromagnetic fields are used to generate plasma to increase the temperature of the bulk propellant. The thermal energy imparted to the propellant gas is then converted into kinetic energy by a nozzle of either solid material or magnetic fields. Low molecular weight gases (e.g. hydrogen, helium, ammonia) are preferred propellants for this kind of system. An electrothermal engine uses a nozzle to convert the heat of a gas into the linear motion of its molecules so it is a true rocket even though the energy producing the heat comes from an external source.

1. Resistojets
2. Arcjets

Resistojets:

A resistojet is a method of spacecraft electric propulsion that provides thrust by heating a (typically non-reactive) fluid. Heating is usually achieved by sending electricity through a resistor consisting of a hot incandescent filament, with the expanded gas expelled through a conventional nozzle. Resistojets have been flown in space since 1965 on board military Vela satellites, however they became used in commercial applications in 1980 with launch of first satellites in the INTELSAT-V program. Nowadays resistojet propulsion is used for orbit insertion, attitude control, and deorbit of LEO satellites, including satellites in the Iridium satellite constellation and do well in situations where energy is much more plentiful than mass, and where propulsion efficiency needs to be reasonably high but low thrust is acceptable.

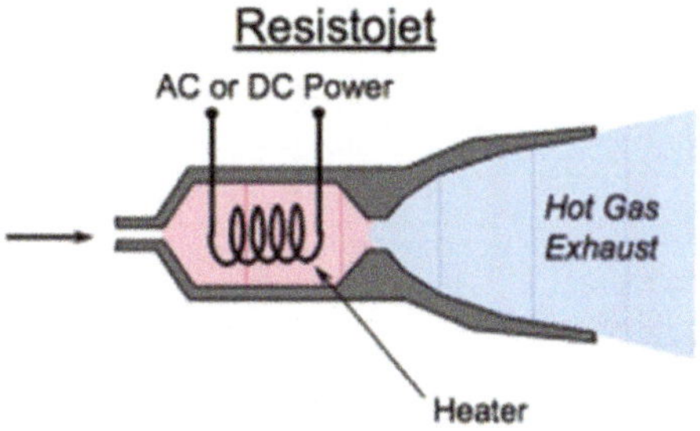

Figure 2: Resistojet

Resistojets operate by passing the gaseous propellant around an electrical heater (which could be the inside of tubes heated radiatively from the outside), then using a conventional nozzle to generate thrust. The heating reduces the gas flow rate from a given upstream pressure through a given nozzle area, thus leading to the familiar increase in specific impulse as$\sqrt{T}$. Nearly any gas could be used (as long as it is compatible with the high-temperature heater), and this may be dictated by considerations such as waste disposal on manned S/C. The most successful application has been based on the superheating of catalytically decomposed hydrazine, which has the advantage of commonality with familiar fuel systems used in hydrazine monopropellant applications. The heaters can operate over the wide pressure range encountered with blowdown systems, and their input voltage is low enough to require no special power conditioning, except for current surge protection. An exception to this would be an S/C power system that would allow more than about 20% voltage variations, in which case a dedicated regulator would become necessary. Operation can continue in a nonsuperheated mode in case of heater failure. The plume is not ionized and poses no unusual S/C interaction problems. Because the molecular mass of the gas (N2/H2/NH3) is relatively high, and because the heating wall is limited by materials (W–Re or something similar) to about 2000 K, the specific impulse (Isp) achieved is only modest, of the order of 300 – 310 s. This is 40% better than that without superheating, and the improvement comes at a very small cost in complexity, if power is available. A favourable situation (for this and also for other EP techniques) occurs in geostationary communications satellites, in which excess power is indeed available most of the time (Sec. V), and this prompted the early commercial introduction of hydrazine resistojets (starting with Intelsat V, 1980) for the north– south station keeping (NSSK) function. A more recent application is for orbit insertion, control, and deorbit of the Iridium low Earth orbit (LEO) constellation. One of the few technical problems posed by these thrusters is the tendency of the hydrazine to produce non-volatile deposits at the hot inlet to the catalytic chamber; this is common to all hydrazine thrusters, but the problem is made more critical by the reduction in flow rate because of the higher Isp. Solutions have included the use of ultrapure hydrazine and thermal shunts to reduce heat flux at the critical points. Ammonia resistojets have also been used for higher specific impulse (lighter gas), at some cost in complexity.

Arcjets:

Arcjets are a form of electrically powered spacecraft propulsion, in which an electrical discharge (arc) is created in a flow of propellant (typically hydrazine or ammonia). This imparts additional energy to the propellant, so that one can extract more work out of each kilogram of propellant, at the expense of increased power consumption and (usually) higher cost. Also, the thrust levels available from typically used arcjet engines are very low compared with chemical engines. When the energy is available, arcjets are well suited to keeping stations in orbit and can replace monopropellant rockets.

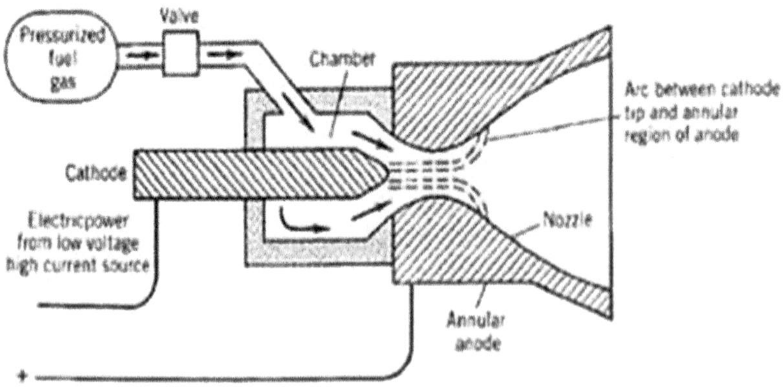

Figure 3: Arcjet

Arcjets are, like resistojets, electrothermal devices, but the wall temperature limitation of the resistojet is overcome here by depositing power internally, in the form of an electric arc, typically between a concentric upstream rod cathode and a downstream anode that also serves as the supersonic nozzle. The flow structure at the throat is extremely non uniform, with the arc core at temperatures of 10,000– 20,000 K, and the buffer layer near the wall at no more than 2000 K. Because of this there is practically no flow through the arc core, which can be thought of as an effective fluid plug; this reduces the flow, without reducing the pressure integral, and leads to the high specific impulse. On the other hand, some intrinsic loss mechanisms are now introduced.
1) Compared with a uniformly heated stream, the nonuniformity reduces by itself the propulsive efficiency because, as in any thermal propulsion device, maximum thrust for a given power and flow is obtained when the heat is added uniformly across the flow.
2) The power invested in ionizing the arc gas is mostly lost because of the small recombination time available (in addition, in molecular gases, there is a substantial dissociation loss as well).
3) There are near-electrode voltage drops, which mainly constitute a local heat loss to the electrodes.

Electrostatic Propulsion:

If the acceleration is caused mainly by the Coulomb force (i.e. application of a static electric field in the direction of the acceleration) the device is considered electrostatic.

1. Electrostatic ion thrusters
2. Colloid ion thrusters

Its key principle is that a voltage difference between two conductors sets up an electrostatic potential difference that can accelerate ions to produce thrust. The ions must, of course, be neutralized--often by electrons emitted from a hot filament. The three main stages of an ion-thruster design are ion production, acceleration, and neutralization.

Electrostatic Ion Thrusters:

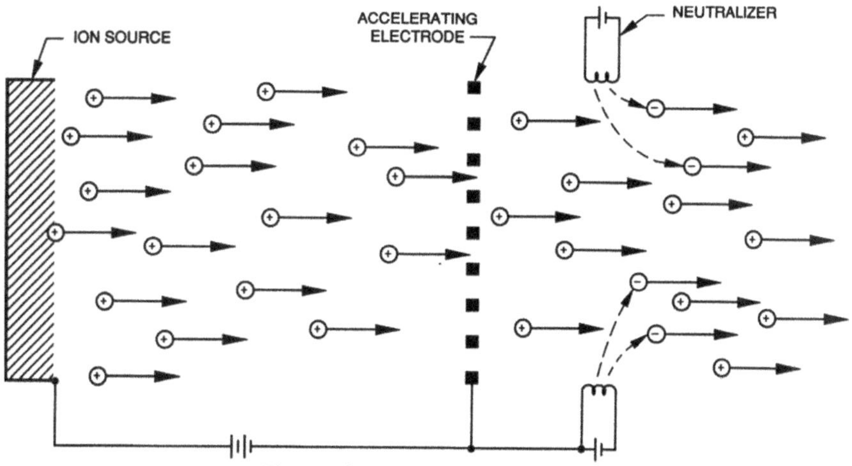

Figure 4: Ion thruster schematic

a) Ion Sources:

In practice, the most amenable propellants for electrostatic thrusters have proven to be cesium, mercury, argon, krypton, and most commonly xenon, and many possible sources of such ions of the requisite efficiency, reliability, and uniformity have been conceived and developed. Of these, only three, the electron bombardment discharge source, the cesium-tungsten surface contact ionization source, and one form of RF discharge source, have survived to application. The essential elements of the bombardment sources are some form of cylindrical discharge chamber containing a centerline cathode that emits electrons, a surrounding anode shell, and a permeating azimuthal and radial magnetic field that constrains the electrons to gyrate within the chamber long enough to ionize the injected propellant gas and to direct it, once ionized, to extractor and accelerator grids downstream. This particular device employs a hollow cathode electron source, wherein is sustained a secondary discharge that facilitates electron emission from the interior walls of the cathode cavity. The magnetic field permeating the entire chamber is provided by three ring magnets, empirically configured to establish a grossly diverging but doubly cusped field pattern that optimizes the discharge for ionization and ion extraction purposes. The magnitude of this field is adjusted in concert with the anode-cathode voltage differential to maximize the ionization efficiency and discharge stability while minimizing the production of doubly charged ions, which would be out of focus in the accelerator gap and thus tend to erode the grids through high-energy sputtering. Typical values for xenon and mercury propellants would be in the regimes of 0.25 T and 30 V, respectively. Slightly different chamber configurations and field values have also been used successfully.

b) Accelerator Grids:

In virtually all classes of ion thruster, the positive ions are extracted from the source and accelerated downstream by a system of grids configured to achieve the desired exhaust velocity with minimum beam impingement. In U.S. bombardment engines, for example, a double grid configuration is usually dished downstream to improve its mechanical and thermal stability against distortion. The upstream grid is maintained at a higher positive potential than required by the desired exhaust speed in order to enhance the ion extraction process and increase the space-charge limited current density that can be sustained. The downstream grid then reduces the exhaust plane potential to the desired value. This 'accel-decel' scheme has the advantages of higher beam density at a given net voltage and of reducing electron back streaming from the neutralized beam downstream. The grid perforations are configured analytically and empirically to focus the ion stream into an array of beamlets that pass through with minimum impingement. In this process, the downstream surface of the discharge plasma in the chamber acts as a third electrode, and since this contour is not independent of the discharge characteristics and applied grid voltages, it can be a source of some instability. Further complications are introduced by the small fractions of double ions or neutrals that find their way into the beam and are henceforth out of focus and free to bombard the grid surfaces.

c) Neutralizers:

If the ion beam emerging from the downstream electrode is not to stall on its own interior potential profile, it must be electrostatically neutralized within a very few units of grid spacing. This is typically achieved by provision of a flux of electrons, usually from another hollow cathode discharge, which fortuitously mix effectively within the ion beam by means of a variety of microscopic and macroscopic internal scattering processes. Once so neutralized, this plasma constitutes a downstream 'virtual electrode' that completes the axial potential pattern.

Colloid Ion Thrusters:

A colloid thruster is a type of thruster which uses electrostatic acceleration of charged liquid droplets for propulsion. It is closely related to electrospray ionization and other hydrodynamic spraying processes. In a colloid thruster charged liquid droplets are produced by an electrospray process and then accelerated by a static electric field. The liquid used for this application tends to be a low volatility ionic liquid. Like other ion thrusters its benefits include high efficiency, thrust density, and specific impulse; however it has very low total thrust, on the order of micronewtons. It provides very fine attitude control or efficient acceleration of small spacecraft over long periods of time.

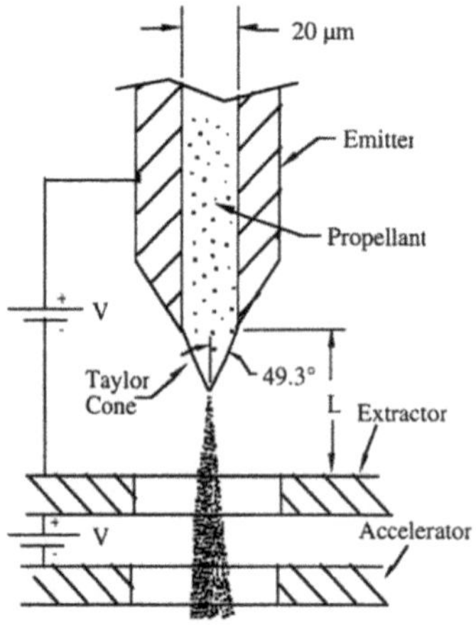

Figure 5: Colloid thruster schematic

The original impulse for the development of this type of thruster in the 1960s was the need for thrusters that would yield higher thrust density and would be efficient at lower specific impulse than ion engines. The principle is analogous to that in FEEP, except that a nonmetallic liquid is used, and submicron-sized charged droplets are extracted rather than individual ions. Intensive NASA-led efforts at that time were frustrated by the inability to increase the charge/mass ratio of the extracted droplets sufficiently, which forced the use of very high voltages (several tens to hundreds of kV) to reach the desired specific impulse range (around 1000 s). There was also difficulty in obtaining sprays with nearly uniform values of the charge/mass ratio, which led to poor beam focusing. With continued development of the charged spray technology for applications such as printing and paint application, some of these limits have been relaxed. Russian work20 has led to the development of 65% efficient colloidal engines in the 1000-s specific impulse range that operate with voltages of 15 – 25 kV. This type of thruster would appear to be a leading contender for micro propulsion applications requiring high D*V* at moderately high thrust /power ratios.

Electromagnetic Propulsion:

If ions are accelerated either by the Lorentz force or by the effect of electromagnetic fields where the electric field is not in the direction of the acceleration, the device is considered electromagnetic.

1. Hall Effect Thruster
2. Magnetoplasmadynamic (MPD) thruster
3. Pulsed Plasma Thruster

Such systems can produce exhaust speeds considerably higher than those of the electrothermal devices, and thrust densities much larger than those of the electrostatic thrusters, but are phenomenologically more complex and analytically less tractable than either of these alternatives. The essence of an electromagnetic thruster is sketched, where some electrically conducting fluid, usually a highly ionized gas, is subjected to an electric field $\boldsymbol{E}$ and a magnetic field $\boldsymbol{B}$, perpendicular to each other and to the fluid velocity $\boldsymbol{u}$. The current density $\boldsymbol{j}$ driven by the electric field interacts with $\boldsymbol{B}$ to provide a streamwise body force $\boldsymbol{f} = \boldsymbol{j} \times \boldsymbol{B}$ that accelerates the fluid along the channel. The process may alternatively be represented from a particulate point of view in terms of the mean trajectories of the current-carrying electrons, which, in attempting to follow the electric field, are turned downstream by the magnetic field, transmitting their streamwise momentum to the heavy particles in the stream by collisions and/or by microscopic polarization fields. It is important to note that in either representation, the working fluid, although highly ionized, is macroscopically neutral, hence not constrained in its mass flow density by space-charge limitations as in the electrostatic accelerators.

Hall Effect Thruster:

In spacecraft propulsion, a Hall thruster is a type of ion thruster in which the propellant is accelerated by an electric field. Hall thrusters trap electrons in a magnetic field and then use the electrons to ionize propellant, efficiently accelerate the ions to produce thrust, and neutralize the ions in the plume. Hall thrusters are sometimes referred to as Hall Effect thrusters or Hall current thrusters. Hall thrusters are often regarded as a moderate specific impulse (1,600 s) space propulsion technology. The Hall Effect thruster has benefited from considerable theoretical and experimental research since the 1960s. 6 kW Hall thruster in operation at the NASA Jet Propulsion Laboratory. Hall thrusters operate on a variety of propellants, the most common being xenon. Other propellants of interest include krypton, argon, bismuth, iodine, magnesium, and zinc. Hall thrusters are able to accelerate their exhaust to speeds between 10–80 km/s (1,000–8,000 s specific impulse), with most models operating between 15–30 km/s (1,500–3,000 s specific impulse). The thrust produced by a Hall thruster varies depending on the power level. Devices operating at 1.35 kW produce about 83 mN of thrust. High power models have demonstrated up to 3 N in the laboratory. Power levels up to 100 kW have been demonstrated by xenon Hall thrusters.

Today's Hall thrusters are sometimes referred to as 'closed-electron-drift' devices, given the azimuthal drift electrons that is common to all present variants of such thrusters. The most common versions are the stationary plasma thruster (SPT) (also termed the 'magnetic layer thruster') and the anode layer thruster (ALT).
The former differs from the latter by its extended channel, the use of insulator chamber walls, and the extent of the quasineutral acceleration region, but both rely on the same basic principles for ionizing and accelerating the propellant.

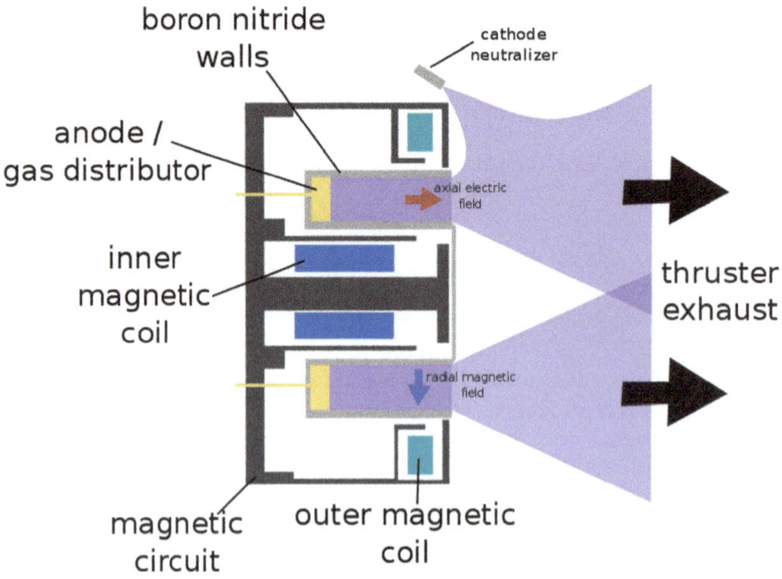

Figure 6: Hall Effect Thruster

Electrons from the cathode enter the chamber and are subjected to an azimuthal drift in the crossed radial magnetic and axial electric fields, wherein they undergo ionizing collisions with the neutral propellant atoms (typically xenon) injected through the anode. While the magnetic field strength is sufficient to lock the electrons in an azimuthal drift, it does not significantly affect the trajectory of the ions, which are directly accelerated by the axial electric field. An axial electron flux equal to that of the ions reaches the anode due to a cross field mobility that often exceeds classical values, and the same flux of electrons is available from the cathode to neutralize the exhausted ions. Quasi-neutrality is thus maintained throughout the chamber and exhaust beam, and consequently no spacecharge limitation is imposed on the acceleration, which allows relatively high thrust densities compared with those of conventional electrostatic propulsion devices. Nominal operating conditions of a common flight module (e.g., the Russian SPT-100) operating with xenon are a 2- 5mg/sec mass flow rate; a 200- to 300-V applied voltage, yielding a plasma exhaust velocity of 16,000 m/sec; and a thrust of 40-80 mN, at efficiencies of about 50%.

Magnetoplasmadynamic (MPD) Thruster:

In MPD thrusters, a current along a conducting bar creates an azimuthal magnetic field that interacts with the current of an arc that runs from the point of the bar to a conducting wall. The resulting Lorentz force has two components:

Pumping: a radially inward force that constricts the flow.
Blowing: a force along the axis that produces the directed thrust.

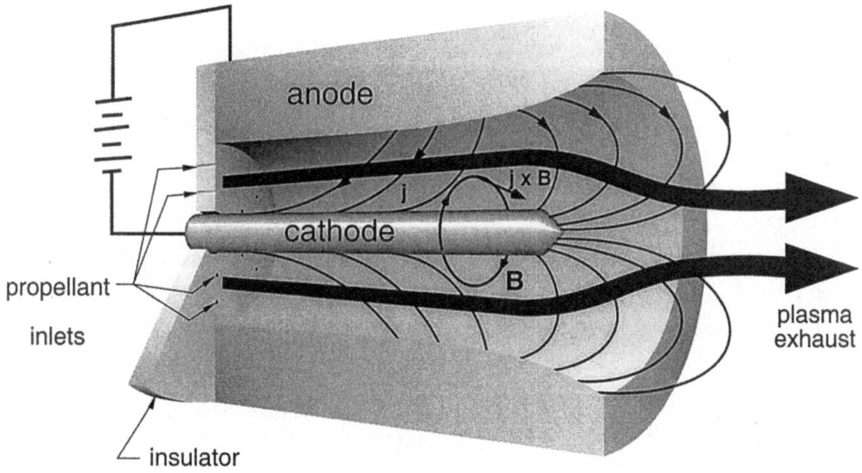

Figure 7: Magnetoplasmadynamic (MPD) Thruster Schematic

The magnetoplasmadynamic (MPD) thruster is characterized by a coaxial geometry constituted by a central cathode, an annular anode, and some form of interelectrode insulator. Gaseous propellants are introduced into the upstream portion of the channel, whereafter they are ionized by passage through an intense, azimuthally uniform electric arc standing in the interelectrode gap. If the arc current is high enough, its associated azimuthal magnetic field is sufficient to exert the desired axial and radial body forces on the propellant flow, directly accelerating it downstream and compressing it toward the centerline into an extremely hot plasma just beyond the cathode tip. Subsequent expansion of this plasma, along with the direct axial acceleration, yields the requisite exhaust velocity.

The MPDT has demonstrated its capability of providing specific impulses in the range of 1500-8000 sec with thrust efficiencies exceeding 40%. High efficiency (above 30%) is typically reached only at high power levels (above 100 kW); consequently, the steady-state version of the MPDT is regarded as a high-power propulsion option. When the thruster is operated below 200 kW, the self-induced magnetic field becomes only marginally sufficient to provide the desired body force, and external fields are frequently added to enhance performance in this range. However, in its megawatt versions, the self-field MPDT has the unique capability, among all developed electric thrusters, of processing very high power levels in a simple, compact, and robust device that can produce thrust densities as high as 10^5 N/m^2. These features have rendered the steady state MPDT particularly attractive for energetic deep-space missions requiring high thrust levels, such as piloted and cargo missions to Mars and the outer planets, as well as for nearer-term orbit raising missions.

Pulsed Plasma Thruster:

A Pulsed Plasma Thruster (PPT), also known as a plasma jet engine, is a form of electric spacecraft propulsion. PPTs are generally considered the simplest form of electric spacecraft propulsion and was the first form of electric propulsion to be flown in space, having flown on two Soviet probes (Zond 2 and Zond 3) starting in 1964. PPTs are generally flown on spacecraft with a surplus of electricity from abundantly available solar energy.

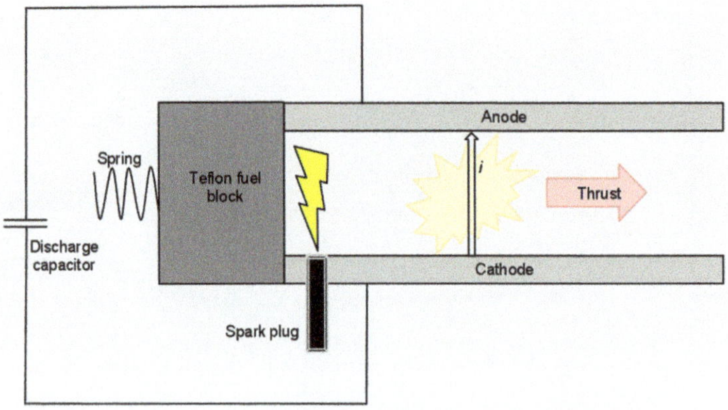

Figure 8: Schematic Layout of a Pulsed Plasma Thruster

Most PPTs use a solid material (normally PTFE, more commonly known as Teflon) for propellant, although a minority use liquid or gaseous propellants. The first stage in PPT operation involves an arc of electricity passing through the fuel, causing ablation and sublimation of the fuel. The heat generated by this arc causes the resultant gas to turn into plasma, thereby creating a charged gas cloud. Due to the force of the ablation, the plasma is propelled at low speed between two charged plates (an anode and cathode). Since the plasma is charged, the fuel effectively completes the circuit between the two plates, allowing a current to flow through the plasma. This flow of electrons generates a strong electromagnetic field which then exerts a Lorentz force on the plasma, accelerating the plasma out of the PPT exhaust at high velocity the pulsing occurs due to the time needed to recharge the plates following each burst of fuel, and the time between each arc. The frequency of pulsing is normally very high and so it generates an almost continuous and smooth thrust. While the thrust is very low, a PPT can operate continuously for extended periods of time, yielding a large final speed. The energy used in each pulse is stored in a capacitor. By varying the time between each capacitor discharge, the thrust and power draw of the PPT can be varied allowing versatile use of the system.

PPTs are very robust due to their inherently simple design (relative to other electric spacecraft propulsion techniques), and draw very little electrical power relative to other comparable thrusters. As an electric propulsion system, PPTs benefit from reduced fuel consumption compared to traditional chemical rockets, reducing launch mass and therefore launch costs, as well as high specific impulse improving performance. However, due to energy losses caused by late time ablation and rapid conductive heat transfer from the propellant to the rest of the spacecraft, propellant efficiency is very low compared to other forms of electric propulsion, at around just 10%.

Electrodynamic Tethers:

As part of a tether propulsion system, crafts can use long, strong conductors (though not all tethers are conductive) to change the orbits of spacecraft. It has the potential to make space travel significantly cheap. It is a simplified, very low-budget magnetic sail. It can be used either to accelerate or brake an orbiting spacecraft. When direct current is pumped through the tether, it exerts a Lorentz force against the magnetic field, and the tether accelerates the spacecraft. In 2012, the company Star Technology and Research was awarded a $1.9 million contract to qualify a tether propulsion system for orbital debris removal.

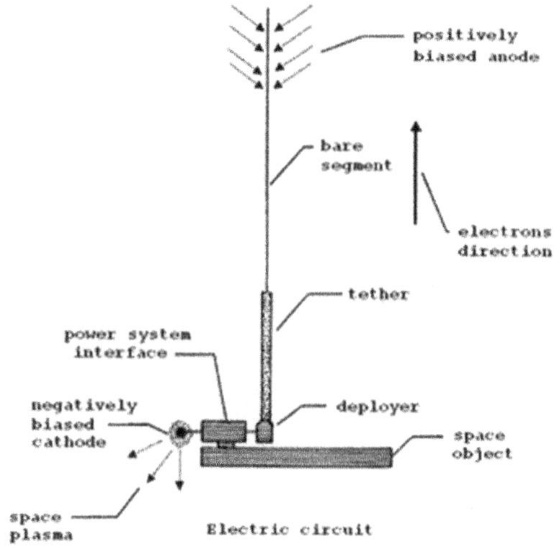

Figure 9: Electrodynamic Tether

Over the years, numerous applications for electrodynamic tethers have been identified for potential use in industry, government, and scientific exploration. The table below is a summary of some of the potential applications proposed thus far. Some of these applications are general concepts, while others are well-defined systems. Many of these concepts overlap into other areas; however, they are simply placed under the most appropriate heading for the purposes of this table. All of the applications mentioned in the table are elaborated upon in the Tethers Handbook. Three fundamental concepts that tethers possess are gravity gradients, momentum exchange, and electrodynamics. Potential tether applications can be seen below:

Table 2: Applications of Electrodynamic Tethers

ELECTRODYNAMICS	
Electrodynamic power generation	Electrodynamic thrust generation
ULF/ELF/VLF communication antenna	Radiation belt remediation
SPACE STATION	
Microgravity laboratory	Shuttle de-orbit from Space Station
Tethered Space Transfer Vehicle (STV) launch	Variable/low gravity laboratory
Attitude stabilization and control	ISS reboost
TRANSPORTATION	
Generalized momentum scavenging spent stages	Internal forces for orbital modification
Satellite boost from orbiter	Tether Assisted Transportation System (TATS)
Tether re-boosting of decaying satellites	Upper stage boost from Orbiter

Future of Electric Propulsion:

More and more companies are beginning to use satellites with electric propulsion to extend the operational life of satellites and reduce launch and operation costs. This produces savings that can be passed along to consumers.

NASA's primary application of ion propulsion will be for main propulsion on long missions that are difficult or impossible to perform using other types of propulsion. The Dawn spacecraft, scheduled for launch in May 2006, will use three NSTAR ion thrusters as main propulsion. The mission will study Ceres and Vesta, two protoplanets located in the asteroid belt that exists between Mars and Jupiter. By studying these protoplanets, which were among the first bodies formed in our solar system, researchers hope to gain valuable information about the solar system's early development.

The Jupiter Icy Moons Orbiter (JIMO) spacecraft will use an array of high-power ion thrusters as main propulsion. JIMO will perform an extensive exploration of Jupiter's icy moons Callisto, Ganymede, and Europa. The spacecraft will investigate each moon's composition, history, and potential for sustaining life.

Research in the area of ion propulsion continues to push the envelope of propulsion technology. Advancements are being made that allow the thrusters to operate at higher power levels, higher speeds, and for longer durations. PPU and PMS technologies are being developed that will allow NASA to build lighter and more compact systems while increasing reliability. As new power sources become available, higher power thrusters will be developed that provide greater speed and more thrust. Supporting technologies such as carbon-based ion optics and ECR discharges may greatly increase ion thruster operational life, enabling longer duration missions or high-power IPS operation. These technologies will allow humankind to explore the farthest reaches of our solar system.

4. Conclusion:

This paper has provided a condensed overview of the electric propulsion field, at a time when it is undergoing a very rapid transition from the laboratory to actual flight application. Inevitably, this means that many of the points made here will become obsolete in a short time. In the domain of interplanetary flight, however, EP offers much more substantial advantages over chemical systems, which extend in several important cases to enabling missions that simply could not be performed by means of any other reasonably projected propulsion technology. These include heavy cargo and/or piloted missions to Mars and the outer planets and many unpiloted probes beyond the solar system and out of the ecliptic plane. Finally, before any such ambitious EP missions can seriously be contemplated, nonsolar alternatives for high-power sources in space must be developed. For mostly political reasons, plans for deployment of nuclear high-power sources in space have so far failed to materialize, and consequently the use of electric thrusters for primary propulsion in energetic missions has had a cyclical history of false starts and disappointments. Indeed, the recent vigorous rejuvenation of the field of electric propulsion can be attributed, at least in part, to a conscious shift in emphasis away from the high-power missions envisaged during the 1960s and 1970s toward the less ambitious but more realistic power-limited small satellites of today. Now that many EP systems have entered the mainstream of astronautic technology, their role in helping to expand human ambition beyond the inner part of the solar system, although still dependent on the hitherto unrealized development of high-power sources, is perhaps on more credible ground.

References:

1. M. Martinez-Sanchez, J. E. Pollard, 1998, Spacecraft Electric Propulsion – An Overview, *Journal of Propulsion and Power,* Volume 14, Issue 5
2. Robert G. Jahn, Edgar Y. Choueiri, 2002, Electric Propulsion, *Encyclopaedia of Physical Science and Technology,* 3rd Edition, Volume 5
3. Jerry M. Seltzman, 2003-2004, Electric Rocket Propulsion, *AE 4451 Propulsion, Georgia Institute of Technology*
4. Roland Florenz, Thomas M. Liu, Alec D. Gallimore, Hani Kamhawi, Daniel L. Brown, Richard R. Hofer, James E. Polk, 2012, Electric Propulsion of a Different Class: The Challenges of Testing for Megawatt Missions, *48th AIAA/ASME/SAE/ASEE Joint Propulsion Conference & Exhibit*
5. www.jpl.nasa.gov/ep/
6. www.nasa.gov/centers/glenn/about/fs21grc.html

CFD and Real Time Analysis of a Symmetric Airfoil

Abhay Sharma[1], Manpreet Singh Sarwara[1], Harsimranjeet Singh[1], Lakshya Swarup[1], Rajeev Kamal Sharma[2]

[1]*UG students, School of Mechanical Engineering, Chitkara University, Punjab.*
[2]*Associate Professor, School of Mechanical Engineering, Chitkara University, Punjab.*
Author Correspondence: UG students, School of Mechanical Engineering, Chitkara University, Punjab.

Abstract

A NACA 0015 airfoil has been designed and fabricated as per NACA specification. The performance of the designed airfoil has been evaluated using CFD software and the same has been verified by studying its behaviour in a lab scale wind tunnel. The effect of angle of attack on the lift force, flow velocity profile, pressure profile, flow separation and wake formation has been studied. The airfoil has been tested at 0°, 5° 10° and 15° angle of attack and it has been found that lift force, velocity and pressure difference between upper and lower surfaces increases with angle of attack. Extent of flow separation and intensity of wake formation also increases with increase in angle of attack.

Keywords: Symmetric airfoil; real time analysis; CFD analysis; angle of attack.

1. Introduction

In the recent past computational fluid dynamics (CFD) has become popular among the researchers, as it fulfils the requirement for faster and accurate methods for predicting the various flow field parameters around the geometries of interest. CFD has found applications in the field of aerospace, automobiles, Marine and industrial processes and components where fluid flow place s a major role in assessing the performance. Flow over the geometries can not only be visualised but certain flow parameters which are difficult, impossible or expensive to measure experimentally can be approximated correctly by using suitable CFD packages.

Study of variation of various hydrodynamic parameters over an airfoil has been an area of research for long time. Before the inception of CFD the variations of hydrodynamic parameters during the flow over an airfoil was estimated by conducting experiments over airfoil in wind tunnels. During the flow over an airfoil, the flow splits and passes over the upper and lower surfaces of the airfoil. The shape of airfoil is such that the flow stretches over the upper surface resulting in decrease of pressure. On the other hand the flow over the lower portion of the airfoil moves with same speed and pressure resulting in an upward force called LIFT. As the flow progresses over the airfoil it experiences the drag force, in the direction of flow. The lift force experience by an airfoil is smaller in comparison to the weight of the airfoil in low Reynolds numbers. With the increase in Reynolds number the lift force also increases, and when lift force is greater than the gravitational force the airfoil starts moving in upward direction. As a Reynolds number of the flow increases the drag force over the airfoil also increases. Karna et. al. (2014) studied the lift and drag forces on NACA 0012 airfoil at zero degree angle of attack using CFD package. The flow was subsonic and it was observed that for such airfoil the lift force was zero at zero degree angle of attack. Variations of static and dynamic pressure over NACA 4412 airfoil has been studied by Parbhakar A. (2013) using CFD analysis at different angles of attack. Maximum dynamic pressure was found to occur at upper surface near and around maximum camber and minimum static pressure was found to occur at and around the same location. It was also concluded that with the increase in the angle of attack the stagnation point moves away from leading edge on the lower surface of the airfoil. Many other researchers have tried to model the flow around airfoil with different geometries using various CFD packages [Shankara MHM. et al (2013), Mevadiya M (2013)], but the CFD outcomes has not been validated with real time analysis.

This paper presents a real time and CFD analysis comparison for NACA 0015 airfoil. NACA 0015 has been designed using NACA generator. Real time analysis has been carried out in 600×270mm wind tunnel. CFD analysis has been done using Fluent 13.0. CFD analysis shows good agreement with real time analysis.

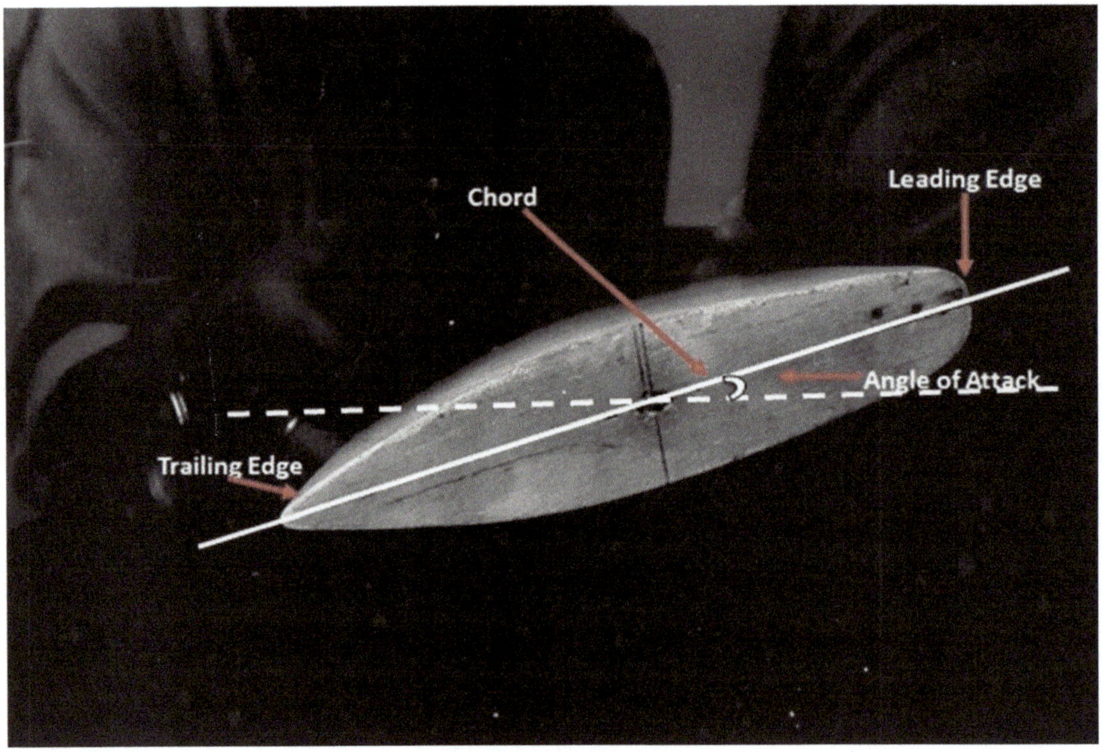

Figure 1 Basic nomenclature of Airfoil.

2. Airfoil Nomenclature

The basic terms of airfoil are leading edge, chord, camber, Angle of attack, trailing edge. In symmetrical airfoil chord and camber lies on each other because of symmetry in shape. Angle of attack is the angle between the incoming air and relative wind and a reference line on the airplane or wing. Figure 1 shows the basic nomenclature of airfoil.

3. Wind tunnel Experimental Setup

Figure 2 shows the wind tunnel (600mm×270mm) set up used for real time flow analysis which has two distinct sections (i) transparent constant cross section for flow visualization and (ii) a divergent outlet. A suction fan has been installed at the end of divergent outlet and a honey comb mesh at the flow entry has been used to decrease the turbulence in the air flow. The real time flow analysis has been carried out at a Reynolds number= 3×10^4. Angle of attack has been varied from 0° to 15° with an increment of 5°. Ambient conditions are pressure equal to 1 atmosphere and temperature equal to ambient temperature. NACA 0015 airfoil profile has been used for real time flow analysis.

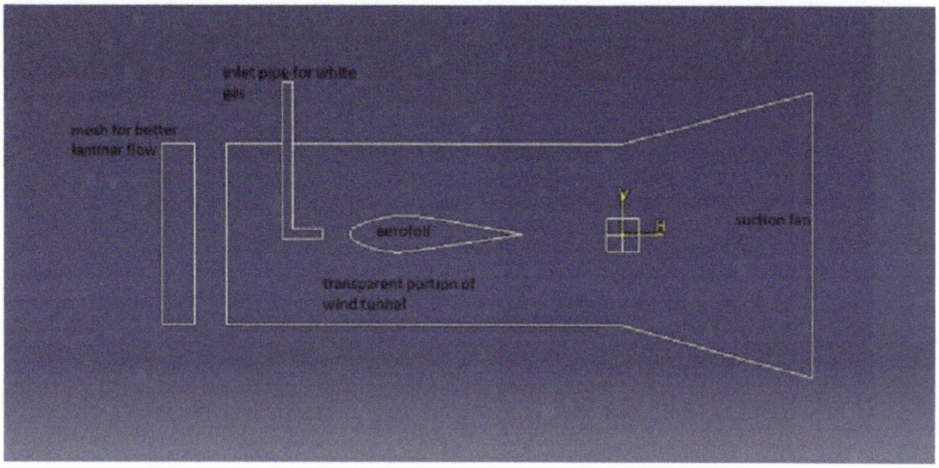

Figure 2 Wind tunnel set up line diagram

4. CFD Analysis

4.1 Modelling of NACA 0015 airfoil.

Figure 3 this shows the NACA 0015 airfoil profile has been generated using NACA Generator [http://airfoiltools.com/airfoil/naca4digit]. CATIA V6 (R2013x) has been used for reproducing 2D model of airfoil as shown in figure 4 below. The dimensions of the airfoil are shown in table 1.

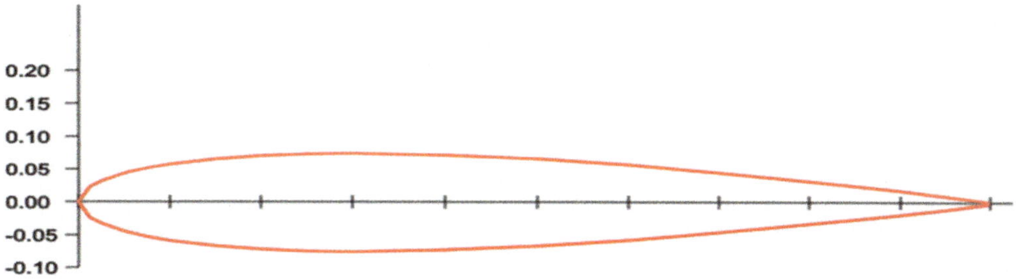

Figure 3 NACA 0015 airfoil

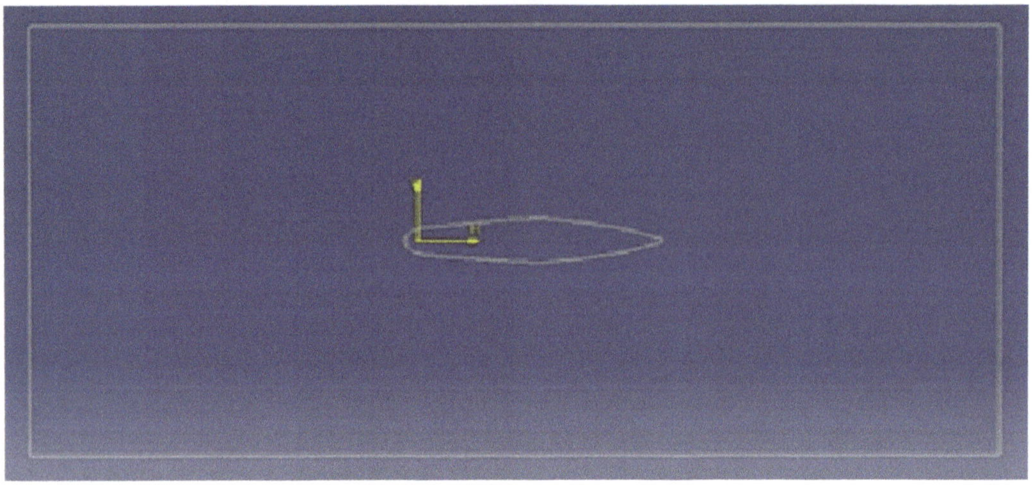

Figure 4 CATIA designed Airfoil

Table 1 Dimensions of NACA 0015 airfoil

S.No.	Parameters	Dimensions
1.	Chord Length	182mm
2.	Span of Airfoil	110mm
3.	Thickness	15%
4.	Leading Edge Radius	4.2%
5.	Lower Flatness	37.4%
6.	Trailing Edge Angle	20.5°
7.	Camber	0%

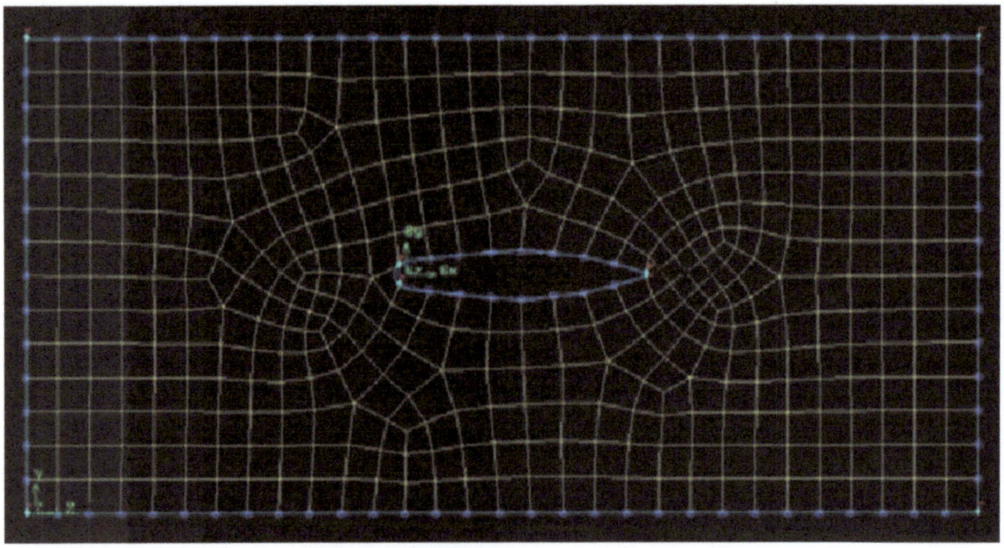

Figure 5 Generated mesh in Gambit

Table 2 Input parameters

S.No.	Parameters	Readings
1.	Pressure(atm)	1atm
2.	Temperature (Kelvin)	300K
3.	Velocity (m/sec)	2m/sec
4.	Density (kg/m^3)	1.2335kg/m^3
5.	Iterations.	1000.
6.	Residual	0.001
7.	Boundary conditions	No slip

4.2 Mesh Generation.

Quadrilateral mesh has been created using Gambit 2.2.30 as shown in figure 5.

4.3 Flow analysis

Fluent 13.0 has been used as a solver for this particular problem. The input parameters have been shown in table 2.

The generated Mesh file with boundary conditions is used in Fluent for further analysis.

5. Results and discussions

5.1 Angle of attack =0°

Figure 6 shows the air flow over NACA 0015 airfoil at AOA = 0 °. The Flow and its visualisation can be improved further by increasing the density of the mesh. Figure 7(a) represents the Velocity Profile and figure 7(b) represents the pressure profile over the NACA 0015 airfoil at AOA = 0°. It can be observed that the velocity and pressure profiles are symmetric about the camber line when AOA=0°. Same can be confirmed from the real time flow around the airfoil (figure 6) as the white smoke used for analysis is flowing along the surfaces of the airfoil. No separation of flow has been observed at the trailing edge of the airfoil.

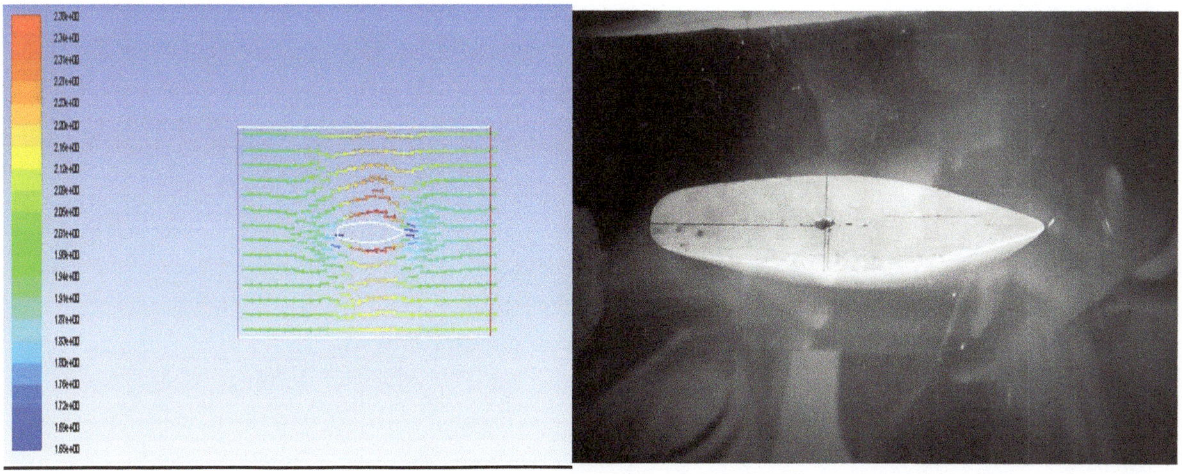

Fig 6 Real time and CFD generated air flow over NACA 0015 airfoil at AOA=0°

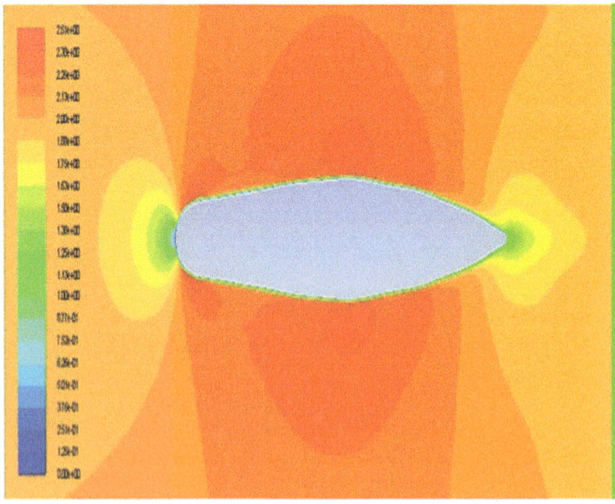

Figure 7(a) Velocity profile over NACA 0015 at AOA=0°

Figure 7(b) Pressure profile over NACA 0015 at AOA=0°

5.2 Angle of attack = 5°

Figure 8 shows the air flow over NACA 0015 airfoil at AOA = 5°. Figure 9(a) represents the Velocity Profile and figure 9(b) represents the pressure profile over the NACA 0015 airfoil at AOA = 5°. The velocity at the upper surface of the airfoil is greater than the velocity at the power surface of the airfoil, whereas the pressure at the lower surface of the airfoil is greater than the pressure at the upper surface of the airfoil. Due to this difference in pressure airfoil experiences a lift force in the upper direction. A very wake formation can be observed at the trailing edge of the airfoil.

5.3 Angle of attack =10°

Figure 10 shows the air flow over NACA 0015 airfoil at AOA = 10°. Figure 11(a) represents the Velocity Profile and figure 11(b) represents the pressure profile over the NACA 0015 airfoil at AOA=10°. A larger difference in velocity and pressure between upper and lower surfaces of the airfoil can be observed in figure 11(a) and figure 11(b) respectively. The velocity at the trailing edge of the airfoil is minimum and this may be because of the flow separation at the trailing edge which leads to the wake formation at the trailing edge of the airfoil. Same can be observed from the real time diagram in figure 10.

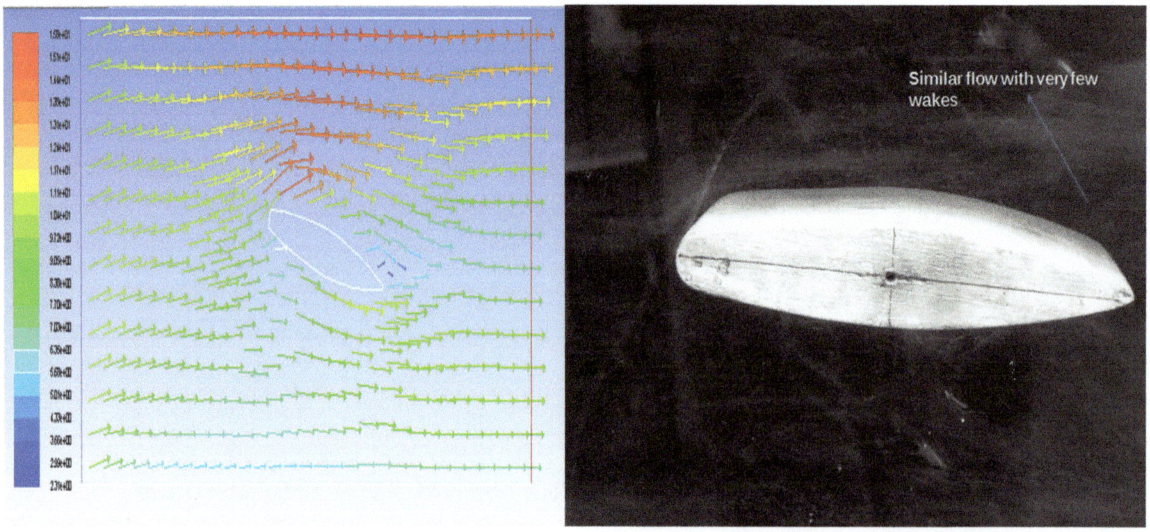

Figure 8 real time and CFD generated air flow over NACA 0015 airfoil at AOA= 5°

Figure 9(a) Velocity Profile over NACA 0015 at AOA= 5°

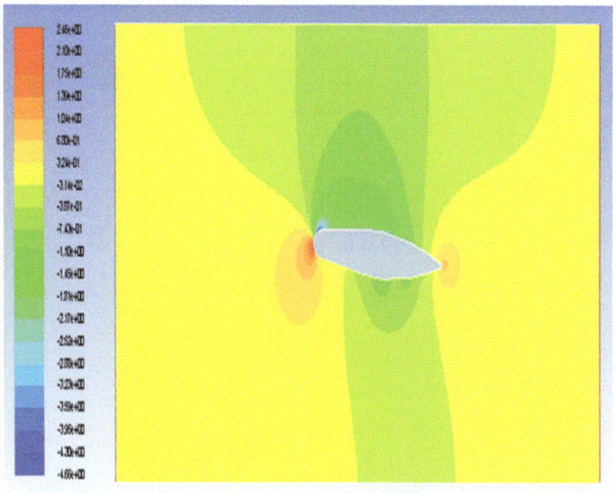

Figure 9(b) Pressure Profile over NACA 0015 at AOA= 5°

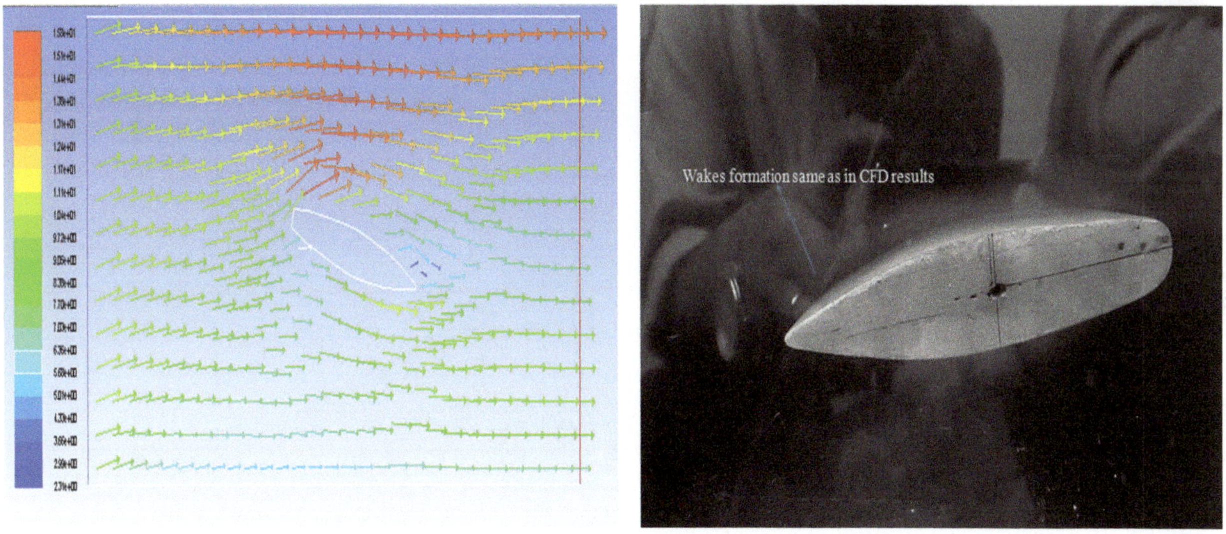

Fig 10 real time and CFD generated air flow over NACA 0015 airfoil at AOA= 10°

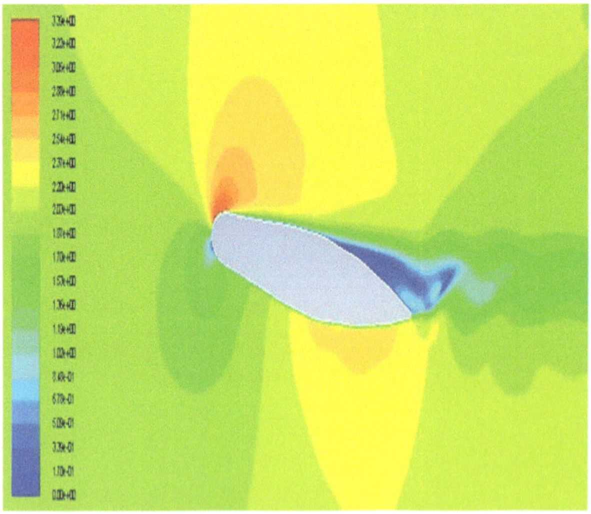

Figure 11(a) Velocity Profile over NACA 0015 at AOA=10°

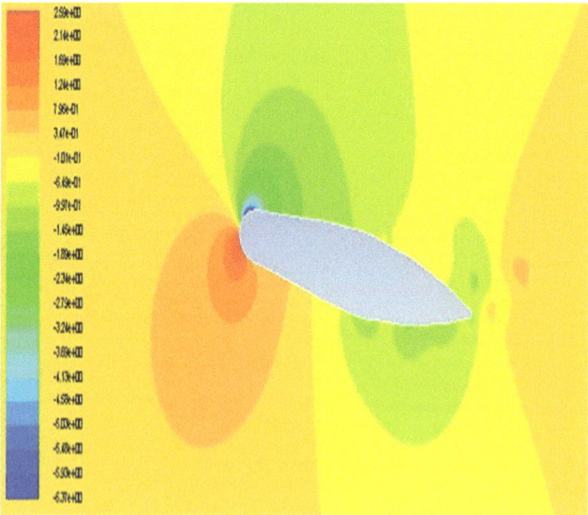

Figure 11(b) Pressure Profile over NACA 0015 at AOA= 10°

5.4 Angle of attack 15°

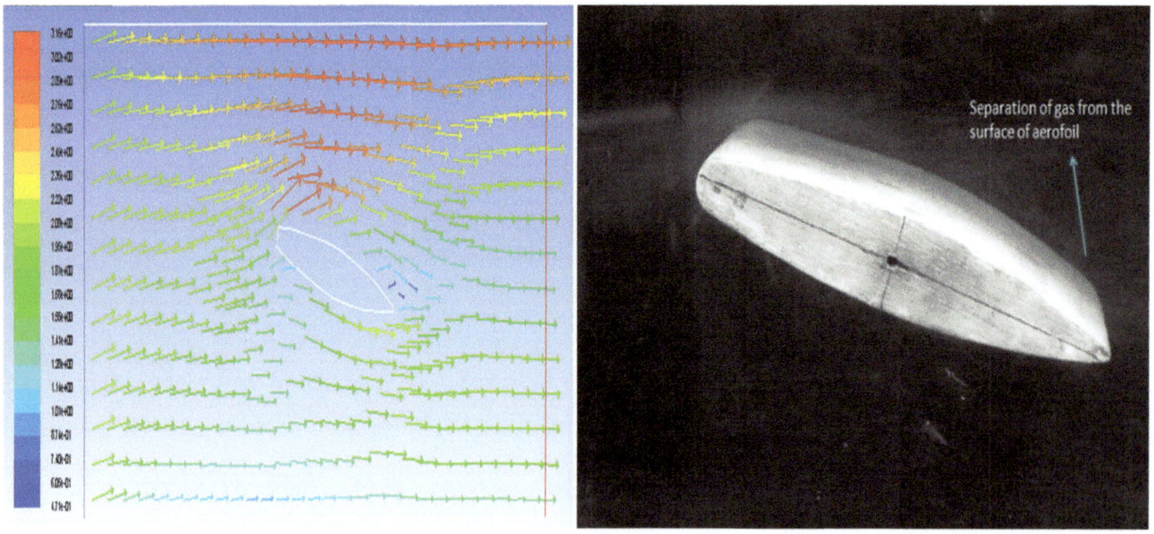

Figure 12 Real time and CFD generated air flow over NACA 0015 airfoil at AOA= 15°

Figure 13(a) Velocity Profile over NACA 0015 at AOA=15°

Figure 11(b) Pressure Profile over NACA 0015 at AOA=15°

Figure 12 shows the air flow over NACA 0015 airfoil at AOA = 15°. Figure 13(a) represents the Velocity Profile and figure 13(b) represents the pressure profile over the NACA 0015 airfoil at AOA = 15°. It can be observed that the velocity is highest at a point on upper surface of the airfoil close to its leading edge and at the same point the pressure is minimum. This increase the lift further but on the same side the drag force also increases as the frontal area of the airfoil increases. The separation of flow can also be observed at the upper surface of the airfoil in figure 12.

6. Conclusion

The behaviour of the NACA 0015 airfoil at different angles of attack has been studied using CFD and the same has been confirmed by the real time flow analysis in the experimental wind tunnel. Flow has been found to depart(flow separation) from the upper surface with the increase in angle of attack. Due to the flow separation a region of wake formation is observed at the trailing edge and the intensity of wake increases with the increases in angle of attack. The lift force will also increase as the pressure difference between the upper and lower surface increases with increase in angle of attack.

References

http://airfoiltools.com/airfoil/naca4digit

Karna SP, Saumil BP, Utsav BP, Ahuja AP, CFD Analysis of an Airfoil, *International Journal of Engineering Research*, 2014,vol. 3, issue 3, pp 154-58.

Mevadiya M, CFD Analysis of Pressure Coefficient for NACA 4412, *International Journal of Engineering Trends and Technology,* 2013, vol. 4, issue 5, pp 2041-43.

Prabhakar A,CFD analysis of static pressure and dynamic pressure for NACA 4412, *International Journal of Engineering Trends and Technology,*2013, vol 4, issue 8, pp 3258-65.

Shankara MHM, Madhukeshwara.N, Kumarappa S, Assessment of boundary layer flow modeling approaches in computational fluid dynamics for compressible external aerodynamics using NACA-64618 Airfoil, *International Journal of Innovative Research in Science, Engineering and Technology,* 2013, vol. 2, issue 7, pp 3137-45.

Author Biography

Abhay Sharma: Abhay Sharma is an enrolled student in School of Mechanical Engineering at Chitkara University, Rajpura, Punjab. His areas of interest are Computational Fluid Dynamics, Designing, Problem Solving Techniques, and Heat Transfer.

Manpreet Singh Sarwara: Manpreet Singh Sarwara is an enrolled student in School of Mechanical Engineering at Chitkara University, Rajpura, Punjab. His areas of interest are Computational Fluid Dynamics, Problem solving Techniques, Thermodynamics, and Fluid Mechanics.

Harsimranjeet Singh Dhaliwal: Harsimranjeet Singh Dhaliwal is an enrolled student in School of Mechanical Engineering at Chitkara University, Rajpura, Punjab. His areas of interest are Computational Fluid Dynamics, Designing, Mathematics, and Thermodynamics.

Lakshya Swarup: Lakshya Swarup is an enrolled student in School of Mechanical Engineering at Chitkara University, Rajpura, Punjab. His areas of interest are Mathematics, Computational Fluid Dynamics, and Thermal Engineering.

Rajeev Kamal Sharma: Rajeev Kamal Sharma is working as Associate Professor in the School of Mechanical Engineering at Chitkara University, Rajpura, Punjab. His areas of interest are fluid mechanics, heat transfer and computational fluid mechanics. He has published research articles in many journals, international conferences and national conferences.

5

A Review on Methods Used for Estimation of Aerodynamic of Damaged Aircraft

Yassir A. Awad Elkarim[1], Mohammed M. Abd Elrahman[2]

[1] *PhD student, yasirabas75@hotmail.com*
[2] *Professor, dr_madbouli@hotmil.com*
Aeronautical Engineering Department, Engineering College, Cairo University
Giza State, Egypt

Abstract

In this study a comprehensive review on methods used by aerospace researchers for the estimation of aerodynamic data for aircraft that sustained damage are illustrated. Design of flight control laws, verification of performance predictions, and the implementation of flight simulations are tasks that require a mathematical model of the aircraft dynamics. The importance of study of damaged aircraft aerodynamic is to generate a good aircraft model to account for various damage effects including the changes in aerodynamic, mass and centre of gravity shift. All aspects of damage are important and minor damage from structural viewpoint may be major damage from a flight dynamic viewpoint. The review mainly focused on: wind tunnels investigations and simulation codes.

Keywords: Aerodynamic, Damaged aircraft, Wind tunnels.

1. Introduction

The control of aircraft has been of paramount concern since the beginning of flight. The dynamic models are characterized by coefficients (aerodynamic derivatives) whose values must be determined from wind tunnels and computational fluid dynamics. To generate a good aircraft model, it is important to have a good estimate of the aerodynamic parameters that closely represents the characteristics of the actual aircraft. In damage events, significant portion of the aircraft's aerodynamic lifting surface may become separated or damage to the aircraft's frame as shown in figure (1), and as a result these may cause the aircraft unstable due to the change of flight envelope.

A flight dynamics model of damaged aircraft is important to develop to account for various damage effects including the changes in aerodynamics, mass, inertia and centre of gravity shift. Understanding the aircraft aerodynamics characteristics during a damage event is critical to developing flight control strategies for stability recovery of damaged aircraft.

Figure1: Wing and horizontal stabilizer damage. [1]

When the aircraft is damaged then several problems may arise and answers are required to questions such as:

a. Can the aircraft complete a mission if it is damaged in a given way?

b. Can aircraft take off again but with reduced capability?

c. Is it safe for the aircraft to take off and fly to another location to be repaired?

d. Will the damage cause the loss of the aircraft?

2. Investigation methods

To know the damage effect on aircraft's flying capabilities the wind tunnels investigations and simulation codes are carried out to calculate the new aerodynamic and stability derivatives so as to design a proper reconfigurable flight control system.

2.1 Wind tunnels investigations:

Hayes [2] at Langley Research Centre on 1968 conducted investigations in a supersonic wind tunnel to determine the effects of simulated wing damage on the static aerodynamic characteristics of a swept wing aircraft model as in figure (2).His model is an ogive-cylinder fuselage with a mid-swept wing and slap vertical tail. Damage was simulated by removing the leading edge section of a wing, the trailing edge section or the complete wing. Removal of the leading edge resulted in an 11% reduction of the total wing area and the lift curve slope is decreased by 10%. However, the lift curve slope decreased by only 13% in case of trailing edge removal while the area resulted in a 17% reduction. Coefficients ($C_L, C_D, C_Y, C_m, C_l, C_n$) as well as parameters ($C_{L\alpha}, L/D, \Delta C_m/\Delta C_L$) were determined for a range of values of angles of attack $\left(-4^0 < \alpha < 22^0\right)$, side slip angles $\left(-5^0 < \beta < 10^0\right)$ and Mach numbers $\left(1.70 \le M \le 2.86\right)$ for constant Reynolds number of $7.38 * 10^6$.

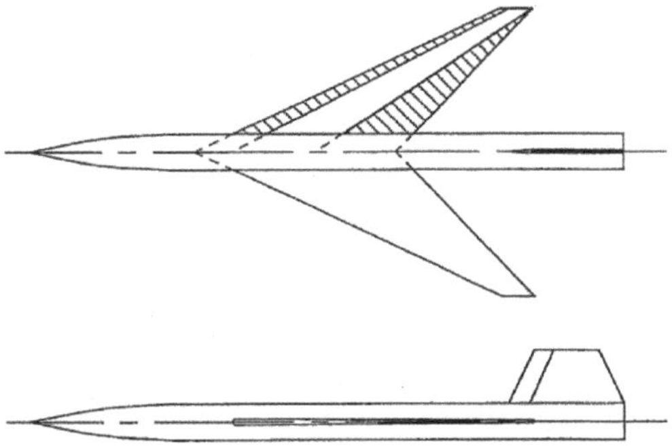

Figure 2: Swept wing aircraft model showing regions of damage (shaded area).

Betzina and Brown [3] measured static aerodynamic characteristics of a McDonnell-Douglas A-4B aircraft with both simulated and actual gunfire damage to the starboard wing. A full scale aircraft was used for experiments which were carried in the NASA-Ames $40 * 80$ ft wind tunnel. A standard fuselage and three different wings were attached to it for the tests. The first wing tested was an undamaged one in which holes had been cut and detachable cover plates installed at various locations. Removal of one or more cover plates gave one of fourteen different simulated damage cases. The other two wings used had been damaged by actual gunfire. Coefficients ($C_L, C_D, C_Y, C_m, C_l, C_n$) were determined for a range of values of angles of attack $\left(-4^0 < \alpha < 26^0\right)$ for the different wings with all other variables held constant.

Spearman [4] summaries transonic wind tunnel tests carried out at the Longley Research Centre using models of undamaged and damaged aircraft to investigate the effects of damage on the aircraft's static aerodynamic characteristics. Three types of aircraft models were used, namely a swept wing aircraft, a delta wing aircraft and a trapezoidal wing aircraft. Damage to the aircraft was simulated by the removal of all or part of a wing, horizontal tail or vertical tail. Figure (4) shows the damage to the trapezoidal wing aircraft.

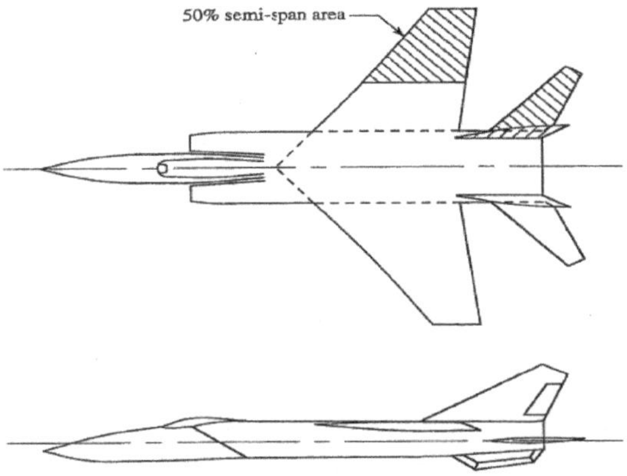

Figure 3: Trapezoidal wing aircraft showing the region of damage (shaded area).

The purpose of the investigation was to determine the probability of completely losing the aircraft, and the extent of damage that the aircraft could be sustained while still completing the mission or at least returning to friendly place. Coefficients (C_L, C_m, C_l, C_n) as well as stability parameters ($C_{n_\beta}, {}^{\Delta C_m}/_{\Delta C_L}$) were determined for a range of values of angles of attack $\left(-2^0 \leq \alpha \leq 4^0\right)$, side slip angles $\left(-4^0 \leq \beta \leq 4^0\right)$ and Mach numbers $\left(0 < M \leq 2\right)$.For the configurations studied in the tests, it was found that losing even half of the wing would not necessarily causes the aircraft to be completely lost, as either enough rudder/aileron power was available or a small side slip could be implemented to offset the rolling moment and the aircraft remained statically stable in the longitudinal plane.

In 1986, Turhal [5] carried out wind tunnel tests using a $1/16$ scale model of an F-16 aircraft to investigate the effect of various types of control surface failure on aircraft's static stability. Three different failure cases were investigated:

a. Fixed deflection of a control surface.
b. Floating left flaperon.
c. Missing left flaperon.

Forces and moments were measured for each failure mode and aerodynamic coefficients ($C_L, C_D, C_Y, C_m, C_l, C_n$) were determined for a range of values of angles of attack, side slip angles and control surfaces deflections. The wind tunnel force and moment data were curve fitted to determine the aerodynamic derivatives as a function of α , β and control surfaces deflections.

Gautam H. Shah [6] at NASA Longley Research Centre conducted in a subsonic wind tunnel investigation to measure the aerodynamic effects of the damage to the lifting and stability/control surfaces of commercial aircraft configurations. The damage is in the form of partial or total loss of the wing, horizontal stabilizer and vertical tail as in figure (4).

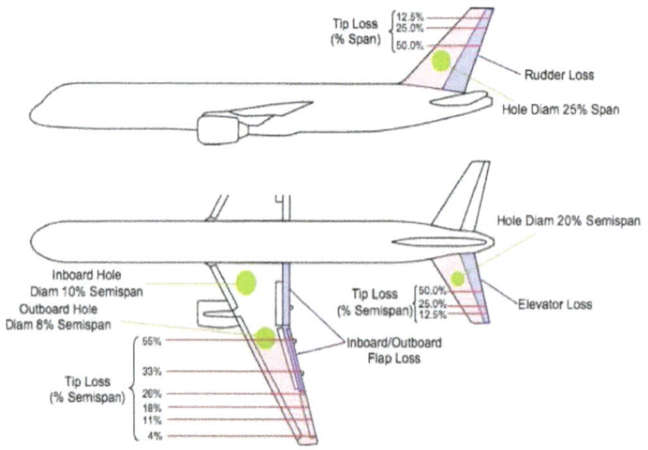

Figure 4: Model sketch and damage conditions tested.

His investigation is focused on:

a. Wing damage:

Figure (5) shows lift coefficient versus angle attack for increasing levels of wing tip loss. The progressive reduction in lift curve slope is seen throughout the angle of attack range up to and beyond the stall at approximately $\alpha = 10^0$ .The effect of tip loss on effective dihedral C_{l_β} is shown in figure (6), as rolling moment coefficient versus sideslip for $\alpha = 0$, where a negative slope indicates static stability. In the low and moderate sideslip region, only a small reduction in stability is noted, though the effect increases for large damage cases at large negative (damaged wing forward) sideslip angles.

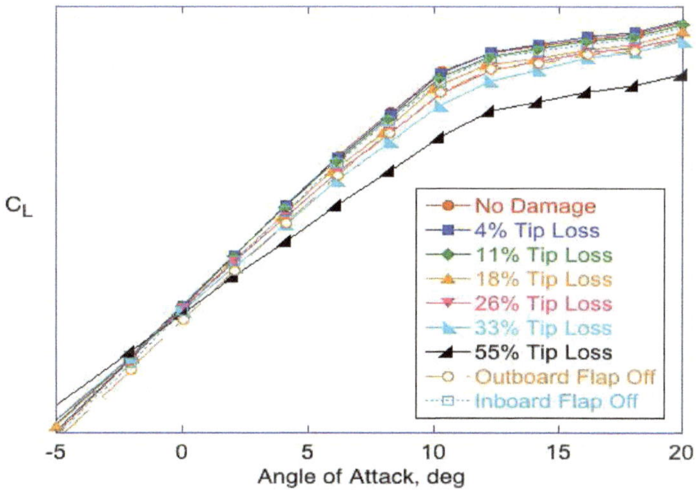

Figure 5: Lift coefficient.

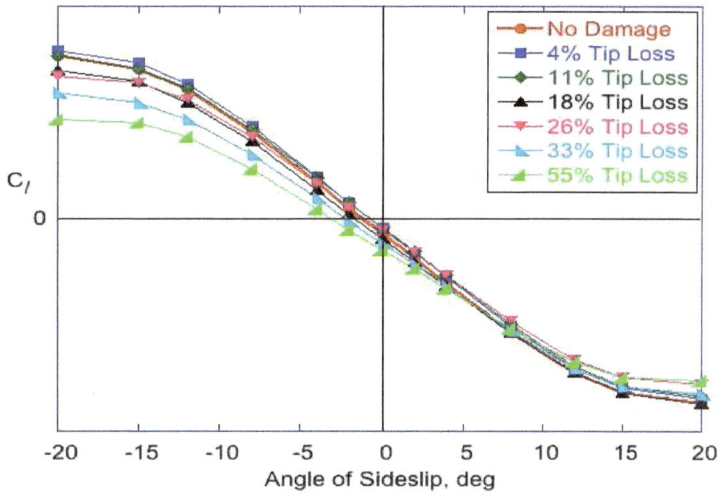

Figure 6: Rolling moment versus sideslip, $\alpha = 0$.

b. Horizontal tail damage:

He showed that for the several damage cases there is a progressive loss in stability (decreasing negative slope) with increasing loss of the stabilizer area as in figure(7). The effect on longitudinal dynamic stability is shown in figure (8) as pitch damping over the angle of attack range for damaged and undamaged cases.

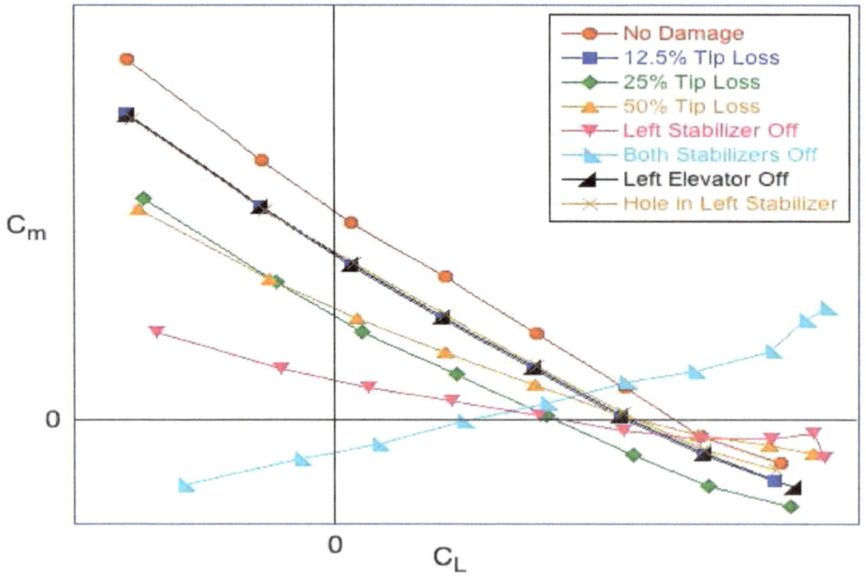

Figure 7: Static longitudinal stability.

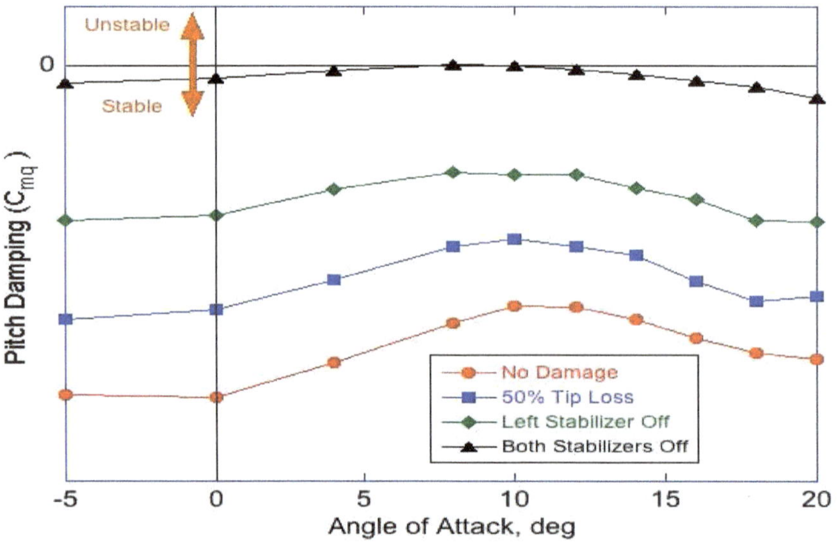

Figure 8: Pitch damping.

c. Vertical tail damage:

Figure (9) shows that at zero angle of attack, a progressive reduction in static directional stability (slope of C_n versus β) with increasing of damage area loss. Also the variation of directional stability over the entire angle of attack range is studied and the $C_{n\beta}$ is computed as the linear slope in yawing moment between side slip angles of $-4^0 \, to \, 4^0$ as in figure (10)

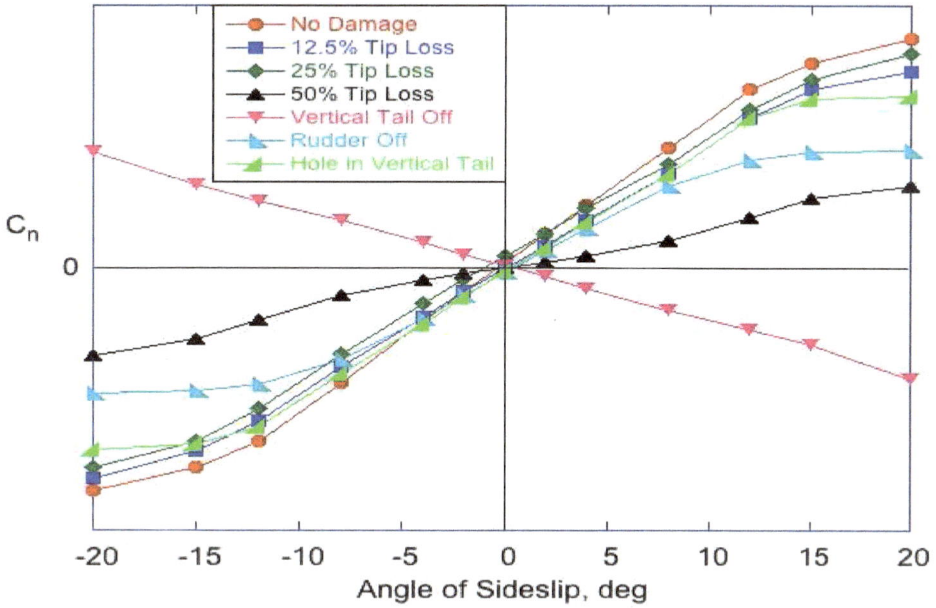

Figure 9: Yawing moment coefficient versus side slip angles at zero angle of attack ($\alpha = 0$).

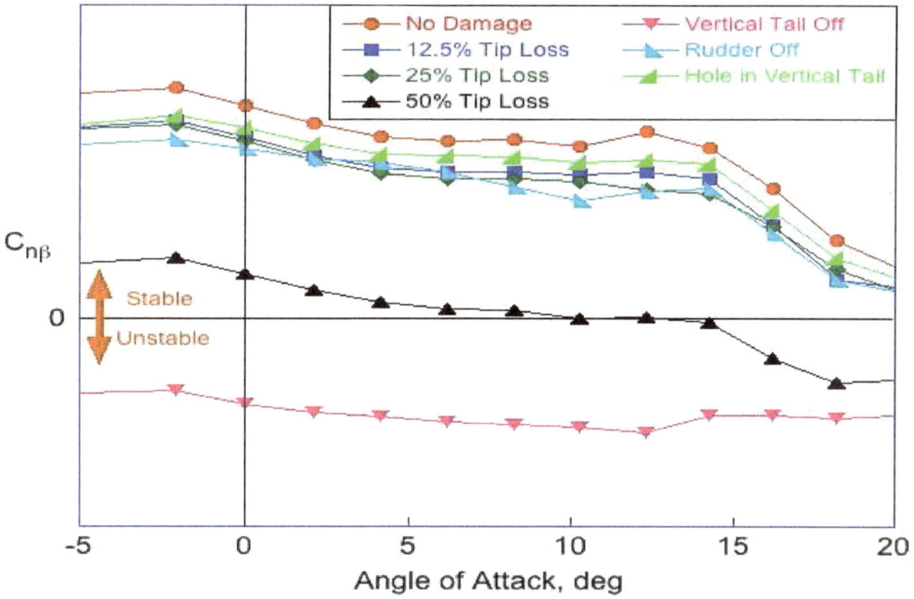

Figure 10: Static directional stability.

Djellal and Ouibrahim [7] on 2009 studied the effect of wing damage on aerodynamic behavior of the aircraft. Three models were used for these investigations. They considered that the damage is only located on the wing, which are from the aerodynamic point of view, the most critical components of the aircraft. The scenarios of damages were expressed in terms of diameters and localizations of the through holes of simulated damage in the wing. Damage size can be expressed in terms of a percentage diameter (d) to the local chord length (c). Wing models with undamaged and damaged cases were investigated by considering moreover the variation of the holes diameter and its position on the span wise and the chordwise.

The models used for the investigations are:

a. Rectangular wing with a profile NACA 64_1-412 and a taper ratio $\lambda=1$.
b. Aircraft wing model (trapezoidal wing) with NASA 23018 at the root section, NASA 23012 at tip section, taper ratio $\lambda=0.5$ and without twist.
c. Triangular plan-form wing with a profile RAE 104, tapered, without sweep and twist.

Figure 11: Damage for three models.

Their investigations are focused on the influence of damage to the aerodynamic behavior and presented as changes in coefficients, ΔC_L and ΔC_D in terms of:

$$\Delta C_L = C_{L damaged} - C_{L undamaged}$$

$$\Delta C_D = C_{D damaged} - C_{D undamaged}$$

For the aircraft wing model three damages of the same size ($40\%c$) centered at mid chord but located at three different wing spanwise locations: tip, mean aerodynamic chord, and root.

The magnitude of the damage effect is sensitive to its position along the span. As shown in figures (12) and (13) the degradation of the C_L and C_D increases from the tip to the root regions.

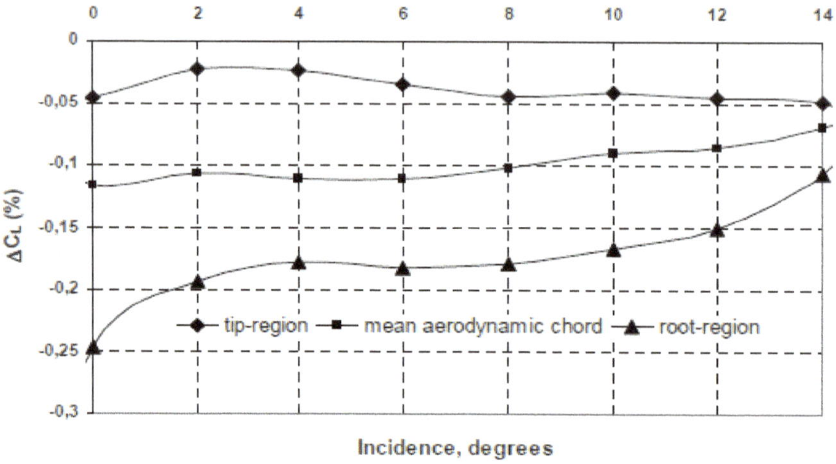

Figure 12: Rate of lift coefficient loss along the span.

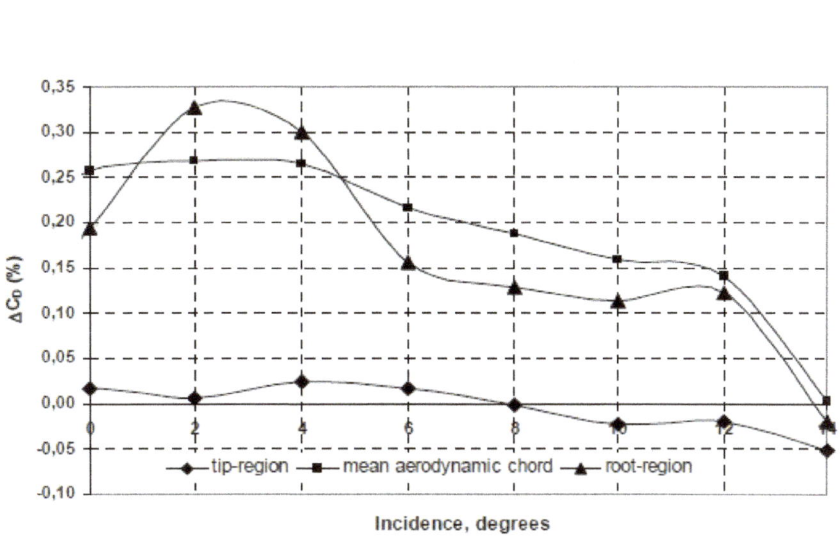

Figure 13: Rate of drag coefficient loss along the span.

2.2 Simulation codes:

Zaiser [8] on Air University used Turhal's data in his studies on the stability characteristics of a combat aircraft with control surface failure. An investigation of the stability characteristics of an aircraft which has sustained damage to a primary control surface was performed. He developed polynomial functions which describe the aircraft static stability derivatives. The polynomials were examined to identify aerodynamic coupling which might be significant. The analysis was performed using wind tunnel data taken on an F-16 model in Turhal's tests. The aircraft stability and control derivatives were developed and analyzed to identify aerodynamic coupling with implications for an aircraft with failed control surfaces. The investigations were conducted at two flight conditions representative of the aircraft at cruise and landing approach velocities.

Nhan, Kalmanje, and John Kaneshige [9] at NASA Ames Research Centre performed aerodynamic modeling to the Generic Transport Model (GTM) to estimate the aerodynamic coefficients, stability and control derivatives. This damage effect is performed using a vortex lattice code. This computational fluid dynamics modeling is capable of rapidly computing the aerodynamic characteristics and control sensitivity of the damaged GTM due to various flight control surface inputs. The damage effects are modeled as partial losses of the left wing, left horizontal stabilizer and vertical tail as shown in figure (14).

Figure 14: Damage to the Generic Transport Model (GTM).

Their studies showed that the effect of wing loss can be seen as a significant source of loss of the lift capability of a damaged aircraft as the lift coefficient can be reduced by as much as 25% for up to 50% span loss of the one of the wings as shown in figure (15) and the changing of the pitching moment due to the wing loss as in figure (16).

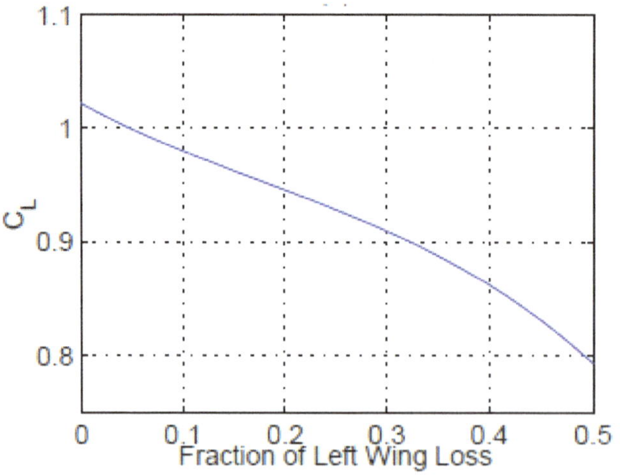

Figure 15: Lift coefficient reduction due to wing loss at $\alpha = 12^0$, $\beta = 0^0$.

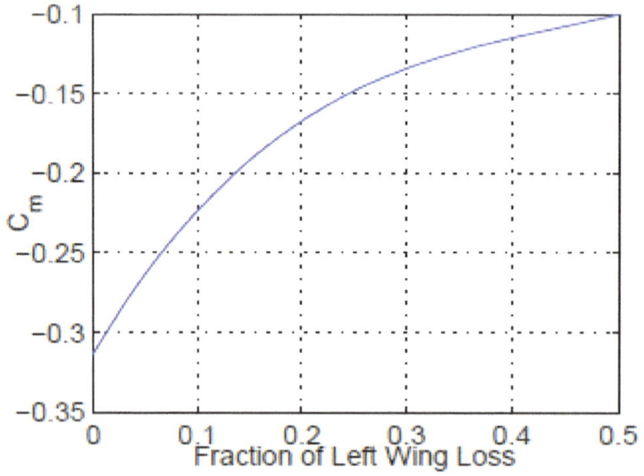

Figure 16: Pitching moment coefficient due to wing loss at $\alpha = 12^0$, $\beta = 0^0$.

Nesrin.S.K, Nespeca.P, T.Marchelli, and M.Sarigul.Klijn [10] represented an approach to predict the flight dynamics and stability derivatives of structurally damaged transport category subsonic aircraft (C-17 Global master model) based on spanwise full loss damage model. In their study they derived the stability and control derivatives from the basic principles and theoretical aerodynamics. The damage is introduced using spanwise parameter dmg which represents the fraction of missing half span of the wing starting from the tip as in figure (17). From the basis of aerodynamic they defined new terms which showed the effect of damage on the basic wing geometry properties. These new definitions are used for the formulas of the stability coefficients and derivatives for the damaged aircraft.

Figure 17: Spanwise losses of structure damage model.

Neal, Shahyar, Harold, Sally and Joseph [11] presented a computational study to assess the utility of the two different NASA unstructured Navier-Stokes flow solvers, USM3D and FUN3D, for capturing the degradation in static stability and aerodynamic performance of transport aircraft configuration due to airframe damage. The damage which studied is the loss in wing, horizontal stabilizer, and vertical tail. Their study demonstrated that high fidelity Navier-Stokes flow solvers could augment flight simulation models with additional aerodynamic data for various airframe damage scenarios.The goal of their study is to evaluate the accuracy and utility of the respective numerical tools by correlating with the damaged aircraft experimental data shown in [6].

Tetrahedral grids were constructed on the full span (no damage) and with damage in wing, horizontal and vertical tail as seen in figures (18, 19 and 20)

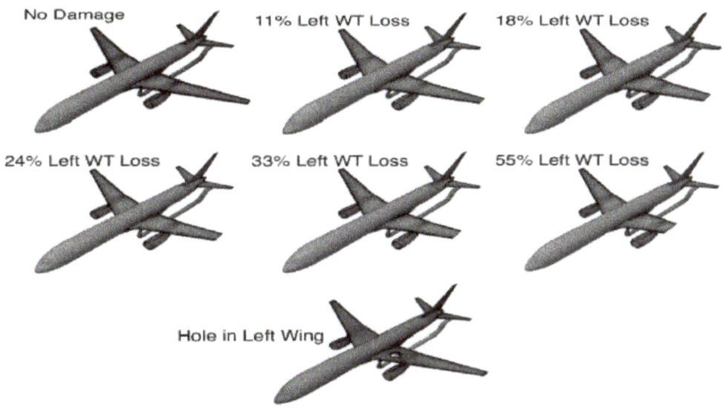

Figure 18: Wing tip damage configurations.

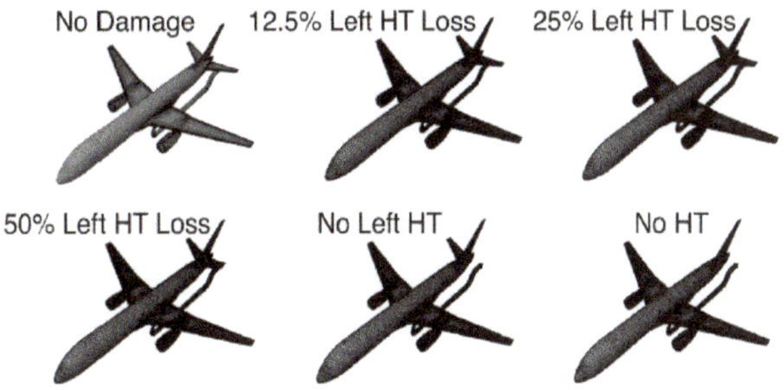

Figure 19: Horizontal tail damage configurations.

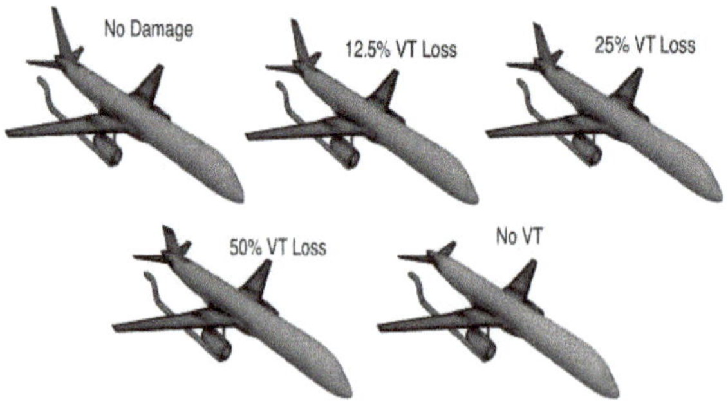

Figure 20: Vertical tail damage configurations.

The effect of wing tip losses in lift coefficient is presented in figure (21) and the overall trend of lift losses with the increasing tip loss is well captured. Also the effect of rolling moment coefficient against sideslip angles at zero angle of attack is illustrated as in figure (22).

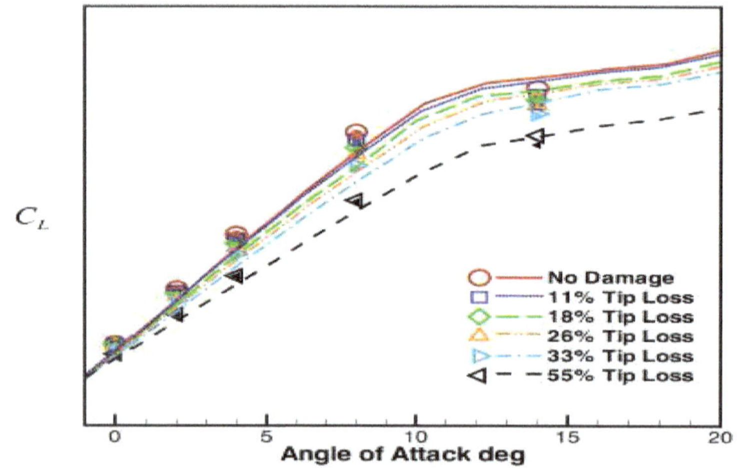

Figure 21: Effect of left wing damage on lift coefficient at $\beta = 0^0$, USM3D (open symbols), FUN3D (solid symbols), experiment (line symbols).

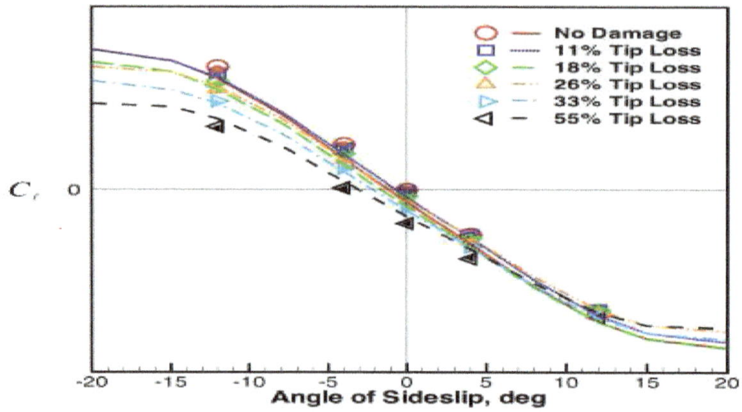

Figure 22: Effect of left wing damage on rolling moment coefficient at $\alpha = 0^0$, USM3D (open symbols), FUN3D (solid symbols), experiment (line symbols).

The effect of horizontal stabilizer damage in longitudinal stability is illustrated in figure (23) in terms of lift and pitching moment coefficients.

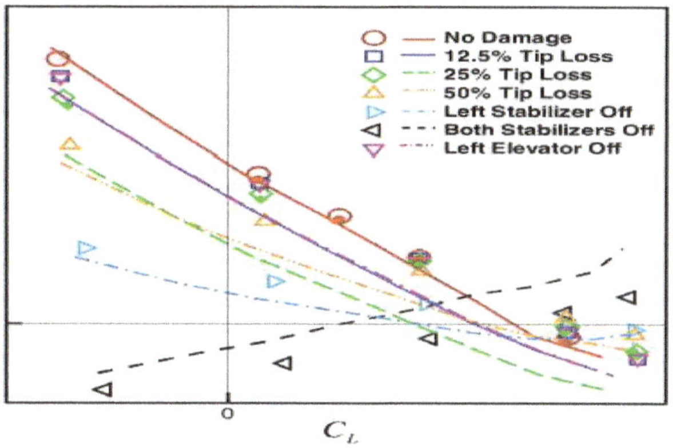

Figure 23: Effect of horizontal stabilizer damage on pitching moment coefficient at $\beta = 0^0$, USM3D (open symbols), FUN3D (solid symbols), experiment (line symbol).

The effect of vertical tail loss in yawing moment coefficient versus angle of sideslip at zero angle of attack is presented in figure (24) and an excellent correlation with the experimental data is observed for the two flow solvers. The static directional static stability (C_{n_β}) is plotted against angle of attack range to yield the stability of the aircraft as in figure (25).

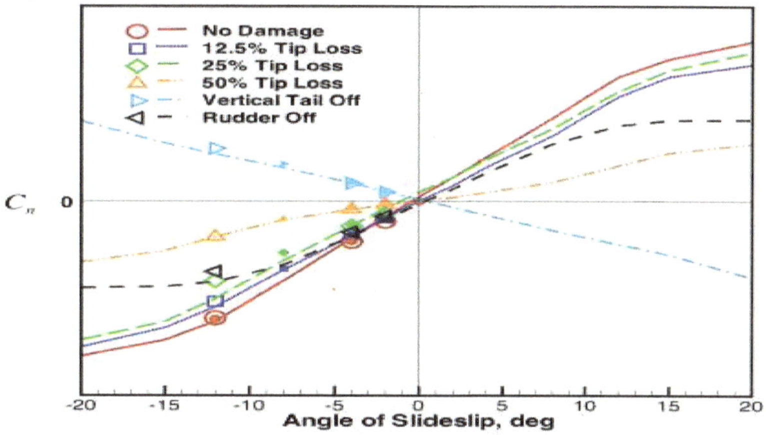

Figure 24: Effect of vertical tail damage on yawing moment coefficient at $\alpha = 0^0$, USM3D (open symbols), FUN3D (solid symbols), experiment (line symbols).

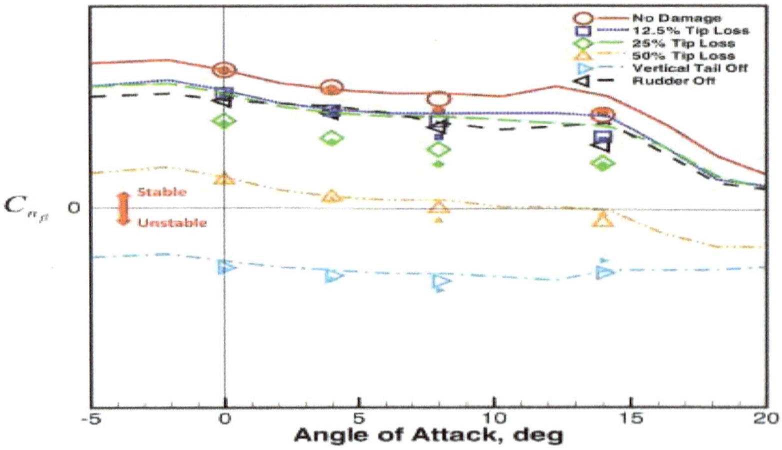

Figure 25: Effect of vertical tail damage on static directional stability, USM3D (open symbols), FUN3D (solid symbols), experiment (line symbols).

B.A.Siddiqui, A.H.Kasim, and A.Z.al-Garani [12] modeled aircraft in Missile DATCOM to predict the lift, drag and pitching moment coefficients and validate the results with [2]. A new technique is proposed for the determination of the aerodynamic derivatives using the data from missile DATCOM and experimental wind tunnel data [2].The model used for the modeling in DATCOM is the same as that used by [2]. A fuselage was of circular cross section having a fineness ratio of 13.8 with an ogive nose. A swept wing was mounted in a mid-wing position. There is no horizontal tail, but a swept vertical tail with a sharp wedge of constant chord at leading edge as seen in figure (26).

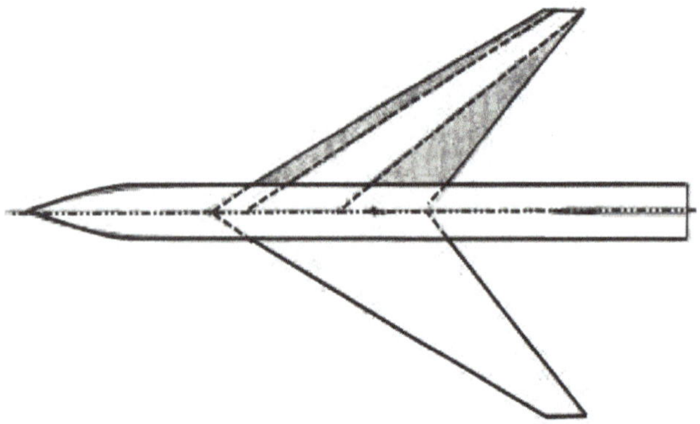

Figure 26: Aircraft model (damage: shaded area).

Damaged aircraft configurations were achieved by:

a. Removal of starboard leading edge portion (forward portion shaded area), this resulted in a 11% reduction in total exposed wing area.
b. Removal of the trailing edge portion of the starboard panel (aft shaded area), this resulted in a 17% reduction in total exposed wing area.
c. Removal of the entire starboard wing panel.

Their steps for this technique are:

a. Get geometric dimensions of the aircraft.
b. Get experimetal aerodynamic data of the complete aircraft.
c. Test a major part of the heathy aircraft in wind tunnel (wing).
d. Model the healthy aircraft in DATCOM to predict the aerodynamics.

e. Improve this DATCOM prediction for the healthy aircraft by wind tunnel study of the wing alone.
f. Model the damaged aircraft in DATCOM.
g. Improve the model in step (f) with wind tunnel data for healthy half wing.
h. Subtract the DATCOM prediction of the healthy aircraft from DATCOM prediction of the damaged aircraft.
i. Add the difference in step (h) and add to healthy complete aircraft aerodynamics in step (b).

A comparison for the various damage cofigurations for the proposed techniques with DATCOM and experimental data are plotted as shown in figures (27, 28 and 29).

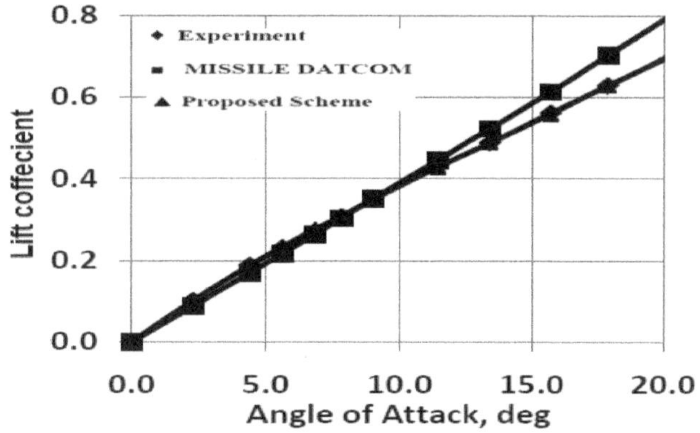

Figure (27): Comparison of lift coefficient for proposed technique and DATCOM predictions with experimental data for wing leading edge damaged aircraft.

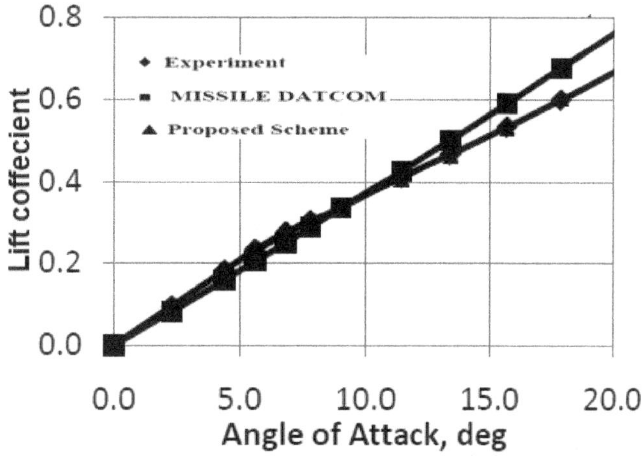

Figure 28: Comparison of lift coefficient for proposed technique and DATCOM predictions with experimental data for wing trailing edge damaged aircraft.

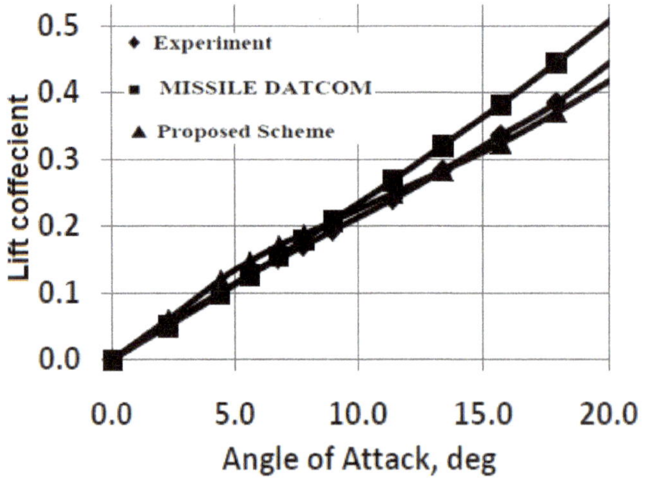

Figure 29: Comparison of lift coefficient for proposed technique and DATCOM predictions with experimental data for wing panel removal damage.

Jinwhan kim, Karthik, and Menon [13] discussed an estimation of aerodynamic model of damaged aircraft using differential vortex lattice method (DVLM) tightly coupled with extended Kalman filters in order to enable the assessment of aircraft performance.Their study is developed as; first, a rapid approach for deriving aerodynamic models of damaged aircraft via DVLM, and second is to use extended Kalman filtering for online estimation of damaged aircraft parameters based on DVLM. The DVLM proposed in their study allows the calculation of the forces on the damaged aircraft using differential circulation (Γ) strength components in the vicinity of the damaged section. And the using of DVLM reduces the dimension of the problem by exploiting the knowledge about the circulation over the aircraft structure before the damage. The problem is formulated in terms of the changes in the circulation on the panels in the neighborhood of the damage.

For the formulation of the DVLM the surface panels are divided into three groups as in figure (30):

a. Group 1 contains undamaged panels that are some distance away from the damaged section, where the effect of damage can be assumed to be negligible or insignificant.
b. Group 2 contains damaged panels which have been taken off from the airframe.
c. Group 3 contains undamaged panels that are in the close neighborhood of the damaged area. The effect of the damage can be felt strongly and clearly over these panels.

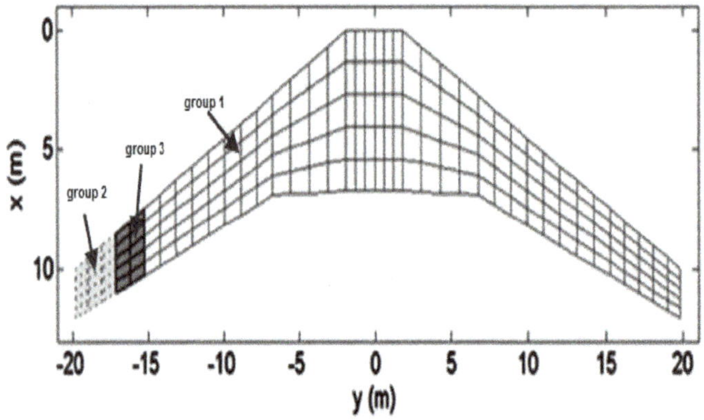

Figure 30: Panels group used in DVLM formulation.

Two cases of damage for the numerical calculation are considered: wing tip damage and a hole in the wing. The pressure distributions is evaluated and integrated to produce aerodynamic forces. These forces can be normalized with respect to dynamic pressure and reference area to obtain the lift, drag and side force coefficients. The formulation of using DVLM is as in figure (31).

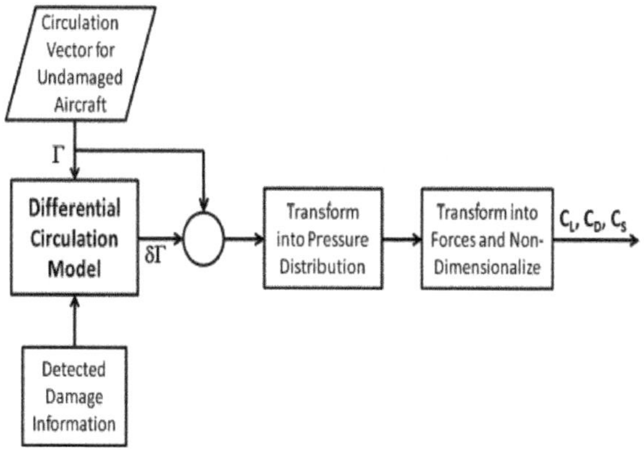

Figure 31: DVLM procedure.

3. Conclusion

The review shows the considerable work done over the last years to estimate the aerodynamic of damaged aircraft. The damage effect aerodynamic modeling has been performed to provide an understanding of the control and stability of an asymmetric damaged aircraft Various damage modeling are used (wing, horizontal and vertical tail) for investigation of damage on aircraft lift, drag and moments coefficients. The most researches at last ten years conducted by mean of computation code rather than wind tunnels because of computers availability.

References

[1] Arruda, M., ''Dynamic Inverse Resilient Control for Damaged Asymmetric Aircraft Modeling and Simulation,'' M.Sc thesis, Aerospace Engineering Dept., Wichita State University, 2007.

[2] Hayes, C., "Effects of Simulated Wing Damage on the Aerodynamic Characteristics of a Swept-Wing Airplane Model," NASA TMX-1550, 1968.

[3] Betzina, M. D., and Brown, D. H., ''Aerodynamic Characteristics of an A-4B Aircraft with Simulated and Actual Gunfire Damage to One Wing,'' NASA TMX-73,119, 1976.

[4] Spearman, M. L., ''Wind Tunnel Studies of the Effects of Simulated Damage on the Aerodynamic Characteristics of Airplane and Missiles,'' NASA TM-84588, 1982.

[5] Turhal, Y. E., ''Investigation of F-16 Control Surface Failures and Optimal Setting of Functional Controls,'' M.Sc thesis, Air Force Institute of Technology, USA, 1986.

[6] Gautam H. Shah, ''Aerodynamic Effects and Modeling of Damage to Transport Aircraft,'' NASA Langley Research Centre, Hampton, 2008.

[7] Djellal, S., and Ouibrahim, A., ''Experimental Investigation on Aerodynamic Performance of Damaged Wing,'' EACWE 5, Florence, Italy, July 2009.

[8] Zaiser, S. M., ''Stability Characteristics of a Combat Aircraft with Control Surface Failure,'' M.Sc thesis, Air university, Ohio, 1989.

[9] Nhan, N., Kalmanje, K., and John, K., ''Dynamic and Adaptive control for Stability Recovery of Damaged Asymmetric Aircraft,'' AIAA, 2006.

[10] Nesrin, S. K., Nespeca, P., Marchelli,T., and Sarigul, K., ''An Approach to Predict Flight Dynamics and Stability Derivatives of Distressed Aircraft,'' AIAA atmospheric flight mechanics conference and exhibit, , Honolulu, Hawaii. August2008

[11] Neal T. F., Shahyar Z., P., Harold L. A., Sally A. V., and Joseph H. M., ''CFD Assessment of Aerodynamic Degradation of a Subsonic Transport Due to aircraft Damage,'' AIAA, 2010.

[12] Bilal, A. S., Ayman, H. K., and Ahmed Z. A., ''Using USAF DATCOM to Predict Nonlinear Aerodynamics of Structurally Impaired Aircraft,'' International Review of Aerospace engineering, Vol. xx.n.x, January 2010.

[13] Jinwhan k., Karthik P., and Menon, P. K., ''Rapid Estimation of Impaired Aircraft Aerodynamic Parameters,'' Journal of Aircraft, Volume 47, July-August2010.

6

A Review on Efficiency of Inlet with Varying Ramp Angle

Karthikeyan.L.M
Department of Aeronautical Engineering,
Techno Global University, India.
lmkarthikeyan@yahoo.in

Abstract

An air inlet can protrude from the aircraft surface or be submerged into the aircraft body. A submerged inlet has in general lower aerodynamic drag than an inlet that protrudes from the surface of an aircraft and is therefore the most preferred design option to the flight industry. An air inlet should ideally not decrease the total pressure of the air which enters and at the same time only give rise to a minimum amount of additional aerodynamic drag. Although angle variation in the air inlet have several affects other than inlet efficiency also the air breathing system of the airplanes. As the varying ramp angle in the airplane, air jet etc, may create the problems on the working of the air inlet. So this review on efficiency of inlet with varying ramp angle is very much needful for the system as well as for the designers, and also it would be very helpful for the future perspective for researches.

Keywords: Ramp Angle, Air Intake, Pressure Distortion.

1. Introduction

The efficiency of air inlet may vary due to the change in angle. When studying air flow over a solid body it is appropriate to divide the analysis of the flow into two parts. Close to the surface of the solid body friction forces play an important part whereas further out into the free stream friction forces can be neglected. The idea is to treat the air flow close to a surface separately. This concept was first suggested in 1904 by a man named Ludwig Prandtl. Due to the friction between the surface and the moving gas, the air flow closest to the surface will tend to adhere. This phenomenon is known as the no-slip condition.

In this paper we would review about the inlets in the airplanes. All types of inlets such as scoop inlet and flush inlet would be reviewed in this thesis. Advantages and disadvantages exist with both design choices. While the scoop inlet has the advantage of avoiding the low energy boundary layer which reduces the efficiency of an air inlet, it has typically the disadvantage of a greater increase of aerodynamic drag compared to a submerged inlet. These criteria cannot be fully met by any air inlet, but design parameters can be changed to come close to an optimum for a specific flight condition. For low drag it is advantageous to use a flushed inlet which is submerged into the surface of the body into which it is placed. The flush type is also advantageous to avoid foreign object damage on the inlet. We would focus on the ramp walls such as Parallel walls, divergent walls and convergent walls. Through this we can evaluate the exact ramp efficiency would be measured as well.

2. Air Inlet – Air Intake System

The air intake is that part of an aircraft structure by means of which the aircraft engine is supplied with air taken from the outside atmosphere. The air flow enters the intake and is required to reach the engine face with optimum levels of total pressure and flow uniformity. These properties are vital to the performance and stability of engine operation. Depending on the type of installation, this stream of air may pass over the aircraft body before entering the intake properly. Figure 1 shows the flow chart of typical air intake system.

2.1 Needs of Air Inlet or Intake System

- A widely used method to increase the thrust generated by the aircraft engine is to increase the air flow rate in the air intake by using auxiliary air intake systems.
- The air flow enters the intake and is required to reach the engine face with optimum levels of total pressure and flow uniformity hence need of an air intake system.
- Deceleration of airflow at high flight mach numbers or aerodynamic compression with help of air intake.

2.2 Air Intake Design Requirement

The airflow first passes through the air intake when approaching the engine, therefore the intake must be designed to meet certain requirements of aircraft engines such as:

- The air intake requires enormous effort properly to control airflow to the engine.
- The intake must be designed to provide the appropriate amount of airflow required by the engine.
- Furthermore this flow when leaving the intake section to enter the compressor should be uniform stable and of high quality.
- Good air intake design is therefore a prerequisite if installed engine performance is to come close to performance figures obtained at the static test bench.
- The engine intake must be a low drag, light weight construction that is carefully and exactly manufactured.
- These above conditions must be met not only during all phases of flight but also on the ground with the aircraft at rest and the engine demand maximum, thrust prior to take off

3. Literature Review

Many studies have been conducted in both the stationary and relative frames of measurement to investigate how fluid flow through a fan or compressor varies due to circumferential inlet total pressure distortions. The literature review resulted in five general publication categories with application to distortion and wake response. The first group of papers concerns inlet flow condition and blade lift response to distortion. The second group of papers concerns blade response to a transient blade incidence angle above the steady-state stalling angle. The third category concerns wake analysis with respect to variations in incidence angle or operating point on the map. The fourth category discusses the phenomena of the suction side total pressure profile appearing as a jet, or pressure excess, in the stationary frame of reference. The fifth category is limited to one paper discussing the effect of inlet total pressure distortion on inlet flow conditions and three dimensional wake responses.

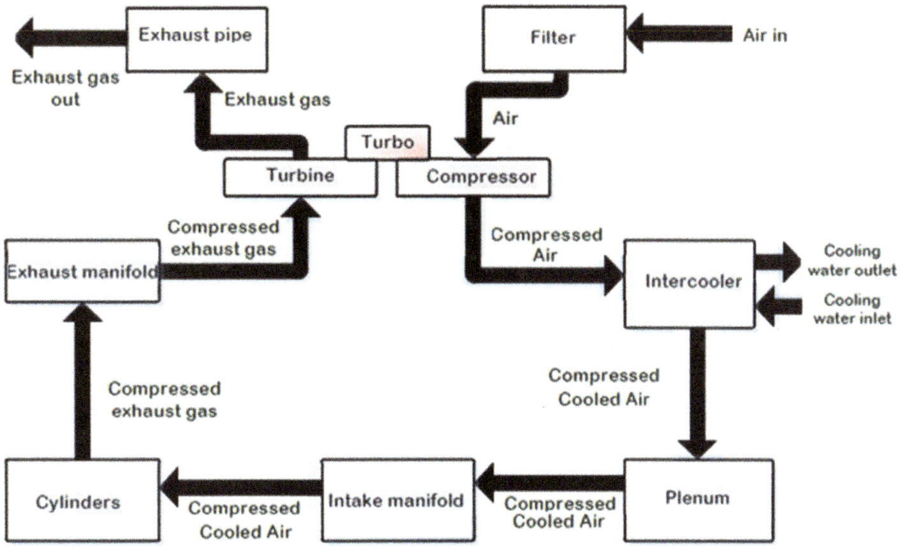

Figure 1: Flow Chart of Intake System

Several hundred documents exist in the literature which describes testing and research performed in an effort to improve compressor response to inlet total pressure distortion. Due to this extreme number of relevant publications, this literature review will emphasize steady-state circumferential pressure distortion in axial-flow fans and compressors.

Kacker and Whitelaw (1971) investigated the turbulence characteristics of two-dimensional wallets. They used static pressure distribution, mean velocity profiles, and wall shear stress to describe the mean properties of air jets. They used turbulence intensity, turbulent shear stress, and a turbulent energy balance to describe air jet properties. They concluded that detailed measurements were required to demonstrate the complexity of wall jet flows and confirmed that a satisfactory prediction of the mean and fluctuating properties was a formidable task.

Walker (1977) reviewed the theoretical relationships of isothermal ventilating air jets. Proper selection of inlet size and design, direction of air movement, air jet throw, entrainment, and spread of ventilation jets were required to predict system performance and to design an effective ventilation system. The review of theory included a free jet issuing from a circular hole and a free jet issuing from an infinitely long slot. He concluded that air jet theory could be used to describe velocity decay, entrainment, and air jet spread for many types of inlets and ventilation configurations.

Boon (1978) tested airflow patterns for various inlet arrangements and ventilation configurations. In most commercial livestock buildings, airspeed at animal level is greatly affected by the rate of ventilation. Temperature, relative humidity, airspeed, ventilation rate, and air movement were measured. Two stable airflow patterns were observed. Temperature variations from floor to ceiling relied on the direction of air movement.

Leonard and McQuitty (1987) investigated design criteria of ventilation inlets for animal housing. They concluded that a minimum Jet Momentum Number of 7.5 × 10–4 was required for developing a stable airflow pattern in isothermal situations. Ogilvie et al. (1990) investigated the effects of air inlets and floor layout on animal-level airspeed. They studied the relationship between jet momentum and airflow pattern near the floor. They found that the location and type of air inlets and the pen layout were two important components of a space ventilation system. They found that Jet Momentum Number (J) and the airflow to floor area ratio (Q/A), both indices of energy input into a room, correlated well with floor airspeed.

Jin and Ogilvie (1992) investigated airflow direction and airspeed at animal level. They showed velocities in the floor region correlated well with inlet configuration, which included inlet type, dimension, and location of the inlet, incoming velocity, direction, and airflow rate.

4. Pressure Distortion Simulation Techniques

Studying about the air flow was first been suggested by Ludwig Prandtl (1904). Inlet flow distortions in fluid machinery are defined as variations of flow properties as a function of space and time and have been of major concern to turbo machinery designers for decades. These deviations from a steady uniform distribution of the flow properties can include variations in swirl, vorticity, turbulence, total and static pressures, velocity, temperature, flow angle, and fluid density. Circumferential distortions are viewed as being more critical than radial distortions due to the production of a disturbance normal to the airfoil motion, and resulting effects on angle of attack and stall margin. Figure 2 shows the inlet total pressure distortions and its configuration.

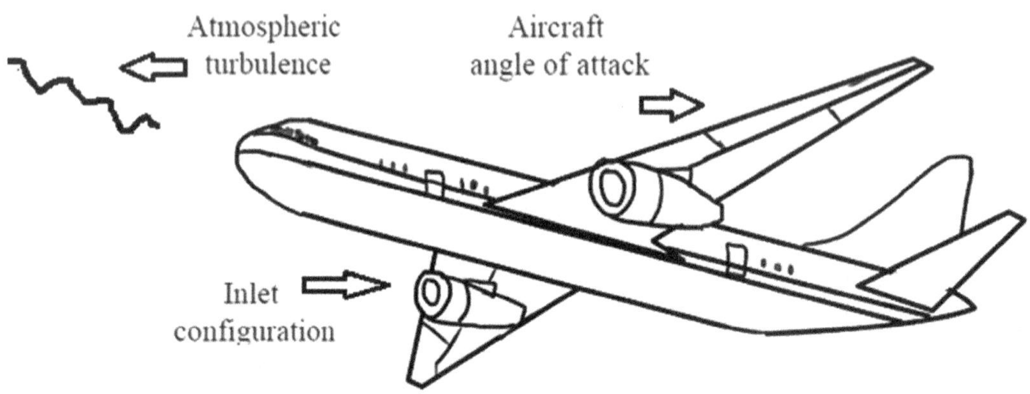

Figure 2: Inlet Total Pressure Distortions and its Configuration

Screens are typically used to impose steady-state total pressure distortions on inlet fans and compressors based on the simplicity of application and cost-effectiveness. The effect of fluid viscosity on the flow is to produce a viscous static pressure drop across the screen grid, as well as a deflection of streamlines to the screen outer edges. It is important to note that a distortion screen installed in front of a rotor will not produce an identical circumferential pressure defect pattern downstream at the aerodynamic interface plane of the rotor inlet due to upstream fluid communication between the screen and rotor. Although ideally the screens would produce "square-wave" pressure patterns at the inlet of the rotor, in reality there are transition regions at the edges of the screen. Therefore, total pressure at the rotor inlet as viewed in the relative frame will never be a "step function" but rather a continuous variation as the blade passes downstream of the screen. Figure 3 indicates the pressure distortion patterns.

5. Distortion and Wake Testing in Axial Flow and Compressors

A number of fan and compressor studies have been performed with regard to variations in inlet flow conditions due to distortion. Soeder and Bobula (47) investigated the effect of steady circumferential total pressure distortion on flow characteristics entering an aircraft engine using classical screens. They found that for a transonic turbofan engine, maximum and minimum flow yaw angles in the absolute frame occurred within the constant intensity distortion sector of the flow field as opposed to at the screen edges. They also found the yaw angle is usually the largest in the hub region for the screen configurations they tested. This yaw angle variation increased in magnitude as the flow approached the engine inlet. Increasing the screen blockage increased the yaw angle variation. They also discovered that the inlet pitch angle variation in the plane of the distortion is much smaller than the yaw angle variation, as would be expected for purely circumferential distortions.

Cousins (7) analyzed the unsteady blade surface pressures due to circumferential inlet total pressure distortion in a low-speed axial-flow compressor rig using on-rotor pressure transducers and a telemetry system. Stationary high-response probes were employed to capture wake pressure variations during rotating stall. He showed that it was feasible to develop a transfer function describing the dynamic blade response using data from on-rotor pressure measurements and Fourier transform techniques.

Gauden (18) investigated the performance and stalling behavior of a low-speed axial-flow compressor subjected to three different circumferential inlet distortion levels. He used steady-state instrumentation in the stationary frame of reference and high response pressure transducers mounted on the blades. He discovered that distortion screens reduced the mass flow rate through the compressor due to their low porosity and precipitated stall at a more open throttle valve setting than for undistorted operation.

Gauden found that with respect to the direction of rotor motion, a sharp increase in axial velocity was observed as the distorted segment was approached, implying a decreased angle of attack at the blade leading edge. This was due to flow blockage created by the distortion screens. As the blades passed circumferentially through the distorted flow region behind the screens, the axial velocity was reduced until the angle of attack was maximized, and then returned to its undistorted value at the trailing edge of the screen. The shape of the circumferential velocity profile remained roughly the same as the flow rate was decreased using a throttle valve.

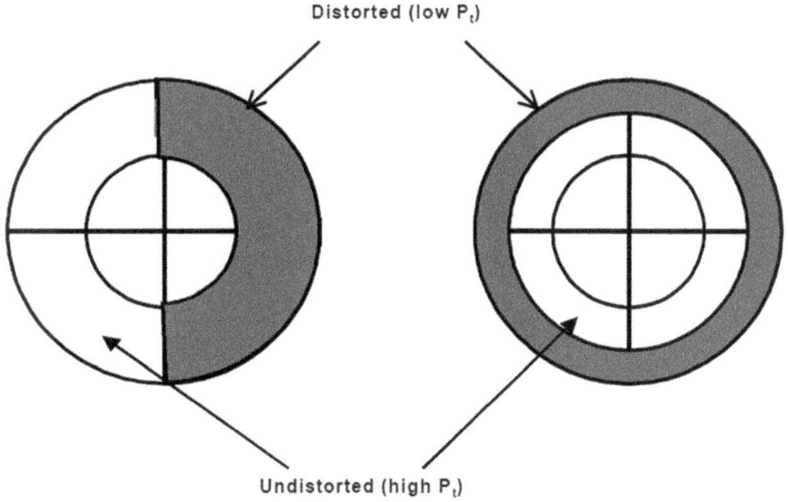

Figure 3: Pressure distortion patterns

Dancy (8) tested the performance and stalling behavior of a low-speed axial-flow compressor with circumferential inlet flow distortion. Similar to the results of Gauden for the same test rig, he discovered that axial velocity increased nearly 10 % as the leading edge of the screen was approached, then fell off sharply and leveled out. An opposite effect occurred at the "departure" end of the screen with respect to rotor rotation. He found that velocities in the undistorted segments of a partially-distorted inlet were higher than the constant flow velocity for a clean inlet. Although the compressor was set to produce the same volumetric flow rate in both cases, the flow rate increase in the undistorted region was insufficient to return the compressor to its original, undistorted volumetric flow rate. As a result, the undistorted compressor always had a higher flow rate. As the back pressure was increased, the flow rates for the distorted and undistorted compressors approached the same value due to a decreased axial velocity and, subsequently, a decreased total pressure drop across the screen. He also found rotating stall for RAF-6 airfoils to originate at the hub of the blade.

The second literature category, blade response to an unsteady incidence that exceeds the steady-state stalling angle, is presented next. Sexton (43) investigated the dynamic stalling characteristics of low-speed axial-flow compressor blades using blade mounted transducers and a multi-channel radio telemetry system. High incidence angles and stalling were induced using a distortion screen mounted in front of the IGV. This screen had a mesh of sufficiently low porosity as to insure incidence angles greater than the steady-state stalling angle. This allowed separation of the blade boundary layer behind the distorted region and reattachment in the undistorted region during each revolution. Sexton then developed a transfer function between the quasi-steady total pressure loss forcing function and the dynamic pressure loss response function. This transfer function in turn described the dynamic response of the rotor blade row flow and made possible the prediction of response to a given inlet distortion for a rotating stall model.

Neal (37) used a multi-channel FM telemetry system in conjunction with miniature blade-mounted transducers to investigate low-speed rotor blade lift response due to circumferential inlet flow distortions. He found that normalized lift for an undistorted compressor decreases as the volumetric flow rate is decreased and angle of attack is correspondingly increased. He explained this unusual behavior by noting that although the coefficient of lift increases with an increase in angle of attack, the decrease in volumetric flow rate causes a decrease in absolute lift. The normalized lift of the distorted compressor first increased and then decreased as flow rate was decreased. He explained this as being due to competing effects of changes in angle of attack, changes in volumetric flow rate, and changes in the level of distortion as the back pressure was increased by closing a downstream discharge valve. He presented an analysis of rotor blade lift and rotor inlet dynamic pressure for a rotor cycle during which the blade experienced a far greater incidence than the steady-state stalling angle.

Neal found that although the blade did eventually stall as it passed well into a distortion; the delay in the inception of stall was significant, with partial stalling of the instrumented rotor blade just prior to the blade's exit from the distorted segment. This corresponded closely to the highest angle of attack on the rotor blade during the rotor cycle. The dynamic stall event was typically characterized by a lift overshoot, which then collapsed and returned to the undistorted lift value after the trailing edge of the screen. This indicated boundary layer reattachment. The phenomenon of a blade experiencing an incidence angle beyond the steady-state stalling angle without stalling was also observed by Melick (35) and Henderson and Horlock (21), and was postulated to be a function of the rotor blade lift response.

Lakshminarayana, et al. (33) studied the effects of rotation and blade incidence on the properties of a low-speed fan rotor wake. They defined a wake semi-width at half the depth on both the pressure and suction sides of the wake and non-dimensionalized them by the rotor blade spacing in the circumferential direction. They found that the wake defect was reduced as rotor rotational speed increased and that the axial velocity wake defect was highest at the hub. Higher rotor speed gave lower axial velocity defect, and higher loading increased the axial velocity defect (or decreased the downstream decay rate of the axial velocity defect). The wake semi-width at a fixed loading did not change with an increase in rotor rotational speed. In addition, they found that wake semi-width was lower at lower loading and the growth was less rapid.

Shreeve and Neuhoff (45) found the wake of a small, transonic single-stage axial compressor to broaden at reduced throttle settings and increased blade speed. Henderson and Shen (20) investigated the influence of unsteady rotor response on a distorted flow field in a low speed axial flow rotor. They found that as the rotor was loaded by decreasing the flow coefficient, the boundary layer thickness, wake defect magnitude, and wake width all increased.

Reynolds, et al. (41) found the wake of an isolated low-speed rotor to be three dimensional in nature with an appreciable radial velocity due to an imbalance in the radial pressure gradient and centrifugal forces. They defined wake semi-widths on the rotor pressure and suction surfaces to obtain a width parameter which was non-dimensionalized by blade spacing.

Ravindranath and Lakshminarayana (39) performed an experimental study of the three-dimensional characteristics of the mean relative frame velocity in the wake of a moderately loaded compressor rotor blade. They defined a non-dimensional semi-width parameter as the sum of the characteristic widths on the pressure and suction surfaces normalized by the semi-blade spacing. They found the effect of blade loading was to sustain the wake asymmetry to a much larger extent downstream of the blade trailing edge. Increased loading also increased the velocity defect magnitude, slowed the decay rate, and induced higher radial velocities. They found the wake width to vary considerably in the radial direction. The width increased towards the hub- and annulus-walls, which was attributed to the complex interaction of the wake, hub-, or annulus-wall boundary layers, and secondary flows (tip vortex in the case of the annulus-wall). They found the static pressure to vary across the wake as well as in the wake near the blade trailing edge due to inviscid effects, which is not reflected in total pressure plots. The static pressure was highest at the center of the wake. They compared the variation of the first two Fourier coefficients of velocity with downstream distance and studied the averaged Fourier coefficients and scatter of the harmonic content in the rotor wake.

Reynolds and Lakshminarayana (40) studied the three-dimensional relative flow characteristics of a lightly loaded low-speed rotor wake. They found that increased loading slowed the decay rates of the axial and tangential mean velocity components and radial velocities in the wake. They found that wake width increased with loading, and that only in the far wake was it acceptable to assume a negligible static pressure variation.
Also, the axial and tangential components of mean velocity were highly asymmetric about the wake centerline, a trait which was more pronounced for increased loading. They found the wake width to increase with radial position about mid-radius and speculated that this may have resulted from large radial transport of mass, momentum, and energy.

Muhlemann (36) performed experiments that showed strong radial variation of the wake with the largest mean velocity defects and wake widths near the hub and tip regions. He also found a more rapid decay of wake width as blade loading was increased, a result verified by Ufer (36).

Kerrebrock, et al. (28) found considerable randomness in absolute exit flow angle at several radii in the wake of a blowdown transonic rotor, even when wake pressure and velocity profiles were quite sharp. They suggested the source of time dependence as random effects from the hub and tip regions influencing the entire blade span.
Shreeve, et al. (46) found impact pressures to be a jet on the suction side of the blades in the wake of a transonic rotor. Shreeve and Neuhoff (44) found the absolute velocity in the wake of a transonic rotor to have a larger magnitude on the suction side of the blades, a phenomena which was more evident at the hub where blade and flow velocities were subsonic. Schmidt and Okiishi (42) found a higher axial velocity downstream of the suction side of the blade when measured in the stationary frame, a phenomena which was, again, more pronounced at the hub region.

Ravindrath and Lakshminarayana (39) found higher total relative velocity and stagnation pressure profiles downstream of the suction side of a low-speed rotor blade, indicating that the phenomena of a suction side jet is not limited to stationary measurements. Cherrett and Bryce (5) studied the unsteady three-dimensional exit flow fields in the stationary frame of a single-stage transonic fan, comparing random stagnation pressure unsteadiness to ensemble-averaged stagnation pressure unsteadiness. They found that in the stationary frame it is possible for stagnation pressures to be higher on the suction, rather than the pressure, surface side of the passage, and for a wake in the relative frame to appear as a jet (pressure excess) in the stationary frame. In the wake region, the stagnation pressure rose rapidly from a pressure trough on the pressure surface side of the rotor wake, to a pressure peak on the suction surface side of the wake. The magnitude of the suction side peak relative to the mean level was two to three times that of the pressure trough. The point in the wake region where the stagnation pressure rose above time averaged values corresponded approximately to the position of maximum random unsteadiness.

A second stagnation pressure trough occurred on the suction surface side of the wake followed by a peak comparable to the strength of that attained within the suction surface side of the rotor wake. Wake unsteadiness was found to be three to four times that elsewhere in the blade passage, with peak unsteadiness at mid-pitch. Ensemble averaged wake stagnation pressures indicated little change in amplitude with increased compressor loading at the same speed, with the exception of the lower blade spans toward the hub where rotor inlet Mach numbers approached unity.

Colpin and Kool (6) studied the propagation of a non-uniform upstream flow field through a low-speed axial-flow compressor stage rotor. Circumferential total pressure distortion was created using a grid which was rotatable with respect to the stationary instrumentation. They found that the distortion was indicated more strongly in flow angles than velocities downstream of the rotor. They found that as a blade passed into a distortion the boundary layer on the suction side thickened due to the increase in loading. Moving further into the distorted region, the axial velocity outside of the wake increased due to increasing blockage caused by the wake. A strong reduction in the relative outlet flow angle corresponded to stronger centrifugation, or radial flow, in the blade wakes. Moving further still into the distorted region, the blade wake reached its maximum thickness indicating boundary layer separation on the suction side while the free stream axial velocity mean value increased to compensate for the additional boundary layer blockage. When reaching the distortion trailing edge, the boundary layer on the blades tended to reattach, inducing a thinner wake. After the blade passed through the distorted region, the investigators noted a reduction in wake circumferential extent and depth, as well as a slight increase in axial velocity. Colpin and Kool also noted that the flow turbulence increased at the inlet and exit of the rotor while in the distorted region. As the suction side boundary layer separated, the turbulence level suddenly grew due to the development of a large wake.

It is interesting to note that Colpin and Kool also experienced incidence angles greater than the quasi-steady stall value. A strong decrease in wake dynamic pressure corresponding to the maximum incidence angle reinforced the observation of blade dynamic stall.

6. Conclusion

The knowledge of inlet efficiency would be outcome of this review. The ramp angle varies continuously due to pressure on ramp that might change the working of the inlet in airplanes. We would also deal with the cooling channel with cooling center what would beneficial for the air cooling for what air inlet is responsible. Cooling center deals with the design of heat exchanges while Cooling channel deals with the inlet, outlet and channel design. However the study has carried in this review focused on inlet efficiency and analytical study of varying ramp angle in combination with air systems inlets.

References

[1] A. Grimminger. Current Practice for Meshing of High Lift Con_gurations with Centaur. Technical report, Airbus, 2009.

[2] Alvin H. Sacks and John R. Spreiter, 1951, Theoretical investigation of submerged inlets at low speed. Technical report, National Advisory Committee for Aeronautics

[3] Anica Bare. CFD Investigation on the Inuence of Wind Tunnel Enclosure and Model Mounting on the Characteristics of the Tails of a Transport Aircraft Con_guration. Technical report, Airbus, 2010.

[4] Aurelien Rigollet, 2009, CFD computations for RAM air inlets. Master's thesis, Ecole Nationale Superieure de Mecanique et d'Aerotechnique

[5] Cherrett and Bryce: study of the unsteady three dimensional exit flow fields in stationary frame.

[6] Colpin and Kool: study of propagation of non uniform upstream flow field through a low speed axial flow compressor
[7] Cousins: analytical view of unsteady blood pressure due to circumferential inlet
[8] Dancy: performance testing and stalling behaviours of low axial flow compressor
[9] David C. Wilcox, 2010, Turbulence Modeling for CFD. DCW Industries, 3rd edition
[10] Dietrich John. A Program to Evaluate the Boundary Layer Flow Field from Navier-Stokes Results in Unstructured Grids. Airbus, 2004.
[11] DNW german-dutch wind tunnels: TWG. http://www.dnw.aero/windtunnels.aspx?id=304&menuid=34&subid=304, May 2011.
[12] Dr. Bruno Stefes. Skill Group Leader of the Fuselage Air Systems Group at Airbus. Verbal communications.
[13] Dr. Christian Bartels. Senior Expert in Aircraft System Aerodynamics at Airbus. Verbal communications.
[14] Dr. Martin Schmid. ECOcents Cooling Channel: CFD Analysis TWG environment. Airbus, 2010. Powerpoint presentation.
[15] Emmet A. Mossman and Lauros M. Randall, 1948, An experimental investigation of the design variables of NACA submerged duct entrances. Technical report, National Advisory Committee for Aeronautics
[16] ESDU. 03006, 2003, Subsonic drag and pressure recovery of rectangular planform ush auxiliary inlets with ducts at angles up to 90 degrees
[17] ESDU. 86002, 2006, Drag and pressure recovery characteristics of auxiliary air inlets at subsonic speeds
[18] Gauden: investigation of the performance and stalling behaviours of a low axial flow pressure
[19] Getting to grips with aircraft performance, January 2002. Airbus.
[20] Henderson and Shen: investigation of the influence of unsteady rotor response on a distorted flow field in a low speed axial flow rotor
[21] Henderson and Horlock: postulated to be a function of the rotor blade lift response
[22] Holger Dettmar. Hybrid-CFD Training, CAD and Mesh generation. Airbus, 2009. Powerpoint presentation.
[23] Institute of Aerodynamics and Flow Technology, Braunschweig. Technical Documentation of the DLR TAU-Code, 2010. Release 2010.2.0.
[24] John D. Anderson, 2008, Introduction to Flight. McGraw-Hill Higher Education, 6th edition
[25] John D. Anderson, 2006, Fundamentals of Aerodynamics. McGraw-Hill Higher Education, 4th edition
[26] John D. Anderson, 1995, Computational Fluid Dynamics. McGraw-Hill Higher Education, international edition
[27] Jiri Blazek, 2006, Computational Fluid Dynamics: Principles and Applications. Elsevier
[28] Kerrebrock, et al.finds considerable randomness in absolute exit flow
[29] Lauros M. Randall Charles W. Frick, Wallace F. Davis and Emmet A. Mossman, 1945, An experimental investigation of NACA submerged duct entrances. Technical report, National Advisory Committee for Aeronautics
[30] Luis Fernando Figueira da Silva Antonio Batista de Jesus Cesar Celis Perez, Sandro Barros Ferreira and Guilherme Lara Oliveira, 2007, Computational Study of Submerged Air Inlet Performance Improvement Using Vortex Generators. Journal of Aircraft, 44(5)
[31] John Seddon and E.L. Goldsmith, 1985, Intake Aerodynamics. Blackwell Publishing Limited
[32] John Dietrich. TAU Code Validation According to the Boundary Layer Flow_eld. Technical report, Airbus, 2007.
[33] Lakshminarayana, the effects of rotation and blade incidence on the properties of a low-speed fan rotor wake
[34] Markus Widhalm. Windtunnel Boundary Conditions. German Aerospace Center, 2011. Powerpoint presentation.

[35] Melick: observation of the phenomenon of a blade experiencing an incidence angle beyond the steady-state stalling angle without stalling

[36] Muhlemann and Ufer: shown experimentally the strong radial variation of the wake with the largest mean and verified simultaneously

[37] Neal: application of multi-channel FM telemetry system in conjunction with miniature blade-mounted transducers to investigate low-speed rotor blade lift response due to circumferential inlet flow distortions

[38] R. P. Dwight. Engine Inow Boundary Conditions for Speci_cation of Mass-Flow. Technical report, DLR, 2005.

[39] Ravindranath and Lakshminarayana: an experimental study of the three dimensional characteristics of the mean relative frame velocity

[40] Reynolds and Lakshminarayana: three dimensional relative flow characteristics of a lightly loaded low speed rotor wake

[41] Reynolds research on the isolated low speed rotor

[42] Schmidt and Okiishi:finds higher axial velocity downstream of suction side of the blade and measured in the stationary frame

[43] Sexton: investigation of dynamic stalling characteristics of low-speed axial-flow compressor blades using blade mounted transducers and a multi-channel radio

[44] Shreeve and Neuhoff: finds absolute velocity in the wake of transonic rotor

[45] Shreeve and Neuhoff: found the wake of a small, transonic single-stage axial compressor to broaden at reduced throttle settings and increased blade speed

[46] Shreeve, et al.find impact pressure to be jet on the suction side

[47] Soeder and Bobula: investigation of the circumferential distortion on flow characteristics

[48] TAU-Code User Guide, 2010. Release 2010.2.0-p1.

[49] Thomas Gerhold, 2007, TAU Training Document - TAU Overview

[50] Thomas Stevens. CFD results comparison of System operating point at Cruise conditions. Airbus, 2011. Powerpoint presentation.

[51] Udo Krause. Research Engineer in Aircraft Aerodynamic Design at Airbus. Verbal communications.

7

CFD ANALYSIS OF LOW SPEED TURBULENT FLOW PAST A DELTA WING

Abhishek Patil [1],Piloo Raghu Kumar[1], Aditya Kodkani[1] and Sekhar Majumdar[2]
[1]Undergraduate Student, [2]Professor
Department of Mechanical Engineering
Nitte Meenakshi Institute of Technology, Karnataka, INDIA

Abstract

Delta wing has been a subject of intense research since decades due to decades due to inherent characteristics of generating increased nonlinear lift due to vortex dominated flows. Lot of work has been carried out in order to understand the vortex dominated flows on the delta wing. The delta wing is a wing platform in the form of a triangle. Aerodynamics of wings with moderate sweep angle is recognized by the aerospace community as a challenging problem. In spite of its potential application in military aircraft, the understanding of the aerodynamics of such wings is far from complete. In order to address this situation, the present work is initiated to compute the 3D turbulent flow field over sharp edged finite wings with a diamond shaped planforms and moderate sweep angle. The detailed flow pattern and surface pressure distribution may further indicate the appropriate kind of flow control during flight operation of such wings. The flow field is computed using an in-house developed CFD code RANS3D.

Keywords: Delta Wing, RANS solver, Turbulent Flow

1. Introduction

Computational Fluid Dynamics (CFD) has evolved today as a potential engineering design tool for both internal and external flow in complex arbitrary shaped configuration. A well-validated CFD code can save a lot of design cycle time, cutting down number of tests on similar geometry or under similar flow conditions. The CFD code RANS3D, solving the Reynolds Averaged Navier Stokes (RANS) equations in curvilinear boundary-fitted coordinates for conservation of mass, moment and other scalars in three dimensional turbulent incompressible flows, has already been developed and extensively validated by the present project supervisor and his research group at the CTFD Division, NAL Bangalore during the last two decades.

The Delta Wing shape has the advantage that the leading edge remains behind the shock wave generated by the nose of the aircraft when flying at supersonic speed. A second advantage is that, as the angle of attack is increased, the leading edge vortices are formed and enhance the lift coefficient of the aircraft. The delta wings therefore stall only at a very high angle of attack. . The major focus of the present work is on numerical simulation of 3D turbulent incompressible flow past an isolated backward swept (Sweep Angle = 620) Delta Wing made of a Biconvex Aerofoil section. The computation uses the pressure based finite volume RANS code RANS3D coupled to eddy viscosity based turbulence model and a multiblock structured boundary orthogonal curvilinear grid. Both the grid generation code MESHGEN and the flow solution code RANS3D have been developed recently at NAL Bangalore by the present supervisor and his research group at NAL. The aerodynamic performance of Biconvex Aerofoil section at Re = 0.36×10^6 is estimated from the predicted distribution of the surface pressure and the wall shear stress, derived from CFD analysis for both 2D and 3D configuration.

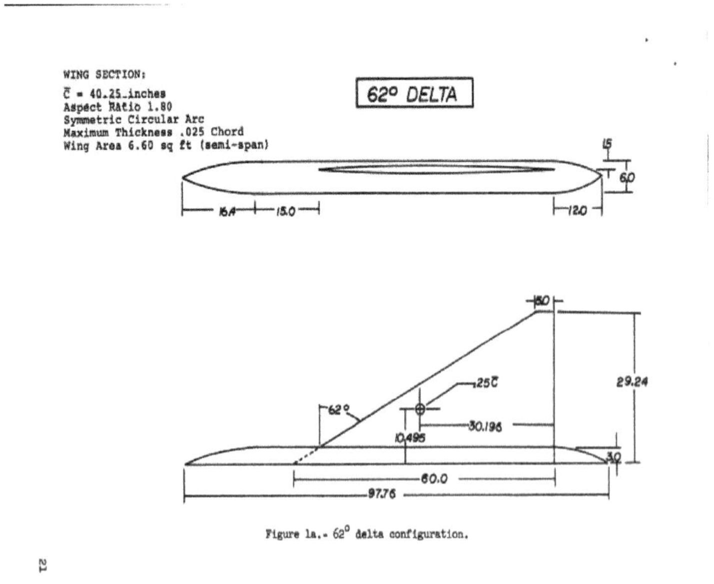

Fig. 1.Configuration for the Delta Wing with a Sweep Angle = 62⁰

2. Finite Volume Method

2.1 *Governing Equations of Fluid Motion*

The conservation of mass and momentum equations for turbulent incompressible flow in an inertial frame of reference are expressed by the continuity equation and the Reynolds Averaged Navier Stokes equations respectively. These equations may be written in the following tensor form using general non-orthogonal curvilinear coordinates where j, k and m are the summing indices; μ and ρ are the fluid viscosity and density respectively; and are the time-averaged pressure and cartesian velocity component respectively; ui is the corresponding fluctuating velocity component due to turbulence. J is the transformation Jacobian between the cartesian and the curvilinear coordinates and and are the relevant geometric coefficients related to the transformation

Momentum Conservation :

$$\frac{\partial \overline{U}_i}{\partial t}+\frac{1}{J}\frac{\partial}{\partial x_j}\left[\left(\overline{U}_i\overline{U}_k\beta_k^j\right)-\frac{\nu}{J}\left(\frac{\partial \overline{U}_i}{\partial x_m}B_m^j+\frac{\partial \overline{U}_k}{\partial x_m}\beta_i^m\beta_k^j\right)+\frac{p}{\rho}\beta_i^j+\overline{u_iu_k}\beta_k^j\right]=0 \qquad (1)$$

Mass Conservation : $$\frac{\partial}{\partial x_j}\left((\overline{U}_k\beta_k^j\right)=0 \qquad (2)$$

2.2 Turbulence Models

The Reynolds Stress tensor is evaluated through appropriate turbulence models. The Linear Eddy Viscosity (LEV) based models, most widely used in RANS computation of complex flows, assume the Reynolds stress tensor components to be directly proportional to the mean strain rates as follows:

$$-\overline{u_iu_k}=\frac{\nu_t}{J}\left(\frac{\partial \overline{U}_i}{\partial x_n}\beta_k^n+\frac{\partial \overline{U}_k}{\partial x_m}\beta_i^m\right)-\frac{1}{3}\delta_{ik}\overline{u_mu_m} \qquad (3)$$

where, δ_{ij} is the Kronecker Delta and the subscript m is a summation index. The eddy viscosity ν_t is evaluated from the relationship with the local turbulence scalars as following:

$$\nu_t = C_\mu E_s T_s \qquad (4)$$

where Es and Ts are appropriate energy scale and time scale respectively defining the local turbulence level, and C_μ is a model constant.

In standard k-ε model, used for high Reynolds number flows, proposed originally by Launder & Spalding [6], the eddy viscosity, in Eq. 3 for Reynolds Stress is computed from a turbulence velocity scale and a turbulence length scale which are predicted at each point in the flow via solution of the following transport equations for turbulent kinetic energy and its dissipation rate.

$$D_t k = P_k - \varepsilon + \nabla \cdot \left(\left(\nu + \frac{\nu_t}{\sigma_k} \right) \nabla k \right) + D \qquad (5)$$

$$D_t \varepsilon = \frac{C_{\varepsilon 1} P_k - C_{\varepsilon 2} \varepsilon}{T} + \nabla \cdot \left(\left(\nu + \frac{\nu_t}{\sigma_\varepsilon} \right) \nabla \varepsilon \right) \qquad (6)$$

Where the eddy viscosity ν_t, the turbulent time scale T and the production term P_k are expressed as following in which the S_{ij} denotes the mean strain rate. D_t represents the total derivative of the relevant flow variable with respect to time representing the convective terms

$$\nu_t = C_\mu k T \; ; T = \frac{k}{\varepsilon} \; ; P_k = 2\nu_t \left| S_{ij} S_{ij} \right| ; S_{ij} = \frac{1}{2}\left(\frac{\partial \overline{U}_i}{\partial x_j} + \frac{\partial \overline{U}_j}{\partial x_i} \right)$$

The model constants are $C_{\varepsilon 1} = 1.44$, $C_{\varepsilon 2} = 1.92$, $C_\mu = 0.09$, and the exchange coefficients for the turbulence scalars are

$\sigma_\varepsilon = 1.3$ and $\sigma_k = 1.0$. However these turbulence equations are not strictly valid in the near wall region which needs some special treatment. In the Standard Wall Function method [6], used in the present computation, the first near wall point is assumed to be in the layer where the logarithmic law of wall is used for computation of the mean velocity and the turbulence scalars.

2.3 Numerical Solution of the Finite Volume Equation

Gauss Divergence theorem is used to transform the governing RANS equations (Eq. 1), into the corresponding finite volume analogue in terms of surface flux balance for each control volume. An implicit predictor-corrector method based on a pressure-velocity solution strategy, similar to the SIMPLE Algorithm of Patankar et al [7] is used for numerical solution of the finite volume equation system. Second order accurate Linear Upwind scheme is used for spatial discretisation of the convective fluxes. Using the relevant geometric factors, appropriate discretisation schemes and linearisation of the source terms, the flux balance equations to be solved for momentum and turbulence scalars are expressed in a generalised implicit manner as follows at the predictor step:

$$A_P \phi_P^{n+1} = \left(\sum A_{nb} - SP \right) \phi_P^{n+1} = \sum A_{nb} \phi_{nb}^{n+1} + SU \qquad (7)$$

where, the coefficient A_{nb} represents the combined effect of convection and diffusion at the four faces of a quadrilateral computational cell ; SU and SP are the components of the linearised source term and is the cell volume. In the corrector step, the continuity equation is also transformed to a similar linearised equation for pressure correction in the form of Eq. 7. The corrections for pressure and velocity field obtained are added to the pressure and the momentum-satisfying velocities at the cell centers and cell faces, obtained at the predictor step. The derivation of Eq. 7 and the decoupled iterative procedure to handle the pressure-velocity link are given in details elsewhere [3-5]. The system of linearised equations (Eq. 7) are solved at each outer iteration level using the Strongly Implicit Procedure of Stone [8].

3. Results and Discussion

The computational and experimental results for turbulent flow around a biconvex delta wing section. Aero foil and also the prediction of three dimensional flow past a finite wing of low aspect ratio of 1.61 with constant biconvex cross section all through the span. The Delta wing of biconvex cross section have been studied mainly for validation of the code "RANS3D" where the present predictions are compared to corresponding measurement data reported in literature [17]. All the computations use the Reynolds Averaged Navier Stokes solver RANS3D for turbulent incompressible flows in complex configuration. The flow solver is provided with six different convective flux discretization schemes and six different turbulence models.

The present 3D computations employ third order accurate quadratic upwind scheme QUICK for discretization of convective flux, coupled to the eddy viscosity based k-ε turbulence model for low Reynolds number flows, proposed by Chien [5]. This turbulence model uses special near-wall exponential damping functions, which allow the eddy viscosity to reduce from its local turbulence-dependent value in the boundary layer to zero on the wall. The objective of these computations is twofold:

(1) To assess the performance and accuracy of the RANS3D code in predicting turbulent incompressible flow past biconvex delta wing sections.

(2) To predict the aerodynamic performance of biconvex delta wing cross section.

The prediction results are compared to available measurement data for the chord wise distribution of surface pressure and also for the variation of aerodynamic coefficients like lift, drag and pitching moment coefficients at different angles of attack. Two dimensional flow patterns around the biconvex delta wing and the corresponding pressure and turbulence energy field for different angle of attacks have also been presented. Flow past a three dimensional wing having biconvex delta wing profile and a rectangular planform with aspect ratio of 1.61 is computed to demonstrate the capability of the 3D flow code The flow results are discussed with adequate physical explanation to justify the prediction results.

The present computation uses H-Grid topology where the wall normal distance of the first near wall grid point is maintained at the order of 1.67x10-5 C for a reasonable resolution of the boundary layers at the operating flow Reynolds number of 0.36 million. The oncoming flow is assumed to be uniform with constant values of the turbulence kinetic energy (k) and its dissipation (ε) decided by the assumption that the turbulence level is 0.1% and the eddy viscosity level is ten times the laminar viscosity of the fluid.

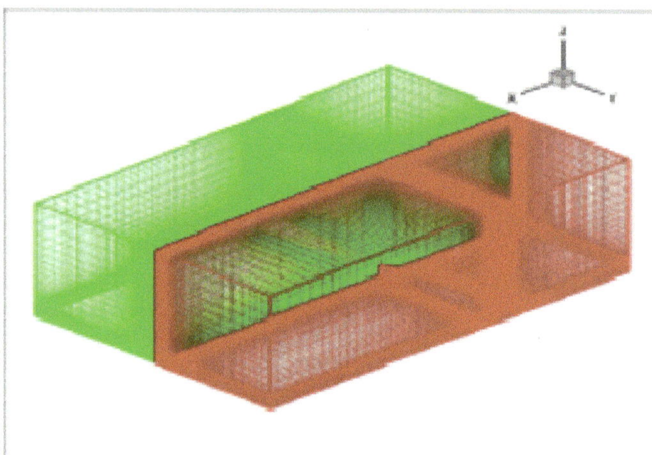

Fig. 2. H grid topology around Aerofoil

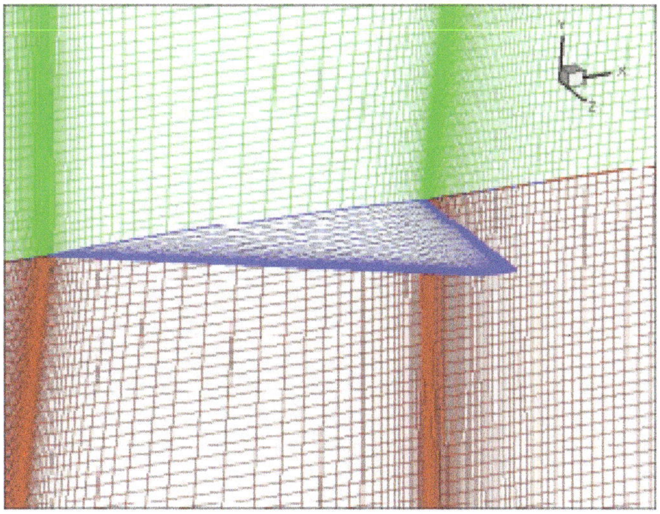

Fig 3. Zoomed view near the aerofoil

3.1 Numerical Convergence

The transport equations are solved in RANS3D to satisfy the convergence criterion at each time step for the time accurate solution. In case of steady state problems however, one single large time-step is used so that the contribution of the unsteady term is reduced to zero and all the equations are solved iteratively up to the prescribed convergence limit. A convergence criterion of 10^{-5} is used in the present study for the maximum

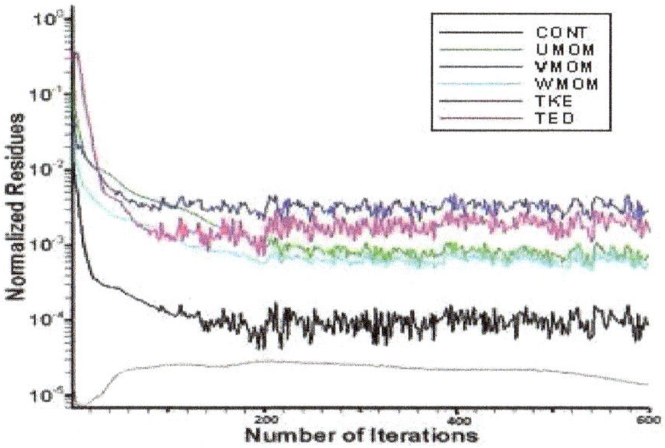

Figure. 4. Numerical convergence history for simulation of flow pas biconvex delta wing

normalized residue of the equations solved and similar behavior of the convergence process has been demonstrated by all the problems studied in the present work. The convergence history is shown in the form of variation of the L2 Norm of the normalized residues (logarithmic scale) of each equation solved, versus the number of iterations for the computation of 3D flow past a Biconvex Delta Wing aero foil. The numerical convergence process for the continuity, for the two relevant momentum equations and for the two transport equations for scalars k and ε are observed to be more or less oscillatory whereas the residues decay by almost four decades in about 532 iterations.

3.2 Flow pattern around the aero foil

Figs.5 shows the flow pattern around the aero foil section. The pictures show the particle traces or streamlines plotted using the Tecplot360 software, based on the computed velocity vectors for the flow past Biconvex Delta Wing at Four different angles of attack. In each case, the flow is observed to hit the aero foil at the so-called stagnation point near the leading edge and divided into two streams – one along the upper and the other along the lower surface of the aero foil. Further the stagnation point moves upwards or downwards from the leading edge as the angle of attack changes to negative or positive values respectively. The flow is observed to remain attached all over the upper and lower surface of the aero foil for all the four angles of attack before the stalling condition.

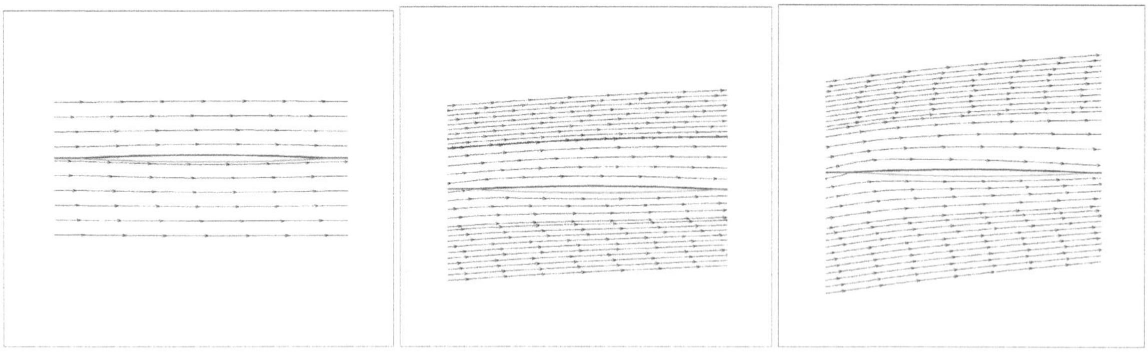

$\alpha=0^0$ $\alpha=4^0$ $\alpha=8^0$

Figure 5(a): Streamline traces with different angles of attack (Re=0.36x10^6)

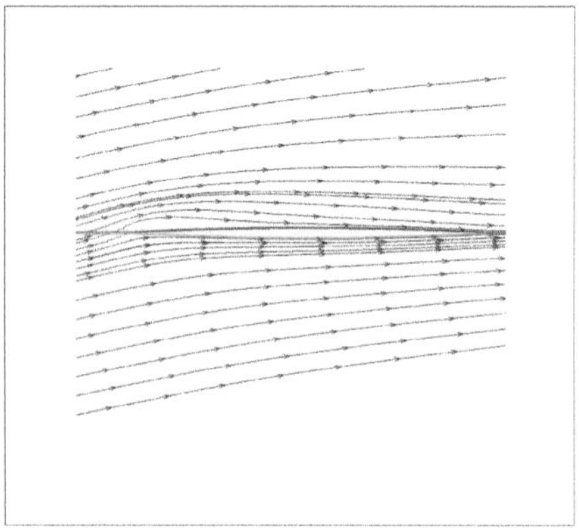

$\alpha=12^0$

Figure 5(b): Streamline traces (Re=0.36x10^6)

3.3 Pressure field around the biconvex delta wing

Fig.6 & 7 shows the contours of constant pressure (Cp) around the aero foil for flow at different angles of attack. The pressure contours for zero angle of attack clearly demonstrate the symmetry of the flow for symmetric aero foil. The highest pressure level at the stagnation point near the leading edge is also observed in the pressure contour plots. The contours also deviate from their symmetric pattern when the angle of attack is non-zero. The flooded color contours also demonstrate clearly the negative pressure value bands on the upper surface and positive pressure values on the lower surface of the aero foil.

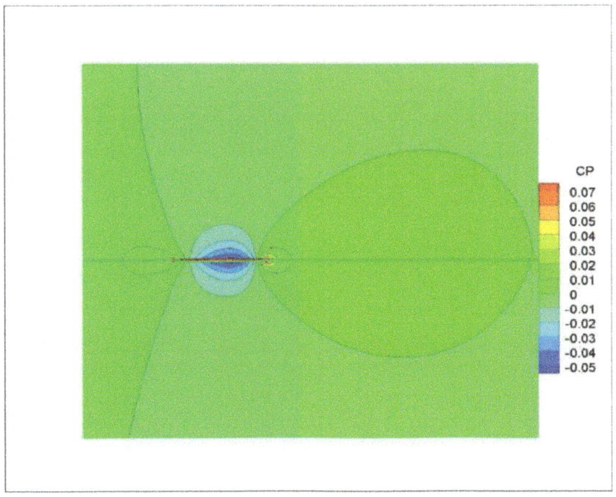

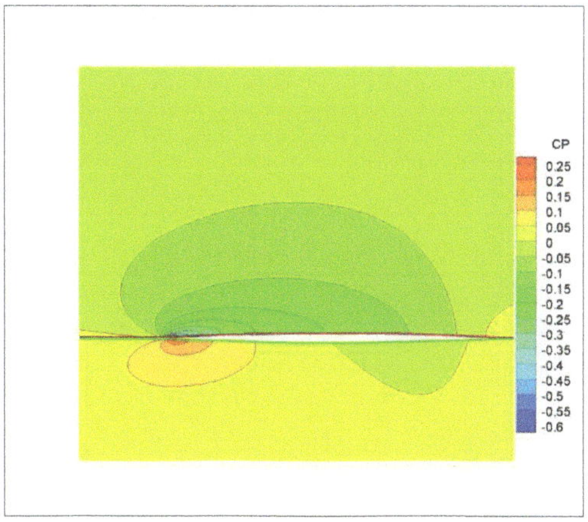

α=0⁰ *α=4⁰*

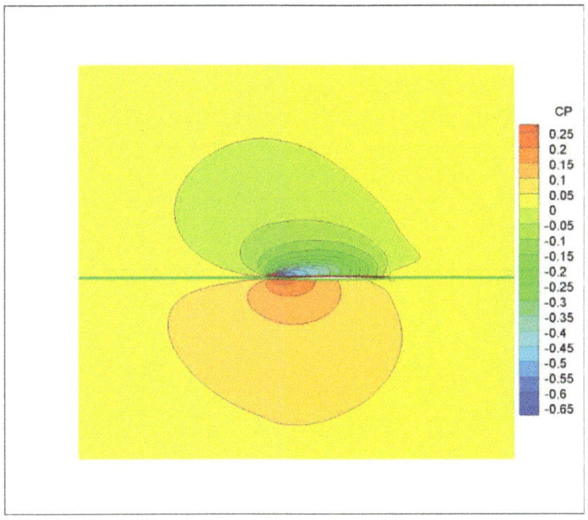

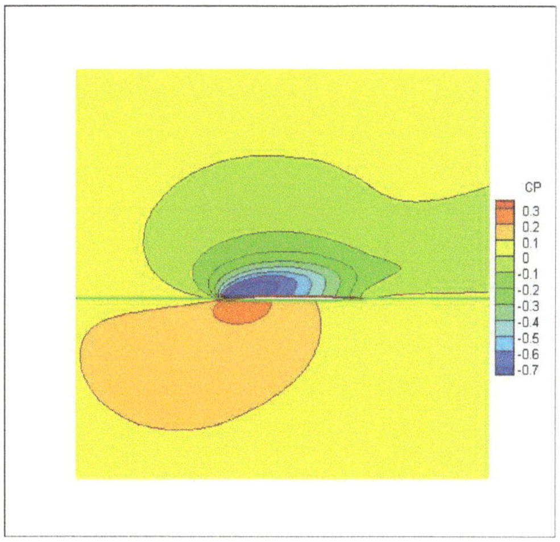

$\alpha=8^0$ *$\alpha=12^0$*

Figure 6: Predicted Pressure contours with different angles of attack (Re=0.36x10^6,,z/Croot =0.27(mid- section))

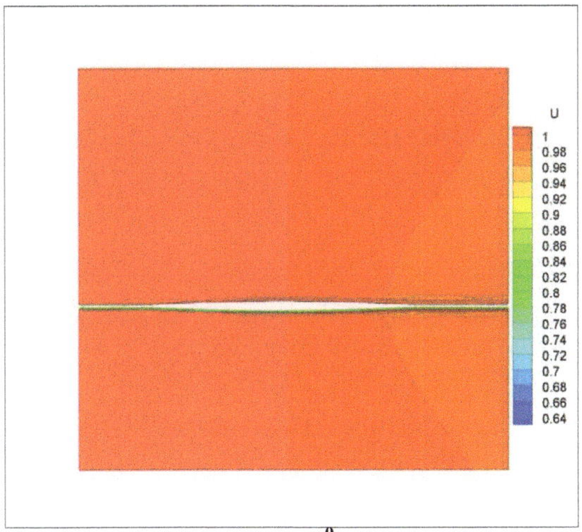

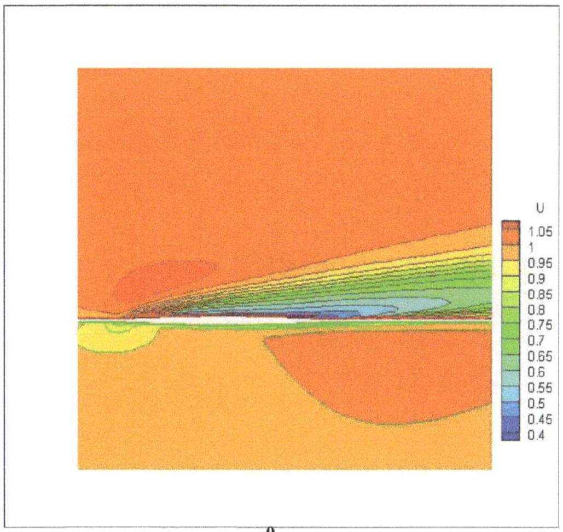

$\alpha=0^0$ $\alpha=4^0$

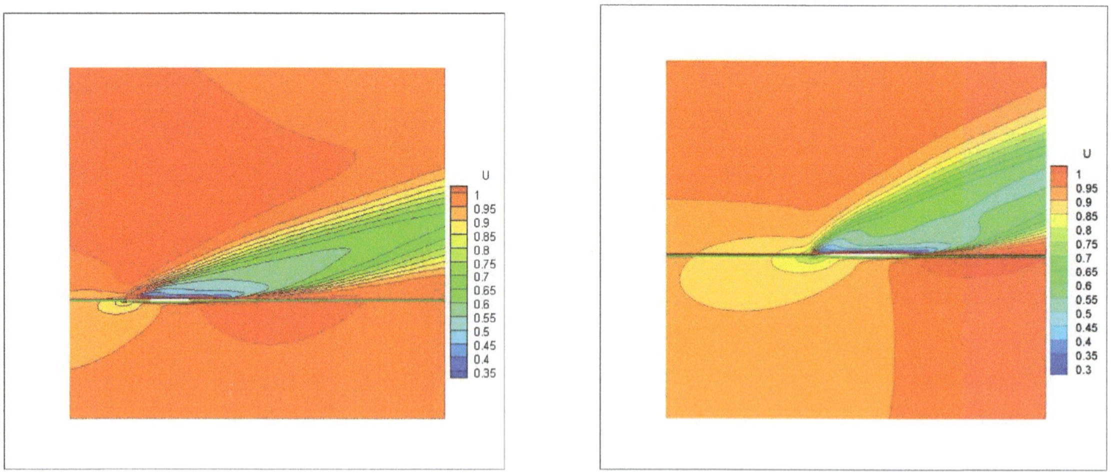

$\alpha=8^0$ $\alpha=12^0$

Figure.7. Predicted pressure contours with different angles of attack (Re=0.36x10^6, z/C=0.45(Tip))

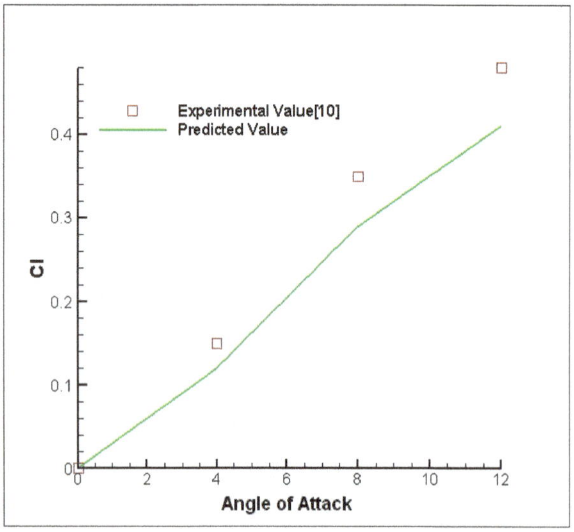

Figure.8. Variation of C_L vs Angle of Attack (α)

4. CONCLUDING REMARKS

1. The computation of flow past a biconvex delta wing cross section with aspect ratio 1.61 shows physically possible flow field, demonstrated by the surface pressure distribution and particle traces on different planes.
2. 3D grid around an isolated backward swept Delta Wing (Sweep Angle = 62^0) is generated by parallel stacking (along spanwise direction z) of the 2D Grid generated by MESHGEN at the root cross section (Biconvex) plane (x-y) of the wing. Beyond the wing tip, the grid planes along the root chord from the upper and lower blocks of the grid are collapsed on a single plane (y=0).

3. The mathematical background and coding of the elliptic grid generator MESHGEN used especially for generating curvilinear boundary-fitted grid for the biconvex wing cross section, has been thoroughly understood in details.

4. The relative performances of different sections are assessed by comparing their surface pressure distribution, skin friction distribution, flow pattern and variation of aerodynamic coefficients

5. ACKNOWLEDGMENT

The authors wish to express their sincere thanks to the Director and Principal NMIT, Bangalore and the Head Mechanical Department, NMIT for their constant encouragement and support in providing the computation facilities of the CFD Laboratory, Mech. Engg. R&D Centre, NMIT.

6. REFERENCES

[1] Fathima, A., Baldawa, N.S., Pal, S., Majumdar, S. (1994), Grid generation for arbitrary 2D configurations using a differential algebraic hybrid method, NAL PD CF9416, April 1994

[2] Majumdar Sekhar, Rajani B.N., Kulkarni D.S., Subrahmanya M.B. (2009), CFD simulation of low speed turbulent flow problems using Unsteady RANS and Large Eddy Simulation approach, Journal of Aerospace Sciences and Technologies, Special Issue, February 2009

[3] Majumdar S., Rajani B.N., Kulkarni D., Mohan S., (2003), RANS computation of low speed turbulent flow in complex configuration, Proc. Symposium on State of the Art and Future Trends of CFD at NAL NAL SP 0301

[4] Majumdar S.,(1998) Pressure based Navier Stokes solver for three-dimensional flow in hydrodynamics and low speed aerodynamics applications, Proc. 3rd Asian CFD Conference, Bangalore, December 7-11, 1998

[5] Majumdar S., Rodi W.,and Zhu J.(1992), Three dimensional finite volume method for incompressible flows with complex boundaries, Journal of Fluids Engineering., ASME, pp. 496-503

[6] Launder B.E., Spalding D.B. (1974), The numerical computation of turbulent flows, Computer Methods and Applied Mechanics and Engineering. vol. 3, p. 269

[7] Patankar S.V. (1980) Numerical Fluid Flow and Heat Transfer, Hemisphere Publishing Corporation, McGraw Hill Book Company

[8] Stone H.L. (1968), Iterative solution of implicit approximations of multidimensional partial differential equations, SIAM Journal of Numerical Analysis, vol. 5, pp. 530-538G.

[9] Chen C. J and Jaw S. Y, (1998), Fundamentals of Turbulence Modeling, Taylor and Francis.

[10] Kulkarni DS, Rajani B N, Majumdar S (2001), Studies on temporal and spatial discretization schemes used in a pressure based RANS algorithm for Incompressible flow, Project Document CF 0108, NAL, Bangalore.

[11] Mansour N. N, Kim J. and Moin P., (1988), "Reynolds-stress and dissipation-rate budgets in a turbulent channel flow", Journal of Fluid Mechanics, vol. 194, pp. 15-44

[12] Patankar S. V (1980), "Numerical Heat transfer and fluid flow", Hemisphere publ. Co

[13] Weber C. and Ducros F., Large Eddy and Reynolds Averaged Navier-Stokes simulations of turbulent flow over an aerofoil. IJCFD,13:327.355,2000

[14] Wilcox, (1993), Turbulence modeling for CFD, DCW Industries Inc., California.

[15] Zhu, J (1990), " A hybrid-differential-algebraic method for three-dimensional grid generation ", International Journal for Numerical Methods in Engineering, vol. 29, pp. 1271-1279,

[16] Majumdar S, (1988), Role of Under-relaxation in momentum interpolation for calculation of flow with non-staggered grids, Numerical Heat Transfer, Vol.13, pp 125-132.

[17] William H. Wentz, Jr. and Michael C. McMahon., " An experimental investigation of the flow fields about Delta and Double-Delta wings at low speeds"

Comparative Studies on Buckling Analysis of Thin Walled Dropped Flange and Rounded edge Channel beams of Aircraft Structures

Sudhir Sastry Y B[1], Y krishna [2], S S Prasad[3], Syed Mohsin A Q[3]

[1]*Dept. of Aeronautical Engg. Institute of Aeronautical Engineering, Hyderabad, India,*
3Dept. of Aeronautical Engg. MLR Institute of Technology, *Hyderabad, India,*
[2] *scientist G, Head, Structural Test Facility, DRDL, Hyderabad, India,*
Corresponding Author, email:sudhirsastry@gmail.com

Abstract

Thin walled members are very important part of aircraft structures. These thin walled structures constitute thin plates fixed on open section beams whose Centroid does not coincide with shear center. The attention here is focused on open section channel beam subjected to buckling loads. Both rounded closed end and rounded open flange end are studied where the main variables are thickness of the sheet, radii of the ends. Finite Element Analysis of these structures for buckling is done using ANSYS .Also, the influences of various shapes and sizes of the flanges are analyzed to get the critical moments. The results are comparisons between cases and with classical solutions are included in this paper which leads to the best types of open section beams for thin walled structures.

Keywords: Thin walled beams, buckling analysis, Finite element analysis

1. Introduction

Marcello pignataro [1] on the basis of the general theory of elastic stability, post buckling analysis of simply supported channels under uniform compression is performed. M.Ma &o. Huges [2] Energy method is developed for analyzing the lateral buckling behavior of the monosymmetric I beams subjected to distributed vertical load, with full allowance for distortion of web. k Magnucki, [4] Thin walled structures where a thin-walled beam is in pure bending subjected to couples m0. The open cross-section profile has two ribs, with cross-section a0 and it is shaped symmetrically towards the plane perpendicular to the bending plane. The ribs are located at the profile ends. The shape of the profile line is searched for. Criterion is the minimal value of the cross-section area a1 of the beam. The problem in this paper is described by means of variational calculus. While numerical calculations are performed using runge–kutta method. Krzysztof Magnucki, [5] aimed at searching for the optimum shape of an open cross-section of a thin-walled beam under strength and stability constraints and presented it through their paper optimum shape of the open cross-section of a thin-walled. B. W. Schafer [6] Open cross-section, thin-walled, cold-formed steel columns have at least three competing buckling modes: local, distortional, and Euler i.e., flexural or flexural-torsional buckling and provided Accurate closed-form methods for their prediction.. A direct strength method is proposed for column design. Nuno Silvestre [7 Thin-Walled Channel Columns and Beams which after giving a brief outline of the Generalized Beam Theory (GBT) fundamentals and linear stability analysis procedure, presented the main concepts and steps involved in the derivation of GBT -based fully analytical formulae are described and discussed. Xiao-ting [8] lateral torsional buckling of thin walled channel section beams restrained by metal sheeting when subjected to an uplift load. N. S. Trahair, [9] simple advanced method of designing steel members against out-of-plane failure. Cheng yu and Schafer, [10] Simulation of cold-formed steel beams in local and distortional buckling with applications to the direct strength A nonlinear finite element model is developed to simulate two series of flexural tests Attard Mario [11] considers the effects of shear deformations. Choi-Ping Poon [12] Jaehong Lee [13] explained Lateral buckling of thin-walled composite beams with monosymmetric. Tomasz kubiak [14] analysis of local buckling of thin-walled beam-columns. Krzysztof Magnucki [15] Elastic buckling of selected flanges Of cold-formed thin-walled beams. Lei zhang, [16] continued their previous studies related to two representative theories with slight differences between them, widely used in investigating the flexural-torsional buckling of thin-walled beams. Foudil Mohria, [17] derived analytical solutions Based on a non-linear stability model, for simply supported beam-column elements with bi-symmetric I sections under combined bending and axial forces. M.m. pastor, [18] Experimental study of open thin-walled section beams subjected to pure bending. Ben young [19] in 2008 reported in local buckling and shift of effective Centroid of slender sections that local buckling is a major consideration in the design of slender sections. E. Magnucka-blandzi [20], reviewed a paper to the buckling and optimal design of cold-formed thin-walled beams with selected problems where cold-formed thin-walled channel beams with open or closed flanges. M. M. Pastor, [21] in 2009 continued their work starting from the set of bending tests of U sections performed. K. Magnucki [22] Poland in the year 2009 devoted a paper to cold-formed thin-walled channel beams with open or closed profile of drop flanges. Geometric properties of two c-sections are described with consideration of warping functions and warping inertia moments. Optimization criterion and the dimensionless objective function as a quality measure are defined. Constraints of feasible solutions are strength, global and local buckling conditions and also geometric condition. Analytical solutions of the problems of global and local buckling for thin-walled beams are presented. P. Borges Dinis, [23] in the year 2010 along with Dinar Camotim of Department of Civil Engineering, ICIST/IST, Technical University of Lisbon, Portugal reported a paper on numerical investigation concerning the post-buckling behavior of cold-formed steel lipped channel beams, subjected to uniform major axis bending. Jerzy Lewinski [24] deals with cold-formed thin-walled beams with the Z-, S- and Clothoid-section. Zbigniew kołakowski [25] A plate Model is adopted for the channel. The structure is assumed to be simply Supported at the ends. Equations of motion of component plates were obtained from Hamilton's principle taking into account all components of inertia forces. In the frame of first order nonlinear approximation, the Dynamic problem of modal interactive buckling is solved by the transition Matrix using the perturbation method and Godunov's orthogonalization, Krzysztof Magnucki [26] The geometric properties of two types of c sections are described by them, the warping functions and the inertia moments the theoretical solution of these beams are numerically found using FEM and experimentally verified. Mariusz Urbaniak [27] Local Dynamic Buckling of C which deals with local dynamic buckling of thin beams subjected to bending. Various shapes of pulse loading with a duration corresponding to the fundamental period of vibration were taken into account. Carine Louise Nilsen, [28] studied how Globally, thin-walled steel sections have been extensively employed as prime load-bearing members. they found that the behavior of thin-walled steel sections, including local buckling, distortional buckling, global buckling and shear buckling Lawrence W. Rehfield [29] in their design methodology for buckling of thin-walled laminated composite beams shoed show buckling by pure bending and by bending- torsion coupled modes can occur and that transition among modes are governed by structural parameters. Syed Muhammad Ibrahim, [30] observed The buckling of thin-walled structures is presented using the 1D finite element based refined beam theory formulation that permits us to obtain *N*-order expansions for the three displacement fields over the section domain. Antonin pistek [31] analytical method for limit load capacity Calculation Of thin walled aircraft structures focused on description and Comparison of different methods for limit load Capacity calculation of thin walled aircraft Structures M.S.Deepak,[32] used cold formed c and z sections because of their easy connections but they suffer from certain buckling modes. Here the structural behavior of c channel lipped beams due to lateral buckling and load carrying capacity is evaluated. Renata Kaspersk [33] Optimization of Cold-Formed Thin-Walled Beams with Open Cross Sections Regarding their work which deals with the problem of optimal design of selected I-type shapes under bending moment. Jung Kwan Seo, [34] paper addresses research gap in LSBs by reporting the development of web opening modeling techniques based on an equivalent reduced web thickness concept and a numerical method for predicting the elastic buckling moments of LSBs with circular web openings. Piotr paczos [35] experimental and numerical (fsm) investigations of Thin-walled beams with double-box flanges reported.

The proposed subject of analytical and numerical investigations is the continuation of a previous research that has been conducted in the Department of Aeronautical Engineering, Institute of Aeronautical Engineering. These investigations will help to improve the current knowledge about the design of thin-walled beams. Finally, the results of the investigation for thin-walled beams with drop flanges and round flanges will be compared in this paper.

2. Thin-walled structures

Three-dimensional structural components in which one dimension is small compared to the other two they include shells, domes, arches, articulated plates, and membrane-type structures.

Thin-walled structures have a high load-carrying capacity, despite their small thickness. They are widely used in the construction industry because of their light weight and economy, particularly for long-span floors in industrial and public buildings and for storage structures for liquids and bulk materials, such as tanks, hoppers, silos, and coal houses. Thin-walled structures are especially suitable for use in constructing Aircrafts, exhibition pavilions, concert halls, and sports arenas because the variety of available shapes permits great architectural expression, the covering of wide areas, and flexibility in the choice of construction materials, including steel, aluminum, reinforced concrete, and laminated plastic.

The design calculations required to estimate the strength, stability, and vibration of thin-walled structures are among the most complex problems encountered in structural design. In general, the calculations are limited to solutions of two-dimensional problems in elasticity or plasticity. Computers have been used to develop methods of calculation based on design schemes, which predict with sufficient accuracy the actual performance conditions and characteristics of the structures, for example, the yield of the supporting structure, the presence of stiffening elements, plastic deformation, variable wall thickness, and the formation of cracks.

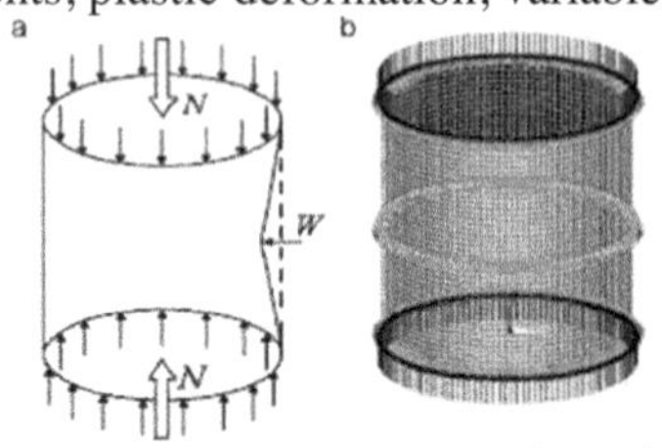

Fig 2.1 Thin cylindrical shell

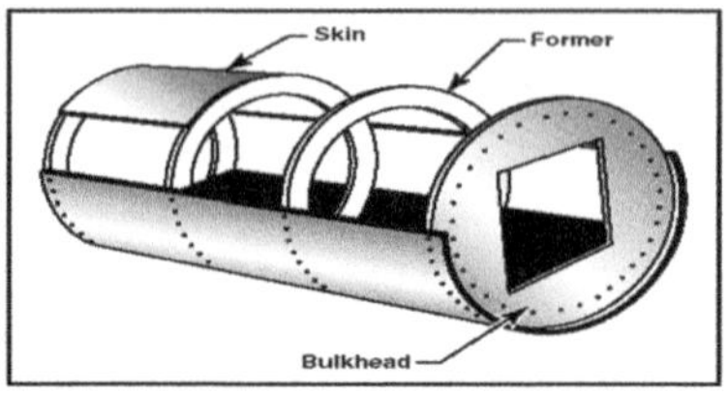

Fig 2.2 Rough fig of aircraft fuselage

2.1 Parts of Air Craft Structures

General the structures of most flight vehicles are thin walled structures (shells) or the semi Monocoque type of aircraft structures. The general function of the aircraft strictures is to resists applied loads (Aerodynamic loads acting on the wing structure) Provides the aerodynamic shape Protects the contents from the environment. The different parts of the air craft structures are Skin, Ribs and Frame, Spar, Stiffeners or Stringers.

2.2 Skin

The functions of the skin are that to reacts the applied torsion and shear forces. It also transmits aerodynamic forces to the longitudinal and transverse supporting members. Skin along with the longitudinal members helps in resisting the applied bending and axial load and also acts with the transverse members in reacting the hoop, or circumferential, load when the structure is pressurized.

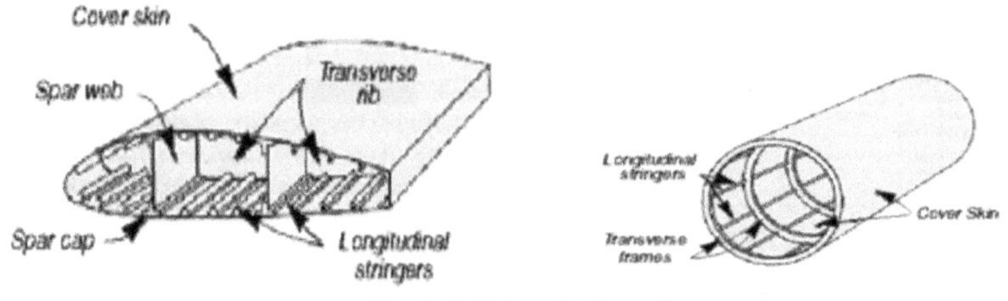

Fig 2.3 Skin of aircraft

2.3 Ribs and Frame

Ribs and frame are the other important part of the aircraft structure. To provide the structural integration of the wings and the fuselage of the air craft. These ribs help the wing and fuselage maintain its aerodynamic profile.

3. Buckling of Thin Walled Structures

In science, buckling is a mathematical instability, leading to a failure mode. Theoretically, buckling is caused by a bifurcation in the solution to the equations of static equilibrium. In practice, buckling is characterized by a sudden failure of a structural member subjected to high compressive stress, where the actual compressive stress at the point of failure is less than the ultimate compressive stresses that the material is capable of withstanding mathematical analysis of buckling makes use of an axial load eccentricity that introduces a moment, which does not form part of the primary forces to which the member is subjected. When load is constantly being applied on a member, such as column, it will ultimately become large enough to cause the member to become unstable. Further load will cause significant and somewhat unpredictable deformations, possibly leading to complete loss of load-carrying capacity. The member is said to have buckled, to have deformed.fig 4.2 shows different stages of buckling.

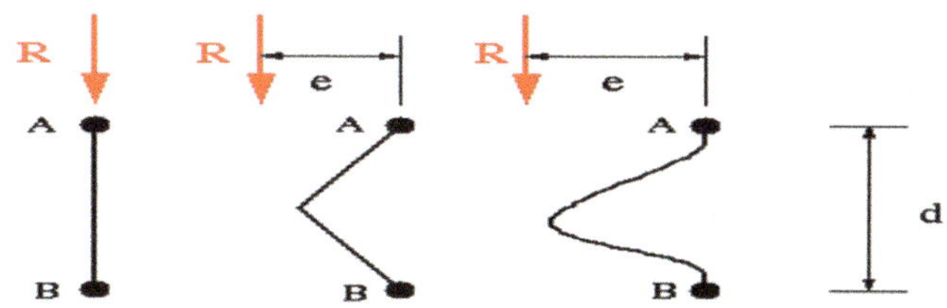

Fig 3.1 Buckling of the column

The ratio of the effective length of a column to the least radius of gyration of its cross section is called the *slenderness ratio*(sometimes expressed with the Greek letter lambda, λ). This ratio affords a means of classifying columns. Slenderness ratio is important for design considerations.

In 1757, mathematician Leonhard Euler derived a formula that gives the maximum axial load that a long, slender, ideal column can carry without buckling. An ideal column is one that is perfectly straight, homogeneous, and free from initial stress. The formula derived by Euler for columns with no consideration for lateral forces is given below. However, if lateral forces are taken into consideration the value of critical load remains approximately the same.

$$F = \frac{\pi^2 EI}{(KL)^2} \tag{1}$$

Where: F = maximum or critical force (vertical load on column), E = modulus of elasticity, I = area moment of inertia, L = unsupported length of column, K = column effective length factor, whose value depends on the conditions of end support of the column, as follows. For both ends pinned (hinged, free to rotate), K = 1.0. For both ends fixed, K = 0.50. For one end fixed and the other end pinned, K = 0.699.... For one end fixed and the other end free to move laterally, K = 2.0. KL is the effective length of the column. The strength of a column may therefore be increased by distributing the material so as to increase the moment of inertia. This can be done without increasing the weight of the column by distributing the material as far from the principal axes of the cross section as possible, while keeping the material thick enough to prevent local buckling. This bears out the well-known fact that a tubular section is much more efficient than a solid section for column service.

Since the moment of inertia of a surface is its area multiplied by the square of a length called the radius of gyration, the above formula may be rearranged as follows. Using the Euler formula for hinged ends, and substituting $A \cdot r^2$ for I, the following formula results. where F/A is the allowable stress of the column, and l / r is the slenderness ratio.

$$\sigma = \frac{F}{A} = \frac{\pi^2 E}{(\ell/r)^2} \tag{2}$$

The above equation is using to calculate crippling stress due to axial loads, but for pure moment the below equation can be used.

$$M_{CR,1} = M_{CR}^{(Glob)} = \frac{\pi E}{L}\sqrt{\frac{J_y J_t}{2(1+\nu)}\left[1 + 2(1+\nu)\frac{\pi^2 J_\omega}{L^2 J_t}\right]} \tag{3}$$

J_y and J_z = principal moments of inertia, J_w =warping moment of inertia

3.1 Flexural-torsional buckling

Occurs in compression members only and it can be described as a combination of bending and twisting of a member and it must be consider for design purposes, since the shape and cross sections are very critical. This mostly occurs in channels, structural tees, double-angle shapes, and equal-leg single angles.

3.2 Lateral-torsional buckling

In wide-flange sections, if the compression flange buckles laterally, the cross section will also twist in torsion, resulting in a failure mode known as lateral-torsional buckling. Typical cross sections requiring check for torsional and flexural torsional buckling

3.3 Plastic buckling

This lower rigidity reduces the buckling strength of the structure and causes premature buckling. This is the opposite effect of the plastic bending in beams, which causes late failure relative to the Euler-Bernoulli beam equation.

3.4 Elastic buckling of thin plates

The flat plate simply supported on all edges and loaded in compression will buckle at an elastic critical stress given by

$$p_{cr} = \frac{K\pi^2 E}{12(1-v^2)}\left(\frac{t}{b}\right)^2 \qquad (4)$$

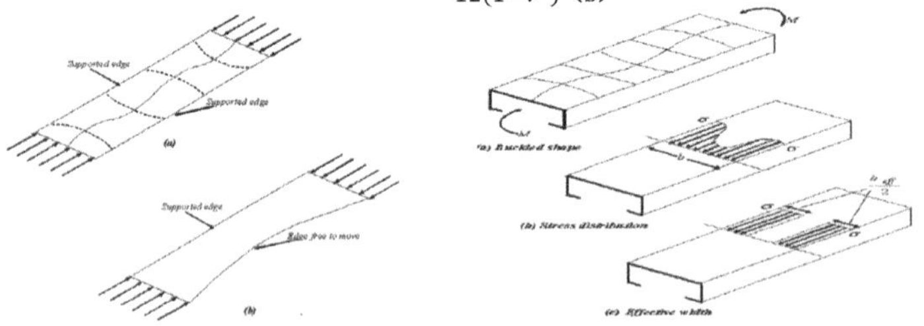

Fig 3.2 buckling of thin plates

3.5 Lipped channel.

The buckling coefficient K1 for the member having a width of B1 in a lipped channel of the type shown above is given by

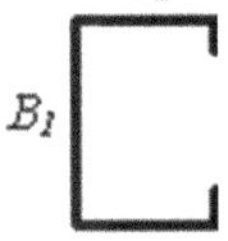

$$K_1 = 7 - \frac{1.8h}{0.15+h} - 1.43h^3 \qquad K_2 = K_2 h^2 \left(\frac{t_1}{t_2}\right)^2 \qquad (5)$$

Where h = B2 / B1, For the member having the width of B2 in the above sketch. Where, t_1 and t_2 are the thicknesses of element width B_1 and B_2 respectively. (Note: normally t_1 and t_2 will be equal). The computed values of K_2 should not be less than 4.0 or 0.425 as the case may be.

3.6 Plain channel (without lips)

The buckling coefficient K1 for the element of width B1 is given by

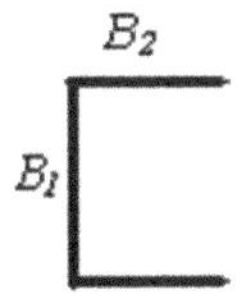

$$K_1 = \frac{2}{\left(1+15h^3\right)^{0.5}} + \frac{2+4.8h}{\left(1+15h^3\right)} \tag{6}$$

4. Lateral Buckling Analysis of Open Section C Channel Beam with drop flanges Using FEA

The beam of the length *L*, the depth *H*, and the wall thickness't' are simply supported and carry two equal moments *M*0. The beams are simply supported and subject to pure bending state. The critical stresses of local and global buckling are defined for the beams

Here the behavior of the channel beam with Circular ends under lateral buckling and finding the critical moment of the beam with change in various parameters such as the length, thickness of the sheet, radius of the round ends, ratio of breadth and height with the help of finite element analysis.

This is being done to achieve more economical design for relatively light loads that use thin walled structures in Aero Industry .and also this type of beams can be useful in providing supporting the structure of wings and fuselage and other parts of the aircraft vehicle.

The geometry of the open section channel beam is as shown in figure 4.6.the channel section consists of web which is 156.86mm,the flange which varies from 19.6mm to 156.86mm,radius varies from 4mm to 20mm,thickness varying from 0.6mm to 3.14mm.

Fig 4.1 Open Section C Channel Beam with drop flanges

4.1Procedure

The development of the FEA model of the beam requires exact properties of the materials to be used for the correct representation of the characteristics in order to achieve the desire results. This is done using Ansys software. At the initial stage we are taking element type of shell 63 having critical's of web 2a as 156.83mm,flange b as 39.21mm ,thickness T 3.14mm ,radius of the circular end R 10mm,length L as 2.8 m. An isotropic material is adopted having E= 71.7 GPa and $v = 0.3$

The ends of the section were fixed and the loads were defined to achieve the solution for problem .the results were recorded. The process was repeated by varying the length, thickness of the sheet, radius of the Circular ends, ratio of breadth and height. The results were tabulated and the graphs were plotted for the results.

The results of the buckling analyses of the channel beam with elongated circular ends section for *Case 1* the selected length is 1600 mm and lateral restraints along the two longitudinal edges of web panel, for this the ratios of flange to web widths *(b/a)* is varied from 0.5 to 2. The curve of the circular end channel is plotted between critical moment and flange to web ratio . As the flange width increases to 156.8 mm, the value of Mcr reduces to 10779 N-m after reaching its peak value of 24917N-m when the flange to web ratio is 1,starting from 7691.3N-m when b/a ratio is 0.25.while the thickness remains at 3.14mm and the radius of the round end is 7.84mm.

L m	a mm	b mm	R mm	T mm	Mcr (N-m)
1.6	78.43	156.8	7.84	3.14	10779
1.6	78.43	78.43	7.84	3.14	24917

1.6	78.43	39.21	7.84	3.14	21990
1.6	78.43	26.14	7.84	3.14	11479
1.6	78.43	19.6	7.84	3.14	7691.3

Table 4.1: showing change in critical moment Mcr with change in b/a ratio.

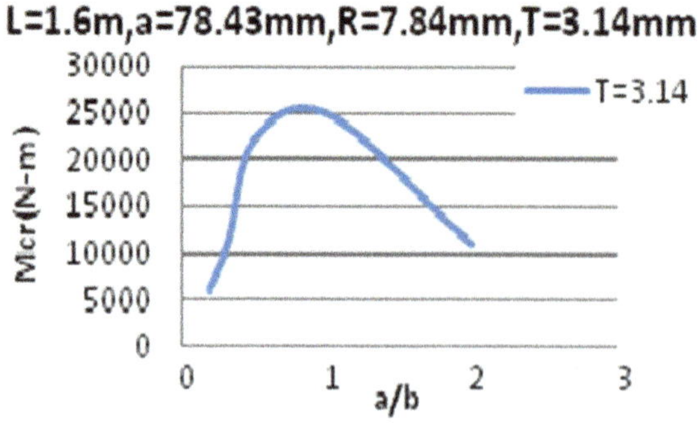

Fig.4.1: showing change in critical moment Mcr with change in b/a ratio

The following are the figures showing each case as tabulated in the above table the figures shown the deformed under the application of forces to do finite element analysis of the defined problem using Ansys software.

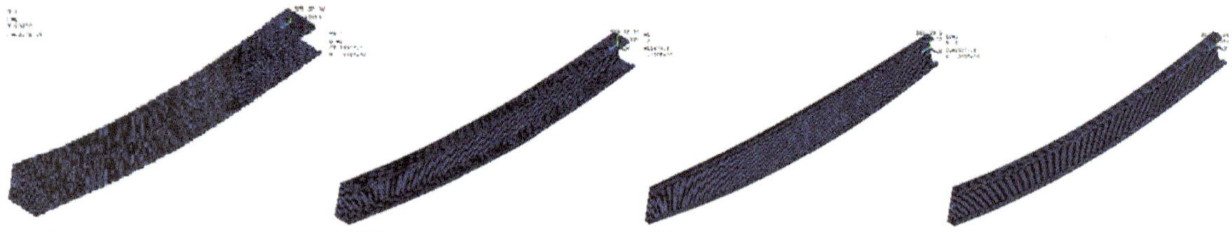

(a) (b) (c) (d)

Fig.4.2: (a) deformed shape l=1.6,a=78.43,b=156.8,R=7.84,t=3.14. (b) Deformed shape l=1.6,a=78.43, b=78.43, R=7.84, t=3.14 , (c) deformed shape l=1.6,a=78.43,b=39.21,R=7.84, t=3.14.(d) deformed shape l=1.6, a= 78.43, b=37.21, R=8, t=3.14.

	T=3.14 mm	T=1.6 mm	T=1.1 mm	T=0.6 mm
R mm	Mcr(N-m)	Mcr (N-m)	Mcr (N-m)	Mcr (N-m)
4	14714	6495	3069.9	566.1
6	17057	7907	3661.5	649.05
8	20396	9532	5136.7	912.44
10	23978	11171	6178	1017.4

Table 4.2:combined table for the L=1600mm,a=78.43 mm,b=37.23 mm and changing radius and thickness.

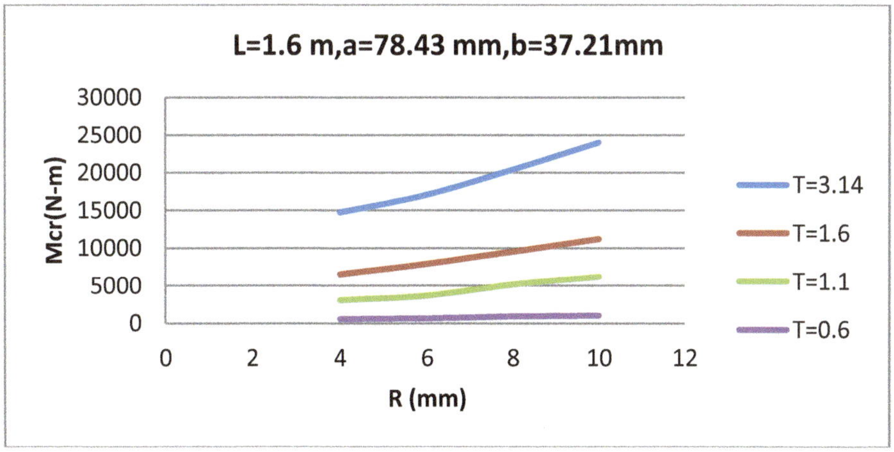

Fig.4.3: showing change in critical moment Mcr with change in radius and thickness.

The results of the buckling analyses of the channel beam with elongated circular ends section for *Case 2* with a length of 1600 mm and lateral restraints along the two longitudinal edges of web panel for the ratios of flange to web widths *(b/a) of* 0.5. The curve for the circular end channel is plotted for critical moment to radius of the circular ends. As the radius is changed from 4 mm to 10 mm, the value of Mcr varies from 14714N-m to 23978N-m the thickness is 3.14mm.

A combined table for all values was formed in order to get more clarity with changing the thickness of the sheet from which the beam is formed and also the graph for this table was formed for the comparison

5. Lateral Buckling Analysis of C Channel Beam with Circular Ends Using FEA

We are studying here the behavior of the channel beam with Circular ends under lateral buckling and finding the critical moment of the beam with change in various parameters such as the length, thickness of the sheet, radius of the Circular ends, ratio of breadth and height with the help of finite element analysis.

This is being done to achieve more economical design for relatively light loads that use thin walled structures in aerospace industry .and also this type of beams can be useful in providing supporting the structure of wings and fuselage and other parts of the aircraft vehicle.

The geometry of the open section channel beam is as shown in figure 1.the channel section consists of web which is 156.86mm,the flange which varies from 19.6mm to 156.86mm,radius varies from 4mm to 20mm,thickness varying from 0.6mm to 3.14mm.

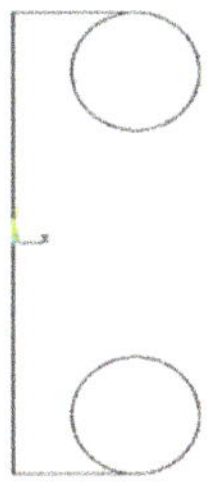

Fig. 5.1 Open Section C Channel Beam with Circular Ended flanges

5.1 Theoretical calculations for Moment of Inertia of C Channel Section with Circular Ends

Total Moment of Inertia= Moment of Inertia of Flanges + Moment of Inertia of Web

(A) Moment of Inertia of Flanges= Moment of Inertia of Bar + Moment of Inertia of Shell since the flange is of circular ends type it is divided into two parts that is of the bar and shell

(a)Moment of Inertia of Bar= $M.I_b = \frac{lt^3}{12} + lt * \bar{x}\bar{y}$ where $\bar{x} = 19.40$ and $\bar{y} = 76.86$

Length of Bar $l = total\ length\ of\ flange - outer\ radius\ of\ shell$

since the length of bar is the distance between the contact point of the bar and the shell and the flange and web.

$l = b - R = b - (r + t)$

$l = 37.81 - (5 + 3.14)$

$l = 29.67\text{mm}$

Now, $M.I_b = \frac{29.67*3.14^3}{12} + 29.67 * 3.14 * 76.86 * 19.405$

$= 138950.85mm^4$

(b)Moment of Inertia of Shell= $M.I_s = \frac{\pi}{64}(D^4 - d^4) + 2\pi(R^2 - r^2)xy$

$= \frac{3.14}{64}(16.28^4 - 10^4) + 2 * 3.14(8.14^2 - 5^2)29.67 * 70.29$

$= 2955.550 + 540375.619$

$=543331.169mm^4$

(c)Moment of Inertia of Flanges= $M.I_f = M.I_b + M.I_s = 138950.85 + 543331.169$

$M.I_f = 682282.019mm^4$

(d)Moment of Inertia of Both Flanges=$M.I_{2f} = 2 * M.I_f = 2 * 682282.019$

$= 1364564.038mm^4$.

(B)Moment of Inertia of Web= $M.I_W = \frac{ta^3}{12} = \frac{3.14*153.72^3}{12} = 950472.442mm^4$

here a is taken as the distance

Total Moment of Inertia= Moment of Inertia of Flanges + Moment of Inertia of Web

$M.I_T = M.I_W + M.I_{2f}$

$= 950472.442mm^4 + 1364564.0384mm^4$.

$= 2315036.476mm^4$.

the values of moment of inertia from the ansys are as follows

$I_{xx} = 0.11264E - 02$, $I_{yy} = 0.12664E - 03$,$I_{zz} = 0.12530E - 02$, $I_{xy} = 0.23717E - 19$

$I_{yz} = 0.0000$, $I_{zx} = 0.0000$, the values of moment of inertia from the calculations M.I=2315036.476mm^4.

The development of the FEA model of the beam requires exact properties of the materials to be used for the correct representation of the characteristics in order to achieve the desire results. This is done using Ansys software. At the initial stage we are taking element type of shell 63 having critical as web 2a as 156.83mm,flange b as 39.21mm ,thickness T 3.14mm ,radius of the circular end R 10mm,length L as 2.8 m. An isotropic material is adopted having E= 71.7 GPa and $v = 0.3$

The ends of the section were fixed and the loads were defined to achieve the solution for problem .the results were recorded. The process was repeated by varying the length, thickness of the sheet, radius of the Circular ends, ratio of breadth and height. The results were tabulated and the graphs were plotted for the results.

The results of the bucking analyses of the channel beam with Circular ends section for *Case A* with a length of 2800 mm and lateral restraints along the two longitudinal edges of web panel for the ratios of flange to web widths *(b/a)* from 0.5 to 2. The curve for the circular end channel is plotted between critical moment Mcr and flange to web ratio b/a. As the flange width increases to 156.8 mm, the value of Mcr reduces to 23438 N-m after reaching its peak value of 46764 N-m when the flange to web ratio is 1,starting from 15221 N-m when b/a ratio is 0.25.while the thickness remains at 3.14mm.

L m	a mm	b mm	R mm	T mm	Mcr(N-m)
2.8	78.43	156.8	10	3.14	23438
2.8	78.43	78.43	10	3.14	46764
2.8	78.43	39.21	10	3.14	31394
2.8	78.43	26.14	10	3.14	20315
2.8	78.43	19.6	10	3.14	15221

Table5.1: showing change in critical moment Mcr with change in b/a ratio.

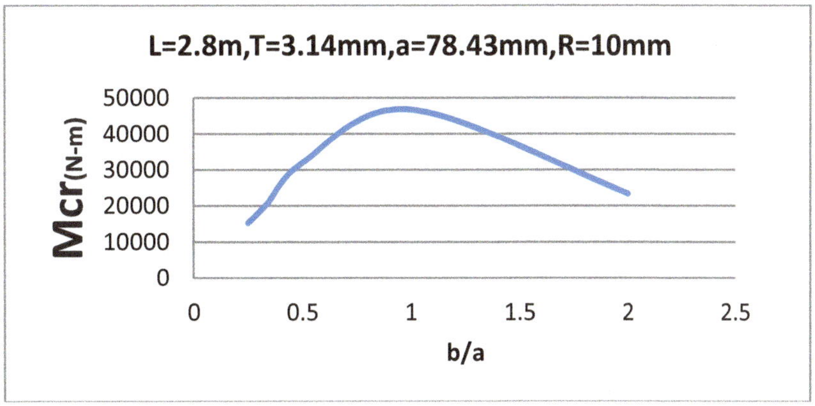

Graph5.1: showing change in critical moment Mcr with change in b/a ratio

The following are the figures showing each case as tabulated in the above table the figures on the left side of the page show the undeformed shape while all the figures on the right side show the deformed and the undeformed shape under the application of forces to do finite element analysis of the defined problem using Ansys software.

The results of the buckling analyses of the channel beam with Circular ends section for *Case B* with a length of 2800 mm and lateral restraints along the two longitudinal edges of web panel for the ratios of flange to web widths *(b/a)* from 0.5. The curve for the circular end channel is plotted between critical moment to radius of the circular end. As the radius is changed from 5mm to 20mm, the value of Mcr is 61988 N-m when the radius is 20 mm, starting from 13570 N-m when radius is 5 mm. while the thickness is 3.14mm.

L	a	b	R	T	Mcr(N-m)
2.8	78.43	37.81	5	3.14	13570
2.8	78.43	37.81	10	3.14	30598
2.8	78.43	37.81	15	3.14	47857
2.8	78.43	37.81	20	3.14	61988

Table5.2: showing change in critical moment Mcr with change in radius of the end ratio.

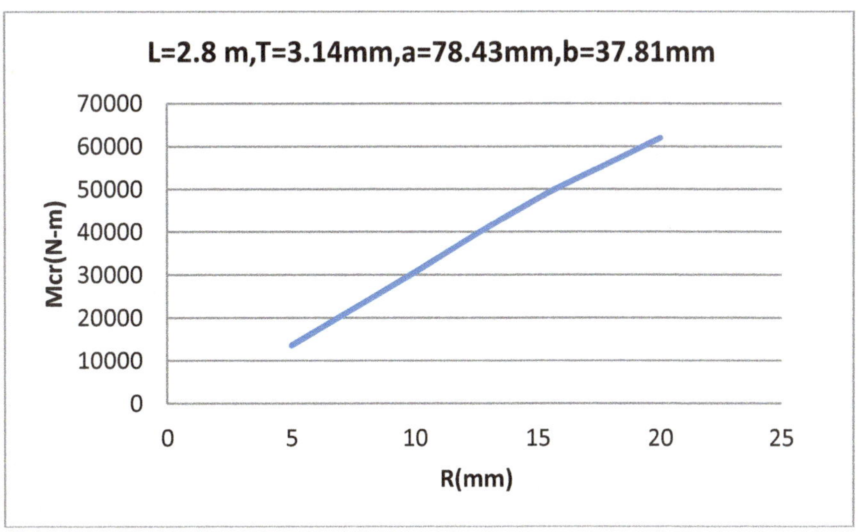

Graph5.2: showing change in critical moment Mcr with change in radius

The following are the figures showing each case as tabulated in the above table the figures on the left side of the page show the undeformed shape while all the figures on the right side show the deformed and the undeformed shape under the application of forces to do finite element analysis of the defined problem using Ansys software.

A combined table for all values was formed in order to get more clarity with changing the thickness of the sheet from which the beam is formed and also the graph for this table was formed for the comparison

	T=3.14	T=1.6	T=1.1	T=0.6
R	Mcr(N-m)	Mcr(N-m)	Mcr(N-m)	Mcr(N-m)
5	13570	6021	3901	783.31
10	30598	12632	5532	903.03
15	47857	17323	6028	978.03
20	61988	20882	6824	1116.2

Table5.3: combined table for the L=2800mm,a=78.43,b=39.21and changing radius and thickness.

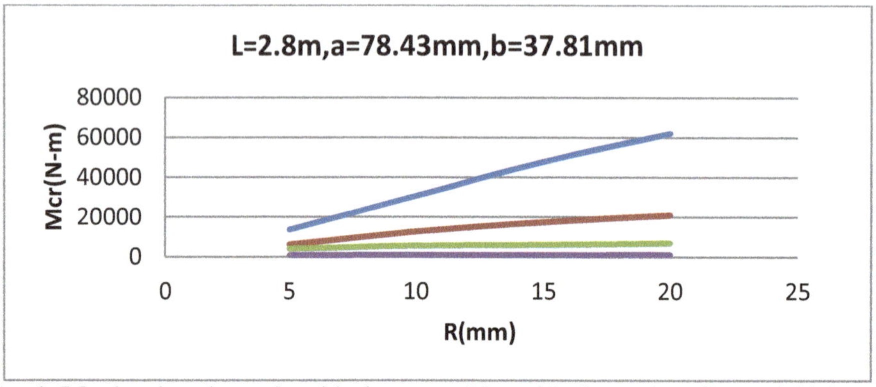

Graph 5.3: showing change in critical moment Mcr with change in radius and thickness

5.2Comparative graph of both designs with same thickness and mass

5.2.1 Same Thickness.

λ	T2=0.6 mm	T2=1.1 mm	T2=1.6 mm	T2=3.14 mm	T=0.6 mm	T=1.1 mm	T=1.6 mm	T=3.14 mm
	Mcr (N-m)							
7.5	207.69	1269.5	3871.4	24992	97.243	577.97	1698.1	10747
10	207.89	1270.4	3868.2	25454	97.634	581.71	1698.9	10779
12.5	207.92	1270.7	3867.3	25089	98.109	581.89	1703.5	10763
15	208.26	1270.2	3861.3	24235	98.33	582.06	1702	10708
17.5	208.52	1269.2	3857.9	23438	98.5	581.9	1701.1	10672

Table 5.4: showing Mcr values of both channels which are having same material physical properties i.e weight length thickness material

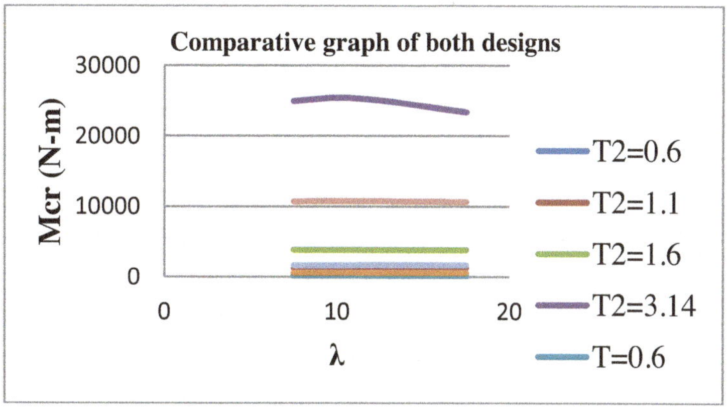

Graph 5.4 Graph showing change in critical moment Mcr with change in λ and thickness
T2 and R2 are the denominations for the c channel with circular ends, T and R are the denominations for c channel with drop flanges

The comparative study of both designs here is done with material of same thickness. Here the comparative parameter is taken as the thickness of the material. The comparison suggests that the case two that is open section C channel section with Circular ends is stronger when compared with same thickness and material of open section C channel with elongated open circular ends.

5.2.2 Same Mass

λ	Mcr (N-m)@T=3.14 mm							
	R2=4 mm	R2=6 mm	R2=7.84 mm	R2=10 mm	R=4 mm	R=6 mm	R=7.84 mm	R=10 mm
17.5	13206	17393	22814	28853	5539.7	5534.7	5540.2	5405.8
15	16475	21090	26853	33118	6792.8	6757	6704.6	6458.5
12.5	21673	26784	32735	38828	8755.2	8653.4	8472.7	7865.6
10	30681	36395	42109	47163	12158	11891	11734	9819.5
7.5	48378	54814	59337	61375	18864	18043	16188	12372

Table5.5: showing Mcr values of both channels which are having same material phisycal properties i.e weight length thickness material

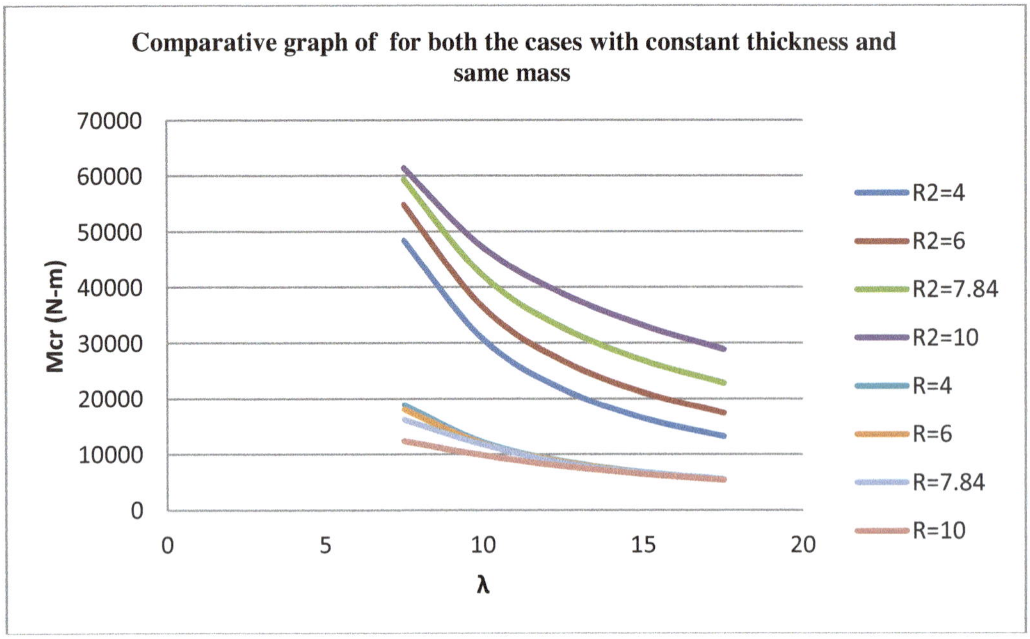

Graph 5.5 Graph showing change in critical moment Mcr with change in λ and thickness

The comparative study of both designs in this section is done is done with material of constant thickness and same mass with varying lengths of the raw plates. Here the comparative parameter is taken as the radius of the material. The comparison suggests that the case two that is open section C channel section with Circular ends is stronger when compared with same thickness and material properties of open section C channel with elongated open circular ends.

5.3 Mass calculations

Dimensions of the plate are length 1.2m, thickness of the plate $3.14*10^{-3}$

Breadth of the plate=$2a+b+2\pi r$=(78.43+37.81+2*3.14*7.84)2* 10^{-3}=330.9540*10^{-3}

the volume of the plate v=l8b8t=1.2*3.14*10^{-3}*330.9540*10^{-3}=1.24710*$10^{-3}m^3$

Mass of the plate= density of the plate * volume of the plate

Density of steel plate=2800kg/m^3

∴mass of the plate m=2800*1.24710*10^{-3} kg =3.491 kg

Similarly, all the masses for plates of different length were calculated and tabulated below.

Length in mts	Mass in Kgs
2.8	8.147072
2.4	6.983204
2	5.819337
1.6	4.65547
1.2	3.491602

Table 5.3 masses of plates at different lengths

6. Conclusion:

The behavior of the channel beam with Circular ends under lateral buckling was examined in order to find critical moment of the beam. The behavior was studied for two types of thin walled C channel beams Elongated Open Circular Ends, Closed Circular Ends.

This behavior of the beams was examined by changing various parameter involved such as the length, thickness of the sheet, radius of the Circular ends, ratio of breadth and height with the help of finite element analysis.

The ends of the section were fixed and the loads were defined to achieve the solution for problem .the results were copied. The process was repeated by varying the length, thickness of the sheet, radius of the round ends, ratio of breadth and height. The results were tabulated and the graphs were plotted for the results for both the beams.

A combined table for all values was formed in order to get more clarity with changing the thickness of the sheet from which the beam is formed and also the graph for this table was formed for the comparison.

Thin walled C channel sections with elongated open Circular end and closed Circular end are analyzed.

Firstly it was concluded that the more the thickness of the material the more is the maximum critical moment of the channel section.

Whereas the maximum critical moment of the channel is more when b/a ratio is around 0.5 to 1

The effect of change in radii of the ends was concluded as the radii of the Circular ends also decided the maximum critical moment in way such that the higher was the value of radius the higher is the maximum critical moment.

It was also concluded that the the effect of change in length can be neglected since there is negligible change in the value of maximum critical moment when compared to the other variables

Therefore the above result can be summarized in the mathematical form in the following way

$$M_{cr}\ \alpha\ t\ ,\ M_{cr}\ \alpha\ r\ ,\ M_{cr}\ \alpha\ \frac{b}{a}\ ,M_{cr}\ \alpha\ l$$

Maximum critical moment is proportional to change in thickness and radius directly and it is maximum with when flange to web ratio is 1.

It is concluded after the comparisons on the grounds of same thickness and mass that the open section c channel with Circular ends has more strength when compared with open section c channel with elongated open circular ends.

REFRENCES

1. Marcello Pignataro Angelo Luongo and Nicola Rizzi "On The Effect Of The Local Overall Interaction On The Postbuckling Of Uniformly Compressed Channels" 01/1985; DOI: 10.1016/0263-8231(85)90009-6
2. M.A.M and O Huges "Lateral distortional buckling of monosymmetric I-beams under distributed vertical load" Thin-Walled Structures ISSN 0263-82311996, vol. 26, n2, pp. 123-145.
3. G.J. Hancock, K.J.R. Rasmussen"Recent Research On Thin-Walled Beam-Columns"Thin-Walled Structures Volume 32, Issues 1–3, September 1998, Pages 3–18 Doi: 10.1016/S0263-8231(98)00025-1
4. K Magnucki, E Magnucka-Blandzi "Variational design of open cross-section thin-walled beam under stability constraints"Thin-Walled Structures (Impact Factor: 1.23). 01/1999; 35(3):185-191. DOI: 10.1016/S0263-8231(99)00031-2
5. Krzysztof Magnucki,Tomasz Monczak,Optimum Shape Of The Open Cross-Section Of A Thin-Walled Beam"Engineering Optimization+A35 01/2000; 32:335-351. Doi: 10.1080/03052150008941303
6. B. W. Schafer "Local, Distortional, And Euler Buckling Of Thin-Walled Columns" Journal Of Structural Engineering / March 2002 / 289
7. Nuno Silvestre And Dinar Camotim "GBT-Based Distortional Buckling Formulae For Thin-Walled Channel Columns And Beams" Sixteenth International Specialty Conference On Cold-Formed Steel Structures Orlando, Florida USA, October 17-18, 2002
8. Xiao-ting Chu, Roger Kettle, Long-yuan Li. "Lateral-torsion buckling analysis of partial-laterally restrained thin-walled channel-section beams."Journal of Constructional Steel Research 60 (2004) 1159-1175.
9. N.S Trahair,G.J Hancoock "Steel member strength by inelastic lateral Buckling"Journal Of Structural Engineering, Vol. 130.No. 1 January 2004,pp.64-69,ASCE 0733-9445(2004)130:1(64)
10. Cheng Yu, B W. Schafer "Simulation of cold-formed steel beams in local and distortional buckling with applications to the direct strength method" Journal of Constructional Steel Research Volume 63, Issue 5,pages 581-590 DOI: 10.1016/j.jcsr.2006.07.008.

11. Emre Erkmen R.Attard Mario" Lateral—Torsional Buckling Analysis Of Thin-Walled BeamsIncluding Shear And Pre-Buckling Deformation Effects" International Journal Of mechanical engineers Issn 0020-7403 2011, vol. 53, n°10, pp. 918-925 [8 page(s) (article)] (34 ref.)
12. Choi Ping , Hamd Raza"Distortional Buckling Of I Beams By Finite Element Analysis" Advances In Structural Engineering 1369-4332 In 2004.
13. Jaehong Lee"Lateral Buckling Analysis Of Thin-Walled Laminated Composite Beams With Monosymmetric Sections" Engineering Structures 28 (2006) 1997–2009.
14. Tomsaz Kubaik "Interactive Buckling In Thin-Walled Beam-Columns With Widthwise Varying Orthotropy" Journal Of Theoretical And Applied Mechanics 44, 1, Pp. 75-90, Warsaw 2006
15. Krzysztof Magnucki,2006" Elastic Buckling Of Selected Flangesof Cold-Formed Thin-Walled Beams"ISSN 1816-1545 Page 116-128
16. Lei Zhanggeng Shu Tong"Elastic Flexural-Torsional Buckling Of Thin-Walled Cantilevers"Thin-Walled Structures - THIN WALL STRUCT 01/2008; 46(1):27-37. DOI: 10.1016/J.Tws.2007.08.011
17. Foudil Mohria, Cherif Bouzerirab, Michel Potier-Ferryc " Lateral Buckling Of Thin-Walled Beam-Column Elements Under Combined Axial And Bending Loads" Thin-Walled Structures 46 (2008) 290–302
18. M.M. Pastor, F. Roure"Open Cross-Section Beams Under Pure Bending. I. Experimental Investigations"' Thin-Walled Structures Volume 46, Issue 5, May 2008, Pages 476–483.
19. Ben Young "Local Buckling And Shift Of Effective Centroid Of Slender Sections" Department Of Civil Engineering,The University Of Hong Kong, Hong Kong 2008 .
20. E. Magnucka-Blandzi, K. Magnucki" Buckling And Optimal Design Of Cold-Formed Thin-Walled Beams: Review Of Selected Problems"Thin-Walled Structures (Impact Factor: 1.23). 01/2011; 49(5):554-561. Doi: 10.1016/J.Tws.2010.09.011.
21. M. M. Pastor F. Roure"Open Cross-Section Beams Under Pure Bending Ii. Finite Element Simulation"Thin-Walled Structures - Thin Wall Struct 01/2009; 47(5):514-521. Doi: 10.1016/J.Tws.2008.10.021
22. Krzysztof Magnucki Piotr Paczos And J.Kasprazak "Elastic Buckling Of Selected Flanges Of Cold-Formed Thin-Walled Beams With Drop Flanges."Journal Structural Engineering ASCE July 2010.
23. P. Borges Dinis Dinar Camotim " Local/Distortional Mode Interaction In Cold-Formed Steel Lipped Channel Beams"Thin-Walled Structures - Thin Wall Struct 01/2010; 48(10):771-785. Doi: 10.1016/J.Tws.2010.01.005.
24. Jerzy Lewinski "Optimization Of Anti-Symmetrical Open Cross-Sections Of Cold-Formed Thin-Walled Beams" Journal Of Theoretical And Applied Mechanics 47, 3, Pp. 553-571, Warsaw 2009
25. Zbigniew Kołakowski " Static And Dynamic Interactive Buckling Regarding Axial Extension Mode Of Thin-Walled Channel" Journal Of Theoretical And Applied Mechanics 48, 3, Pp. 703-714, Warsaw 2010
26. Krzysztof Magnucki "Elastic Buckling Of Selected FlangesOf Cold-Formed Thin-Walled Beams" ISSN 1816-1545 Physical And Mathematical Modeling And Information Technology 2006, No. 3, 116-128
27. Mariusz Urbaniak Tomasz Kubiak" Local Dynamic Buckling Of C shape Profile Subjected To Bending" Mechanics And Mechanical Engineering Vol. 15, No. 2 (2011) 133-148 °C Technical University Of Lodz
28. Carine Louise Nilsen, Md Azree Othuman Mydin And Mahyuddin Ramli "Performance Of Lightweight Thin-Walled Steel Sections: Theoretical And Mathematical Considerations" Advances In Applied Science Research, 2012, 3 (5):2847-2859
29. Lawrence W. Rehfield And Ulrich Mueller "Design Methodology For Buckling Of Thin-Walled Laminated Composite Beams" Department Of Mechanical And Aeronautical Engineering university Of California USA
30. Syed Muhammad Ibrahim, Erasmo Carrera, Marco Petrolo, Enrico Zappino "Buckling Of Thin-Walled Beams By A Refined Theory " J Zhejiang Univ-Sci A (Appl Phys & Eng) 2012 13(10):747-759
31. Prof. Ing. Antonin Pistek, Csc " Analytical Method For Limit Load Capacity Calculation of Thin Walled Aircraft Structures" 28th International Congress Of The Aeronautical Sciences 2012
32. M.S.Deepak,Dr R.kandasamy, Dr K.Thenmozhi " Investigation On Lateral Torsional Buckling Performance Of Cold-Formed Steel C-Channel Sections" International Journal For Emerging Trends In Engineering And Development Issue 2 Vol4 May 2012 ISSN2249-6149.
33. Renata Kasperska,Krzysztof Magnucki, Marian Ostwald " Bicriteria Optimization Of Cold-Formed Thin-Walled Beams With Open Cross Sections"6th World Congresses Of Structural And Multidisciplinary Optimization Rio De Janeiro, 30 May - 03 June 2005, Brazil
34. Jung Kwan Seo, Mahen Mahendrana And Jeom Kee Paikb "Numerical Method For Predicting The Elastic Lateral Distortional Buckling Moment Of A Mono-Symmetric Beam With Web Openings" Thin-Walled structures, VOL 49, ISSUE 6, 2011 DOI 10.1016/J.Tws.2011.01.003
35. Piotr Paczos In 2013 "Experimental And Numerical (Fsm) Investigations Of Thin-Walled Beams With Double-Box Flanges
36. http://aerospaceengineeringblog.com/aircraft-structures/

9

A REVIEW PAPER OF PARAMETRIC ANALYSIS AND RESEARCH DIRECTIONS IN WIND TURBINE BLADES

TUKESH THAKUR

PG SCHOLAR (DEPARTMENT OF MECHANICAL ENGINEERING, MATS UNIVERSITY, RAIPUR)

Abstract

This review paper describe the problems associated with adverse aerodynamic loads will grow more critical. Electricity production from wind energy has grown at a fast pace over the last few years. The size of individual wind turbines has also increased significantly and it is unclear if this trend can b sustained in the future for structural reasons, especially regarding rotor blade components. A research programme is being undertaken to investigate these issues, and finite element modelling will be extensively used to examine the static and dynamic limitations of wind turbine blades. As an initial step in this research, a flexible full blade model was created and is presented in this
paper. The wind energy technical community has begun to seriously consider the potential of aerodynamic control methodologies for mitigating adverse aerodynamic loading. Spatial and temporal attributes of the structures and processes present in these flow fields hold important implications for active aerodynamic control methodologies currently being contemplated for wind turbine applications. The current work uses complementary experimental and computational methodologies, to isolate and characterize key attributes of blade flow fields associated with axisymmetric and yawed turbine operation. During axisymmetric operation, a highly three-dimensional, shear layer dominated flow field yields rotational augmentation of both mean and standard deviation levels of aerodynamic forces.

Keywords – Wind Turbine Blades, Process Automation, blade variables.

1. Introduction

The energy problem we are facing today is articulated around two main drivers: supply and greenhouse gas emissions. Renewable energy sources are an inevitable part of the solution, and wind energy is, at the moment, the fastest growing installed production technology. Commercial wind turbines have developed consistently in size over the last thirty years, largely for economic reasons in an attempt to reduce the electricity production cost, typically measured in p/kW.h. This is due to the fact that wind speed – and hence wind power capture – increases with altitude and that reducing the number of individual turbine units helps to reduce the overall cost of a wind farm, especially offshore. As shown in Figure 1, the largest current machine has a rated output of 5 MW and a rotor diameter of 124 m and so the question arises as to what the ultimate limits on size might be. In all cases, wind turbine blades have pure strength requirements. A static case can, for instance, be calculated on the basis of a 50-year return period gust, while fatigue strength for a 25-year lifetime implies cycle numbers of the order of 107. Another crucial requirement relates to the blade stiffness, since at all times a minimum clearance must be ensured between the blade tip and the turbine tower. The increase in diameter also makes the rotor and blade mass related requirements more severe. Not only do the blade root and the rotor hub need to sustain the static loads (the 62m world-largest blade weighs ~18T), but the nacelle structure, tower and foundations also need to sustain the whole machine dynamics. For a complete discussion of design requirements, interested readers can refer to Burton *et al.* (Burton, 2001).

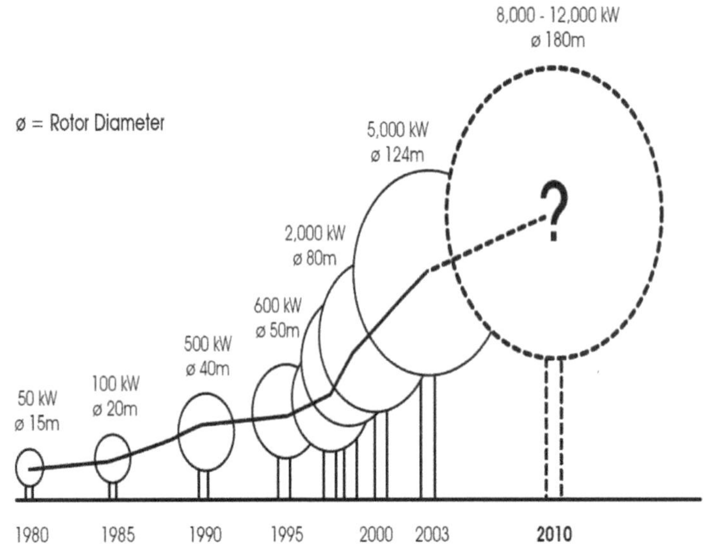

Figure 1. Wind turbine size development (source: European Commission, 2005)

Wind turbine aerodynamics remains a challenging and crucial research area for wind energy. Clearly, steady-state aerodynamic performance is essential to turbine energy capture, since blade aerodynamic forces produce mechanical energy that is subsequently converted to electrical energy. However, more recent inquiry has focused on adverse time varying aerodynamic loads that wind turbines frequently suffer during routine service. These undesirable aerodynamic loads impose excessive stresses on turbine blades and gear boxes, and appreciably shorten machine service life. At present, the wind energy technical community is contemplating the utility of various aerodynamic control methodologies for mitigating adverse aerodynamic loading.

Wind turbine blade aerodynamic phenomena can be broadly categorized according to the operating state of the machine, and two particular aerodynamic phenomena assume crucial importance. At zero and low rotor yaw angles, rotational augmentation determines blade aerodynamic response. At moderate to high yaw angles, dynamic stall dominates blade aerodynamics. As described herein, the spatial and temporal attributes of the structures and processes present

Parametric blade model strategy development

➢ **Generic blade topology**

Although the aerodynamic surfaces can be complex because based on non-planar surfaces, complete blade geometry can be defined by way of a fairly simple topology adequate with our goal of obtaining a parametric model. It is based on the definition of 2D shells for the aerodynamic surfaces and the internal shear-web composites and 3D bricks for volumes of glue used to connect the sub-components. As shown in Figure 2, the outer shape can merely be defined by consecutive sections (e.g. circle at the root and specific aerofoil profiles) and a list of distances from root to tip where each section is present. The aerodynamic shape can then be finally defined by an interpolation process creating a smooth surface through the successive curves.

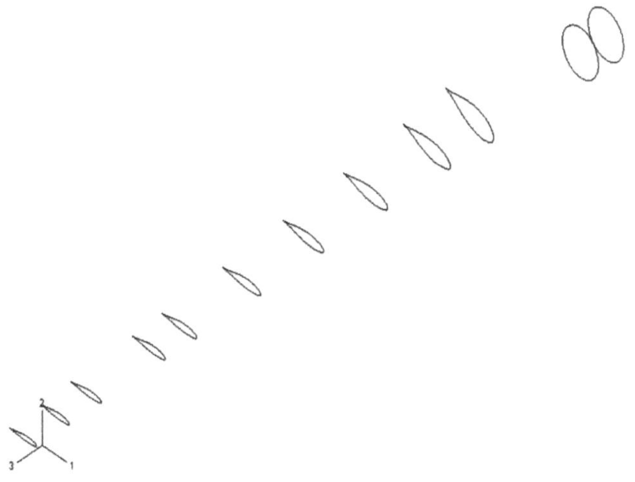

Figure 2. Outer shell topology.

Similarly, the internal structure, typically shear webs, can be parametrically defined by stipulating the distances from root where they start and end as well as the transverse ratios (defined as percentage of chord length) where they will connect to the outer skins. Finally, a glue layer can be inserted between shear-webs and skins, defined by its thickness. The shear-webs and glue volumes can be seen in Figure 3. In this topology, it should be noted that the shear-webs may not necessarily start and end at places defined by aerofoil sections. Similarly, the various regions that receive specific composite layups may not be defined according to the longitudinal split of the blade defined by the aerofoil sections. Finally, even though Figure 2 shows a particular type of shear-web shape, only minor modifications should be needed to the model to reflect other types (e.g. I beam or closed square section). The model adopted should respect these independencies. Because of the nature of our research project, a fully parameterised and extremely flexible modelling strategy is aimed at, so that the blade external geometry, internal structural shape, materials used, loading, analysis type and post-processing operations can be adjusted independently. In turn, the model development time will be optimised and a massive proportion of the Abaqus usage time will be spent producing the results. These specifications lead to the use of a Python script as the analysis definition core. All aspects of an analysis process can be automated in the script, and the full features of the Python programming language confer numerous options for parameterisation.

Examples of parametric analysis

To illustrate the modelling strategy adopted, four parametric analyses are presented below. For each of these studies, once the sweep is set-up and the script execution started, no more user interaction is necessary to produce or compile the complete set of results presented. In these analyses, the blades are simply loaded at the tip in the direction of the wind, which is also the most critical loading direction given the section orientation of a wind turbine blade. The results presented do not constitute significant advances in wind turbine blade engineering.

➢ Mesh convergence analysis

A typical but necessary optimisation problem when conducting FEA work is that of the mesh resolution convergence. How refined does the mesh need to be? How many degrees of freedom can be solved in the time frame available? A mesh convergence analysis was conducted on a typical blade model. Five meshes of increasing resolution were produced. The meshes are shown in Figure 3 as viewed from the root, looking inside the blade – note that this view point means that
the mesh elements mostly look stretched, even though they have in fact good proportions.

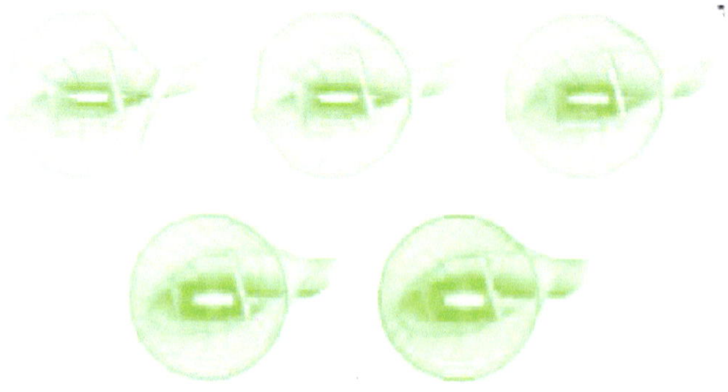

Figure 3. Meshes obtained in mesh size sweep analysis.

These meshes were analysed consecutively and the results presented in Figure 4 were obtained. With the smallest 0.03 m mesh size parameter, the mesh refinement process has converged for general stiffness analyses, as shown by the slope of the tip deflection curve on figure 4. This mesh will be used in the following sections.

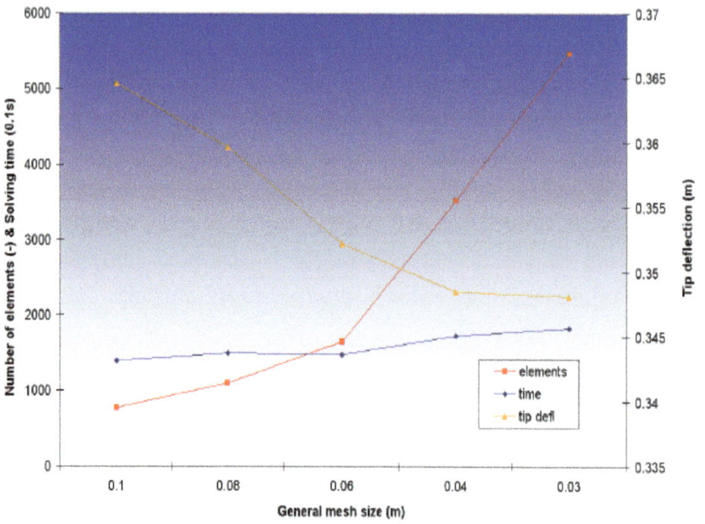

Figure 4. Mesh size sweep results.

➢ Shear-web transverse placement

The transverse placement of the shear-webs within the aerofoil sections naturally influences the structural properties of the assembly. Due to the twist angle variations along the blade, a bending loading state always introduces torsion in some sections of the blade. Such torsion modifies the angle of attack of the aerodynamic surfaces, in turn causing a modification in loading – such a phenomenon is generally described as aero-elastic. The influence of the shear-web placement on the blade mass, bending stiffness and bending-torsion coupling is studied in this example. Six different geometries are examined as shown in Figure 5. In each one, the shear-webs are moved in opposite directions, from being very close to each other to being near the leading and trailing edges. The various geometries are analysed and the results are presented in Figure 6. Note that, as the shear-web spacing increases, the blade mass grows significantly. This is because in the particular layup used, the composite stack applied in the outer skin in-between the two shear-webs is significantly heavier than that applied near the leading and trailing edges. In terms of blade stiffness, it is found that the increase in shear-web spacing also brings higher stiffness values, both in terms of linear and angular displacements. Although the analysis results presented here are valid, they raise the question of cross-coupling between the various design parameters, and it emerges that parameter sweeps with cross coupling of several independent parameters would be necessary in this case. Alternatively, and perhaps more comprehensively, optimization graphs could be produced from the script where iso-parameter contours can be interpolated, improving the understanding of the interdependency between various design parameters.

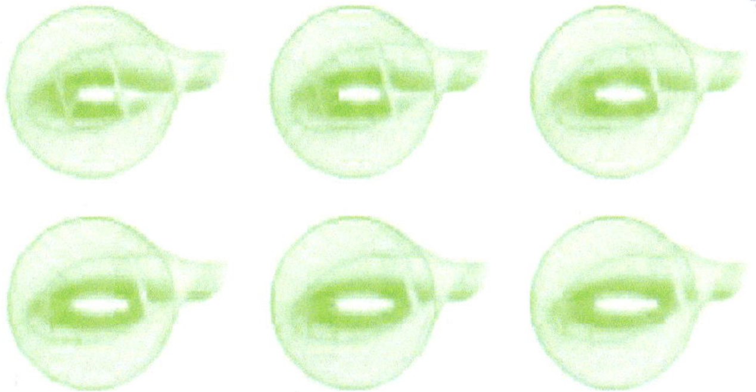

Figure 5. Meshes obtained in shear-web transverse placement analysis.

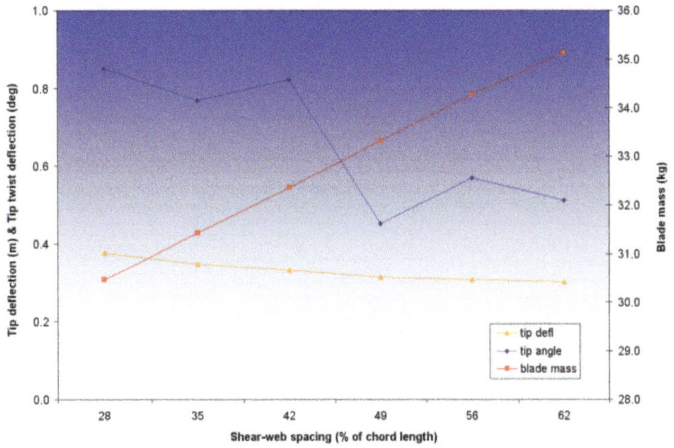

Figure 6. Shear-web placement sweep results.

➢ **Aerofoil shape**

The parametric blade geometry definition also enables a study into the effect of the aerofoil shape on the structural properties of the finished component. Here, we have compared the baseline model used so far with two variations that use different profile series: the Wortmann FX84W and the Althaus AH93W (see Selig, 2006 for full coordinate listing). The blade produced with the Wortmann aerofoils uses a round root section, while a specific root profile, wider but a lot thinner, is provided in the Althaus aerofoil series – note that such a profile would bring significant construction constraints in the case of a pitch operated blade. The aerofoil coordinates are contained in plain text files (one file for each blade section), referred to from within the Python script. Figure 7 shows the meshes obtained for the three aerofoil types. These meshes show once again the validity of the mathematical geometry and mesh generation and its adaptability to substantially different designs.

Figure 7. Meshes obtained with three different aerofoil series.

The static analysis results obtained for the three aerofoil series are displayed in Figure 8. It can be seen that a change from the baseline aerofoils to the Wortmann series, all other parameters constant, would be beneficial in terms of blade mass but detrimental in terms of stiffness. A change for the Althaus series seems very beneficial in mass terms and only a little detrimental in terms of stiffness. These results variations are mostly due to the relatively stiff and heavy sections near the root and the maximum chord area – the Wortmann series has a significantly thinner maximum chord profile while the root section of the Althaus blade has a much smaller circumference, but is still stiff enough compared to the rest of the blade, perhaps providing a better compromise.

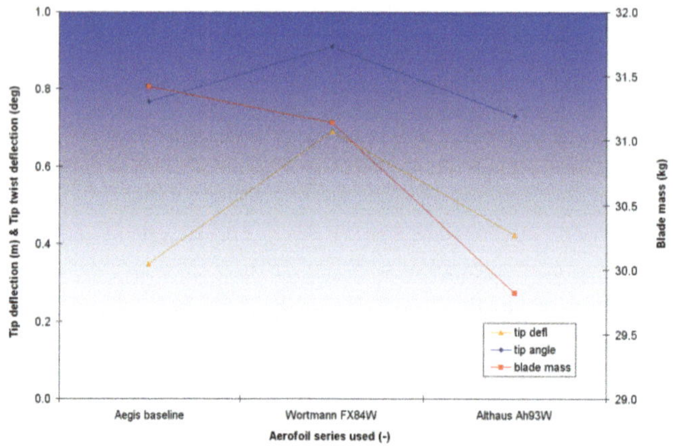

Figure 8. Aerofoil series sweep results.

Rotational Augmentation

Augmentation of rotating blade aerodynamic properties, including stall delay and lift enhancement, was first observed for airplane propellers and qualitatively explained in terms of centrifugal and Coriolis accelerations.[1] Some time later, analytical modeling quantitatively accounted for key elements of the rotating blade flow field.[2]

Subsequent rotational augmentation research was carried out for helicopter rotors. Analytical modeling determined that rotationally induced cross flows played an important role in blade lift production.[3] Experimental research [4] suggested that centrifugal forces are important in the presence of flow separation, but of limited influence otherwise.

Prior research concerning rotational augmentation of airplane propeller and helicopter rotor aerodynamics aided later work aimed at wind turbines. However, wind turbines extract energy from flowing air, while propellers and rotors inject energy. This key distinction pointed out the need for rotational augmentation research specific to wind turbines.

Early wind turbine field testing aimed at rotational augmentation affirmed the importance of blade geometry with respect to rotational influences.[5] A wind tunnel experiment showed that blade geometry coupled with blade rotation maintained blade lift under conditions in which lift otherwise would significantly decline.[6] Subsequent wind tunnel research determined that rotational augmentation was most active at the inboard portion of the turbine blade.[7] Concurrent analytical modeling of rotational augmentation has furnished better comprehension of the aerodynamics underlying this phenomenon, and provided foundational predictive capabilities for design and analysis.[8-13]

More recently, analysis of turbine aerodynamics measurements acquired during controlled wind tunnel experiments, in conjunction with validated computational results, have provided key physical insights regarding rotationally augmented flow fields. [14-17]

2. EXPERIMENTAL PROCEDURES

All surface pressure data used in the current work were acquired during wind tunnel testing of the NREL UAE (National Renewable Energy Laboratory Unsteady Aerodynamics Experiment) horizontal axis wind turbine. This machine was documented in detail for earlier field testing phases of the experiment. Subsequently, several UAE configurations were tested in the NASA Ames 80 ft x 120 ft wind tunnel, all of which are described by Hand, et al.

Wind Turbine

Data analyzed in the current work were acquired from a two bladed upwind rotor, 10.1 m in diameter, with zero cone angle. The rotor turned counterclockwise (viewed from upwind) at a constant 71.6 RPM, was stall regulated, and had a maximum rated power of 19.8 kW. A cylindrical tower 0.4 m in diameter supported the turbine at a hub height of 12.2 m (test section centerline), with 1.32 m rotor overhang. This UAE configuration is depicted in Figure 10.

Figure 10. NREL Unsteady Aerodynamics Experiment wind turbine in NASA Ames 80 ft x 120 ft wind tunnel.

Blade Geometry

The blades used throughout the NASA Ames wind tunnel test were both twisted and tapered. Blade taper distribution is apparent in Figure 11, with chord tapering from 0.737 m at 0.25R to 0.356 m at the tip. Blade section twist decreased from 22.1° at 0.25R to 0.0° at the tip. Between 0.25R and the tip, blade cross-section was uniform, corresponding to the S809 airfoil. The airfoil section at 0.25R was joined to the pitch shaft section at 0.12R using linear segments to yield an uninterrupted transition between these two disparate contours. The blade pitched about an axis located 0.30c aft of the leading edge, and centered between the blade upper and lower surfaces at that chord location. Design procedures, constraints, and measures of merit for this blade have been documented in detail.

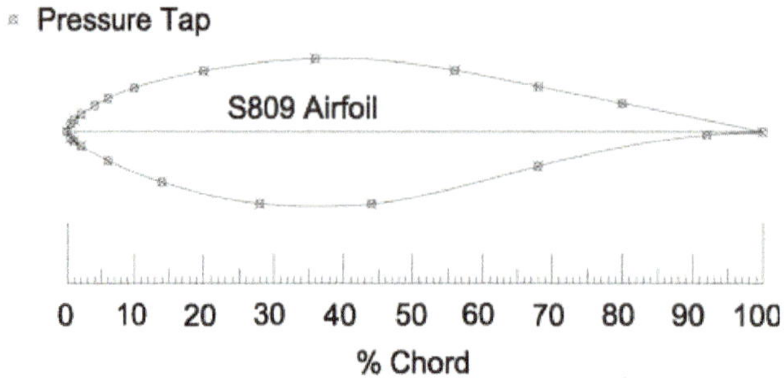

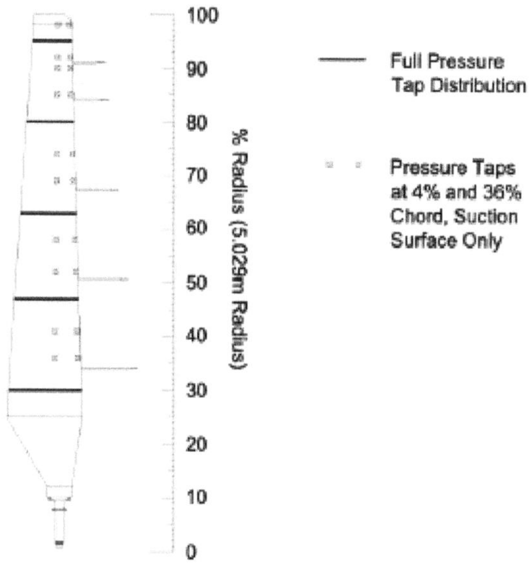

Figure 11. Blade cross-section and planform, showing pressure tap locations.

Data Acquisition Protocol

Data acquisition procedures were designed to decouple the principal physical phenomena that govern blade aerodynamics, and thus were structured in an incremental manner. First, blade aerodynamics was characterized for nonrotating conditions. Then, data were acquired with the blades rotating at zero yaw, for a range of wind speeds. Finally, with the rotor still turning, yawed conditions were characterized for a matrix of wind speeds and yaw angles.

To isolate the effects of rotational influences, a baseline was needed that excluded rotational effects. This was accomplished by halting turbine blade rotation and acquiring pressure data for the stationary (parked) blade. To achieve stationary blade conditions, the instrumented turbine blade was first fixed at the 12 o'clock azimuth position. Then, for $U\infty = 20$ m/s, turbine blade pitch was incremented in 5° steps, from 90° to -15° and -15° to 90°. Blade pitch was incremented in both negative and positive directions to disclose any hysteresis in blade aerodynamic force response. At each pitch angle, one 13 second data record was acquired, and cp data from each record were integrated over the sectional chord to obtain time records of Cn. Finally, these 13 second time records of cp and Cn were time averaged to yield mean values for each blade pitch angle. Prior to acquiring data at each pitch angle, appropriate delays were included to allow time for pitching, blade flow field equilibration, and wake relaxation.

To quantify the effects of rotational influences, data were collected with the rotor turning at 71.6 RPM. Turbine blade plane of rotation was maintained orthogonal to the test section centerline, yielding $\gamma = 0°$. Blade pitch angle was held constant at 3.0°. $U\infty$ was varied between 5 m/s and 25 m/s, in nominal 1 m/s increments. At each $U\infty$, a data record of 30 seconds duration was acquired, corresponding to 36 blade rotations.

Inflow Angle and Force Nondimensionalization

Derivation of angle of attack and lift coefficient for rotating blades using measurements on or near the blade remains a challenging and essential area of inquiry. However, in the current work, these dependencies were excluded in order to simplify physical relationships and concentrate analyses on the blade flow field. This was accomplished by analyzing measured LFA and Cn in lieu of derived α and Cl.

Normalization of aerodynamic forces also presents challenges. In the current work, normal force was nondimensionalized by local dynamic pressure. Local dynamic pressure was computed as the difference between test section static pressure (p_∞) and local total pressure. Local total pressure was sensed at each full pressure tap distribution as the highest pressure in the tap distribution. This methodology for quantifying dynamic pressure has been analyzed previously, and found to induce average errors in dynamic pressure of approximately 1.0 percent.

3. RESULTS AND DISCUSSIONS

Results and discussion below are divided according to turbine operating state into two sections, axisymmetric operation and yawed operation. As summarized above, these two operating states produced flow fields dominated by rotational augmentation and by dynamic stall, respectively. Within each of these two sections, results and discussion follow the same progression. First, inflow relationships are described. Then, typical data for mean and standard deviation of normal force are presented.

4. Future developments

The future work in this research will be articulated over three main drivers. First, the blade model described in this paper needs to be developed with known and potential improvements. The blade loading, as described above, will be fully parameterised and implemented within the blade model script. This will imply that the aerofoil performance data is accessed from external files produced by the panel code tool. Later, this can be developed into an aero-elastic analysis tool by creating a feedback loop from the calculated blade deflections at various sections to the generation of an updated load field. At this stage, fully dynamic blade analyses can be conducted and the structural dynamics of whole wind turbine machines be investigated. Also, the main composite mesh elements, currently described by 2D shells, would perhaps more accurately be modelled by 3D continuum shells. This investigation will involve a comprehensive model calibration, which can be conducted at two levels:

• On a sub-component level, in collaboration with materials research specialists within our Supergen consortium, bespoke tests of manufactured components will be carried out and reproduced computationally.

• On the whole blade level, full blade certification test data from our industrial partner collaborators will also enable crucial model validation and calibration. And of course substantial work will be devoted in parallel to address the project questions regarding future generations of wind turbines. The current blade technology will first be replicated and potential improvements will be investigated. To this end, the material data such that collected in the Optimat Blades Project (Knowledge Centre Wind Turbine Materials and Constructions , 2006), in which the second author was an active partner, will be an invaluable knowledge source.

5. CONCLUSION

Using complementary experimental and computational methodologies, key attributes of blade flow fields were isolated and characterized for axisymmetric and yawed turbine operation. These characteristics are likely to have strong implications for control of turbine blade flow fields.

For axisymmetric operation, elevated inflow angles drive Cn to mean and standard deviation levels that are significantly augmented relative to two-dimensional non-rotating conditions. Both mean and standard deviation levels are substantial, and are caused by a highly three-dimensional, shear layer dominated flow field.

In this paper, a generic model was presented to generate and analyse the structural behaviour of large composite wind turbine blades. The model is programmed in the analysis process is entirely automated and parameterised. Typical potential applications to wind turbine blade structural engineering issues were presented and the model showed good coherence and extreme adaptability. Special considerations were then presented regarding the generation of accurate aerodynamic loading procedures, and future developments of the model and of the current research were described.

6. REFERENCES

[1] Bechly, M. E., and P. D. Clausen, "Structural Design of a Composite Wind Turbine Blade Using Finite Element Analysis," Computers & Structures, vol. 63, pp. 639-646, 1997.

[2] Burton, T., D. Sharpe, N. Jenkins, and E. Bossanyi, "Wind Energy Handbook," John Wiley & Sons Ltd, 2001.

[3] de Goeij, W. C., M. J. L. van Tooren, and A. Beukers, "Implementation of bending-torsion coupling in the design of a wind turbine rotor blade," Applied Energy, vol. 63, pp. 191-207, 1999.

[4] McCroskey, W. J., "Measurements of Boundary Layer Transition, Separation and Streamline Direction on Rotating Blades," NASA TN D-6321, Apr. 1971.

[5] Madsen, H., and H. Christensen, "On the Relative Importance of Rotational, Unsteady and Three-Dimensional Effects on the HAWT Rotor Aerodynamics," *Wind Engineering*, v. 14, n. 6, 1990, pp. 405-415.

[6] Ronsten, G., "Static Pressure Measurements on a Rotating and a Non-Rotating 2.375 m Wind Turbine Blade. Comparison with 2D Calculations," *J. of Wind Engineering and Industrial Aerodynamics*, v. 39, 1992, pp. 105-118.

[7] Corten, G., "Inviscid Stall Model," Proceedings of the European Wind Energy Conference, July 2001, pp. 466-469.

[8] Schreck, S., and Robinson, M., "Rotational Augmentation of Horizontal Axis Wind Turbine Blade Aerodynamic Response," *Wind Energy*, v. 5, n. 2/3, pp. 133-150, Apr.-Sep. 2002.

[9] Shipley, D., M. Miller, and M. Robinson, "Dynamic Stall Occurrence on a Horizontal Axis Wind Turbine Blade," NREL/TP-442-6912, Jul. 1995. Golden, CO: National Renewable Energy Laboratory.

[10] Robinson, M., R. Galbraith, D. Shipley, and M. Miller, "Unsteady Aerodynamics of Wind Turbines," AIAA 95-0526, 33rd Aerospace Sciences Meeting and Exhibit, Jan 1995.

Vibration Analysis of Dragonfly wing section in Gliding Mode at Low Reynolds Numbers

Husain Mehdi[1], Sayed Fahad Anwer[2], Akhlaque Ahmad[2]

[1]Assitant Professor, MED, Meerut Institute of Technology, Meerut (U.P), 250002/ INDIA

[2]Department of Mechanical Engineering ZHCET, AMU Aligarh, 202002/ INDIA

Abstract

The dragonfly wings are highly corrugated, due to light in weight and good corrugation it increases the aerodynamic performance and strength of the wing. When the wings interact with the air, it is subjected to aerodynamic forces acting on the surface of the wing and the inertial force due to the acceleration or deceleration of the wing mass. The interaction between these inertial and aerodynamic forces resulted in wing deformation. We are interesting to calculate deformation and natural frequency of the dragon fly wing at different Reynolds number and different angle of attack. A dragonfly insect has been chosen because MAVs (Micro air Vehicles) and Dragonfly works almost same Reynolds Number i.e. Re- 10^2 to 10^4. In this work, Numerical study of Vibration Analysis for a Pleated Insect 2D Airfoil at Ultra Low Reynolds Numbers is carried out in gliding mode. The dynamics of a pleated insect wing subjected to aerodynamic loading is studied for different Reynolds Number ranging from 100 to 1000 at different angle of attack ranging from 0^0 to 15^0 by using ANSYS-14 multi physics solver. The result from the CFD solver will be fed in the form of lift and drag forces are then fed into the ANSYS Workbench solver and vibration analysis is performed.

Keywords: Lift Coefficient, Drag Coefficient. Angle of Attack, Reynolds No, Deflection, Natural Frequency.

1 Introduction

Aeshna genus dragon fly can glide up to 30s without any significant loss of altitude [1]. A smaller dragon flies found there gliding periods lasting 0.5s, covering a distance of approximately 1m and thus achieving gliding speed of 2.5ms^{-1}. Thus the typical range of Reynolds number in gliding flight can vary from 10^2 [2]. compared aerodynamic performance of cross-sections at different positions along the span of a wing of an Aeschna cyanea to develop the pleated models and its corresponding profiled airfoil at a chord Reynolds number of 10 000, and the results showed that the pleated airfoils generated higher lift than the profiled airfoils [3]. The orientation of the leading edge does not have an effect in enhancing the lift production [3] [4]. As with the early flow visualization experiments, trapped vortices present in the folds that serve to change the effective profile of the airfoil [3]. Dragon flies have corrugated wings where the pleated configuration varies along the span-wise and chord-wise and also studied the aerodynamic performance of a pleated wing section based on the wing of Aeshna cyanea has been performed at ultra-low Reynolds numbers (5000-10000) corresponding to the gliding flight of these dragonflies in two dimension using computational fluid dynamics. The simulations demonstrate that the pleated airfoil produces comparable and at times higher lift than the profiled airfoil, with a drag comparable to that of its profiled counterpart [5]. The effect of the Reynolds number on the gliding ratio is that at Re 1000 and at angle of attack (here after, AOA) 15^0, the largest gliding ratios are obtained. Flow invariably for all Reynolds number, minimum Drag coefficient is obtained at AOA 15^0 [6].

Micro air vehicles (MAVs offer great potentials for various civil and military applications in revolutionizing sensing and information gathering, such as surveillance, inspection, targeting, or biochemical sensing tasks at remote and hazardous locations. Numerous vehicle concepts, including fixed, rotary, and flapping wings, have been proposed [7][8]. However, the advantage of a flapping wing in MAV compared with fixed or rotary-wing aircraft of the same scale is emphasized by its maneuverability and smooth transition from hovering to forward flight signature [9][10].

The structural deformation during flapping can significantly influence the flow behavior around the wing and, consequently, brings an important on its aerodynamic performance [11]. Physiological studies have shown that insect wing deformation is passive in which the wing is deformed by both inertial acceleration and deceleration of the wing mass, by the aerodynamic force from surrounding air, or by the combination of both [12]. The overall bending patterns of a hawkmoth wing are quite similar when flapped (single degree-of-freedom flap rotation) in air and in helium, despite an 85% reduction in fluid density [11]. The wing deformation leads to substantial power economy for lift production. Moreover, they noticed that the leading-edge vortex (LEV) in the full- kinematic model remains attached to the wing during the entire flapping cycle, which may have contributed to the aerodynamic power efficiency of the wing [13]. The quantitative data on the time-varying camber and twist of the wing in free-flying drone flies [14]. A further study was conducted where they removed the camber and spanwise twist one by one and they observed that the lift in the free-flying drone flies is principally caused by camber deformation [15].

In this work, we investigate Modal Analysis for spatio-temporal dynamics of a cut section of Aeshna Cyanea's wing at ultra low Reynolds numbers (100 to 1000) at different angle of attacks ranging from 0° to 15°. To the best of author's knowledge, The Modal Analysis of 2D pleated airfoil is not studied. These parameter ranges are relevant for both dragon flies and micro-aerial vehicles. Past experimental studies have found no intrinsic three-dimensional effects at these low Reynolds numbers. Thus, 2D simulations are implemented in this study to encompass a relatively wide range of the parameter space necessary to draw some general conclusions regarding pleated airfoils. The pleated airfoil implemented in the numerical simulation corresponds to a cross-section located at the mid-section of the forewing of a dragonfly (Aeshna cyanea) [3][4]. The specific profile chosen for the numerical simulations corresponds to 'Profile 2', which was digitally extracted from the paper [3][5]. From the three pleated geometries to select from the paper of [3], 'Profile 2' was chosen due to its horizontal leading edge, thus eliminating the issue that the orientation of the leading edge has an influence on the aerodynamic performance [4]. For the purposes of reducing the resolution requirements in the simulation, the sharp edges of the pleats were rounded out slightly without affecting the basic geometry of the pleats and the overall shape of the airfoil as shown in Figure-2. The shape-forming and stiffness-adjusting mechanisms of a cell through its inner filled liquid have inspired some air-supported/liquid-supported inflatable structural systems [16] [17].

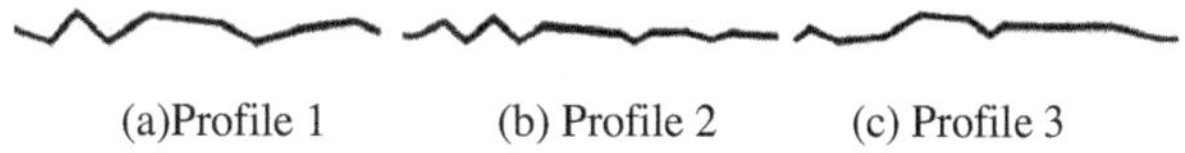

(a)Profile 1 (b) Profile 2 (c) Profile 3

Figure 1: An Illustration of different cut section profiles used by Kessel (2006) and Vargas et al (2008)

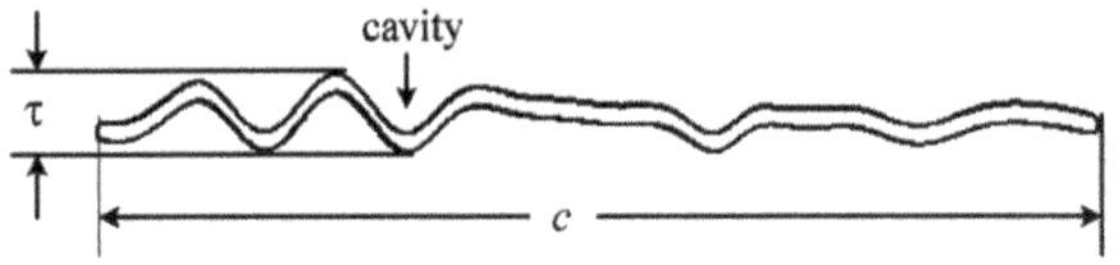

Figure 2: The two-dimensional airfoils used in the numerical simulation. Pleated airfoil representing a cross-section of the forewing of a dragonfly (Aeshna cyanea) having $\tau/c = 7.531\%$

2 Governing Equation

2.1 *Fluid Flow*

The equations governing the flow in the numerical solver are the time-dependent, viscous incompressible Navier–Stokes equations. The non-dimensional momentum and continuity equations are as follows:

$$\frac{\partial u_i}{\partial x_i} = 0 \tag{1}$$

$$\frac{\partial u_i}{\partial t} + \frac{\partial u_i u_j}{\partial x_j} = -\frac{\partial p}{\partial x_i} + \frac{1}{Re}\frac{\partial^2 u_i}{\partial x_j \partial x_j} \tag{2}$$

The equations are non-dimensionalized with the appropriate length and velocity scales, in this case the airfoil chord and free stream velocity. Here Re corresponds to the Reynolds number which is defined as below:

$$Re = \frac{\rho u_\infty c}{\mu} \tag{3}$$

Using the same flow parameters and geometrical dimensions as Kesel (2000), this allowed for validation and a critical analysis of the numerical results. The key quantities examined are the lift and drag coefficients which are defined as

$$C_L = \frac{F_L}{\frac{1}{2}\rho u_\infty^2 c}, \tag{4}$$

$$C_D = \frac{F_D}{\frac{1}{2}\rho u_\infty^2 c} \tag{5}$$

$$\text{Gliding Ratio} = \frac{C_L}{C_D} \tag{6}$$

2.2 *Structure Analysis:*

The interaction of the fluid and the structure at a mesh interface causes the acoustic pressure to exert a force applied to the structure and the structural motions produce an effective "fluid load." The governing finite element matrix equations then become

$$[M_S]\{\ddot{U}\} + [K_S]\{U\} = \{F_S\} + [R]\{P\} \tag{7}$$

$$[M_f]\{\ddot{P}\} + [K_f]\{P\} = \{F_f\} - \rho o\,[R]^T\{\ddot{U}\} \tag{8}$$

[R] is a "coupling" matrix that represents the effective surface area associated with each node on the fluid-structure interface (FSI). The coupling matrix [R] also takes into account the direction of the normal vector defined for each pair of coincident fluid and structural element faces that comprises the interface surface. The positive direction of the normal vector, as the program uses it, is defined to be outward from the fluid mesh and in towards the structure. Both the structural and fluid load quantities that are produced at the fluid-structure interface are functions of unknown nodal degrees of freedom. Placing these unknown "load" quantities on the left hand side of the equations and combining the two equations into a single equation produces the following.

2.3 *Modal Analysis*

2.3.1 *Damping Matrices*

Damping may be introduced into a transient, harmonic, or damped modal analysis as well as a response spectrum. The type of damping allowed depends on the analysis as described in the subsequent sections.

2.3.2 *Transient Analysis and Damped Modal Analysis*

The damping matrix, [C], may be used in transient and damped modal analyses as well as substructure generation. In its most general form, the damping matrix is composed of the following components.

$$[C]=\alpha[M]+(\beta+\frac{2}{\Omega}g)[K]+\sum_{i=1}^{N_{ma}}\alpha_i^m[M_i]+\sum_{j=1}^{N_m}[(\beta_j^m+\frac{2}{\Omega}g_j+\frac{1}{\Omega}g_j^E)[K_j]]+\sum_{k=1}^{N_e}[C_k]+\sum_{m=1}^{N_v}\frac{1}{\Omega}[C_m]+\sum_{l=1}^{N_g}[G_l] \quad (9)$$

where: [C] = structure damping matrix, α = mass matrix multiplier, [M] = structure mass matrix, β = stiffness matrix multiplier, [K] = structure stiffness matrix, N_{ma} = number of materials, α_i^m = mass matrix multiplier for material i, $[M_i]$ = portion of structure mass matrix based on material i, N_{mb} = number of materials, $[K_j]$ = portion of structure stiffness matrix based on material j, N_e = number of elements with specified damping, $[C_k]$ = element damping matrix, Ng = number of elements with Coriolis or gyroscopic damping, $[G_l]$ = element Coriolis or gyroscopic damping matrix, β_j^m = stiffness matrix multiplier for material.

2.3.3 *Harmonic Analysis*

The damping matrix ([C]) used in harmonic analyses is composed of the following components.

$$[C]=\alpha[M]+(\beta+\frac{2}{\Omega}g)[K]+\sum_{i=1}^{N_{ma}}\alpha_i^m[M_i]+\sum_{j=1}^{N_m}[(\beta_j^m+\frac{2}{\Omega}g_j+\frac{1}{\Omega}g_j^E)[K_j]]+\sum_{k=1}^{N_e}[C_k]+\sum_{m=1}^{N_v}\frac{1}{\Omega}[C_m]+\sum_{l=1}^{N_g}[G_l] \quad (10)$$

The input exciting frequency, Ω, is defined in the range between Ω_B and Ω_E via

$\Omega_B = 2\pi f_B$, $\Omega_E = 2\pi f_E$, f_B = beginning frequency, f_E = end frequency

Substituting equation (10) into the harmonic response equation of motion and rearranging terms yields

$$[[K]+i\{2g[K]+\sum(2g_j+g_j^E)[K_j]+\sum[C_m]\}+i\Omega[\alpha[M]+\sum\alpha_i^m[M_i]+\beta[K]+\sum\beta_j^m[K_j]+\sum[C_k]+\sum[G_j]]](u_1+iu_2)=F_1-\Omega^2[M] \quad (11)$$

The complex stiffness matrix in the first row of the equation consists of the normal stiffness matrix augmented by the structural damping terms given by g, g_i, g_i^E, and $[C_m]$ which produce an imaginary contribution. Structural damping is independent of the forcing frequency, Ω, and produces a damping force proportional to displacement (or strain). The terms g, g_i, and g_i^E are damping ratios (i.e., the ratio between actual damping and critical damping, not to be confused with modal damping).

The second row consists of the usual viscous damping terms and is linearly dependent on the forcing frequency, Ω, and produces forces proportional to velocity.

2.3.4 *Mode-Superposition Analysis*

The damping matrix is not explicitly computed, but rather the damping is defined directly in terms of a damping ratio ξd. The damping ratio is the ratio between actual damping and critical damping. The damping ratio ζ_i^d for mode i is the combination of

$$\zeta_i^d = \zeta + \zeta_i^m + \frac{\alpha}{2\omega_i} + \frac{\beta}{2}\omega_i \tag{12}$$

ξ = constant modal damping ratio, ζ_i^m = modal damping ratio for mode shape i (see below), ω_i = circular natural frequency associated with mode shape i = $2\pi f_i$, f_i = natural frequency associated with mode shape i, α = mass matrix multiplier

The modal damping ratio ζ_i^m can be defined for each mode directly (undamped modal analyses only).

Alternatively, for the case where multiple materials are present whose damping ratios are different, an effective mode-dependent damping ratio ζ_i^m can be defined in the modal analysis if material-dependent damping is defined and the element results are calculated. This effective damping ratio is computed from the ratio of the strain energy in each material in each mode using,

$$\zeta_i^m = \frac{\sum_{j=1}^{N_m} \beta^m{}_j E_j^s}{\sum_{j=1}^{N_m} E_j^s} \tag{13}$$

3 Numerical Methodology

Ansys Fluent 14.0 is used to solve the above equations. The solver used is based on collocated methodology and finite volume discretization technique is used.

$$\iiint \frac{\partial \rho \emptyset}{\partial t} dV + \oint \rho \emptyset \vec{v} . d\vec{A} = \oint \Gamma_\phi \nabla \emptyset . d\vec{A} + \iiint S_\emptyset \, dV \tag{14}$$

$$\frac{\partial\rho\emptyset}{\partial t}V + \sum_f^{N_{faces}} \rho_f \vec{v}_f \, \phi_f . \vec{A}_f = \sum_f^{N_{faces}} \Gamma_\phi \, \nabla\phi_f . \vec{A}_f + S_\phi V$$

The gradients are calculated using Green-Gauss Cell-Based methodology proposed by [21] [22].

$$\overline{\emptyset_f} = \frac{1}{N_f} \sum_n^{N_f} \overline{\emptyset}_n \quad (15)$$

where N_f is the number of nodes on the face Convection terms in the momentum equation are discretized using the second order upwind methodology given by in the second order up wind quantities at cell faces are computed using a multidimensional linear reconstruction approach [23]. The result from the CFD solver will be fed in the form of lift and drag forces are then fed into the ANSYS Workbench solver and one way Fluid Structure Interaction analysis is performed.

4 Grid Design:

The pleated cross section of modeled dragon fly is taken from the paper [5]. The length of pleated plate is assumed to be of unit length by giving X coordinate=1, Height is also given to pleated section in the Y direction and it is 0.05. The pleated section is covered with a circular region of diameter 20C where C is set to 1.This circular region is then divided into 2 parts along semi-circle parallel to Y axis.Then 150 nodes are taken on each semi circle.60 nodes are also taken on airfoil's upper and lower part.Then command of triangular meshing is giving and the required meshed profile is obtained.

Figure 3:- Mesh of the problem

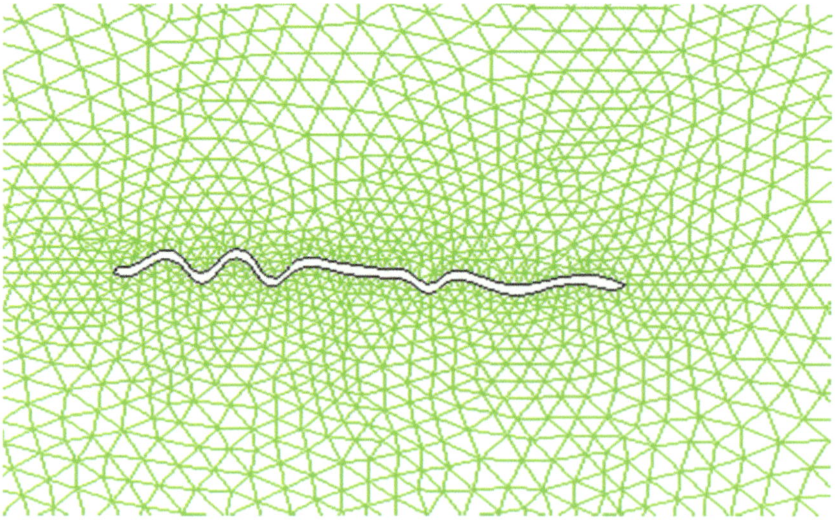

Figure 4:- The zoomed view of the mesh

Grid information:

- 195834 triangular cells
- 196878 nodes,
- Inflow and out flow boundaries are pressure far field
- Inner meshed region is fluid.

Face area statistics:

- Minimum face area (m^2) : 4.428917 e-003
- Maximum face area (m^2) : 2.160249 e-001

5 Boundary Condition

A constant velocity u= 0.14607 m/sec is imposed on the left side of grid, and the right side set as an outflow region where the gradient values are set to zero. The components are taken in accordance with the angle of attack. Pressure on the both sides was taken as atmospheric i.e. $P=P_{atm}$

Velocity inlet boundary conditions are used to define the velocity and scalar properties of the flow at inlet boundaries.

Out flow boundary condition are used to model flow exits where the details of the flow velocity and pressure are not known prior to solution of the flow problem. They are appropriate where the exit flow is close to a fully developed condition, as the out flow boundary condition assumes a zero normal gradient for all flow variables except pressure.

6 Results and Discussions

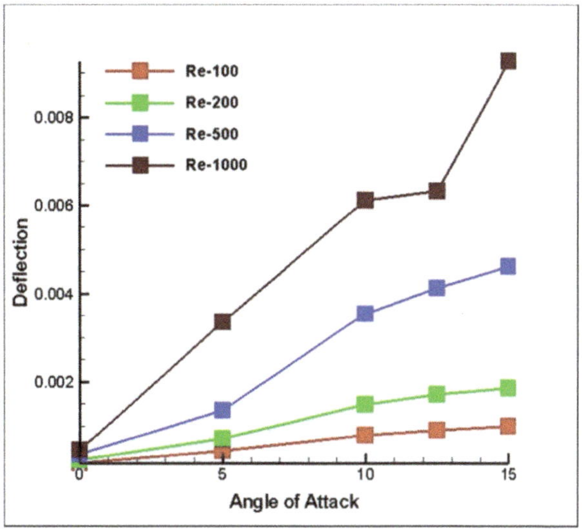

(a)

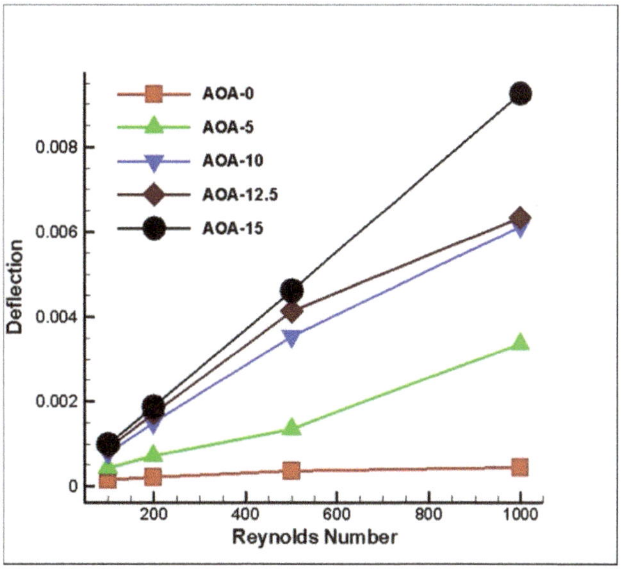

(b)

Figure-5:- (a) Variation of deflection and AOA, (b) Variation of deflection and Re

The material density of wing ρ is taken as 1,200 kg/ [18], and confirmed by Vincent and Wegst [19] and we used a Poisson's ratio ν of 0.49 [18]. Combes and Daniel [11] also used a Poisson's ratio of 0.49.

ANSYS 14 was utilized to solve this model of a dragonfly wing, and airfoil was assumed to be fully clamped along the wing root. Figure-5 shows the variation of deflection and angle of attack. It is seen to be that deflection is increased with increasing the Reynolds Number with their angle of attack. The maximum deflection occur in Re-1000, angle of attack 15^0 i.e. $9.2597 \text{x} 10^{-3}$ mm and minimum value of deflection occurs in Re-100 with angle of attack 0^0 i.e.$1.4618 \text{ x } 10^{-4}$ mm. we have calculated all six modes of vibration, according to figure-6 the minimum value of natural frequency is 0.6985 and the maximum natural frequency is 34.353 Hz which are bending and torsional mode respectively.

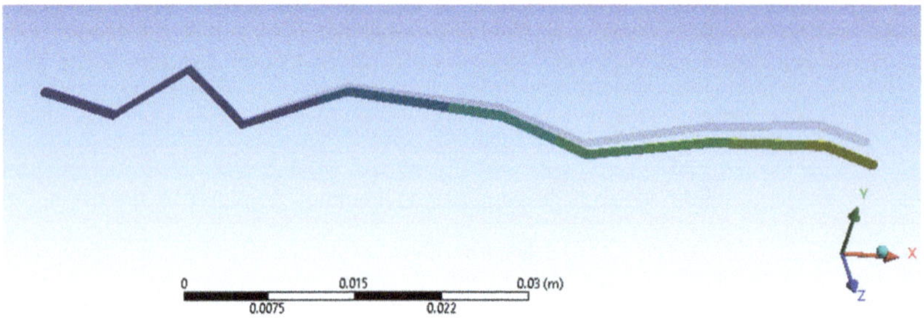

(a)

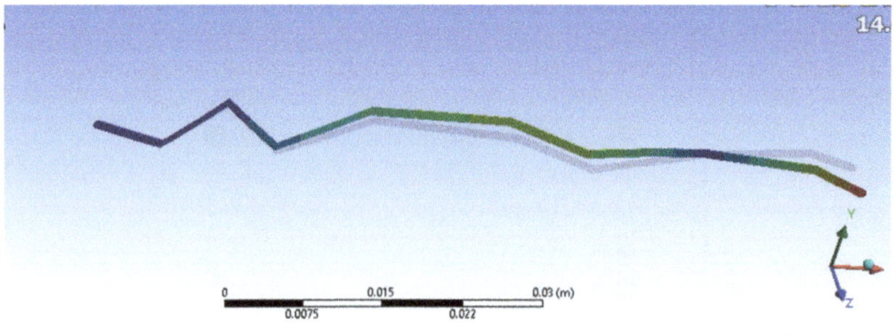

(b)

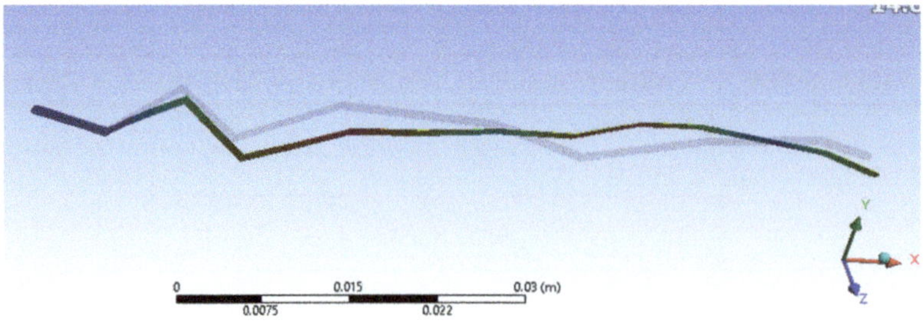

(c)

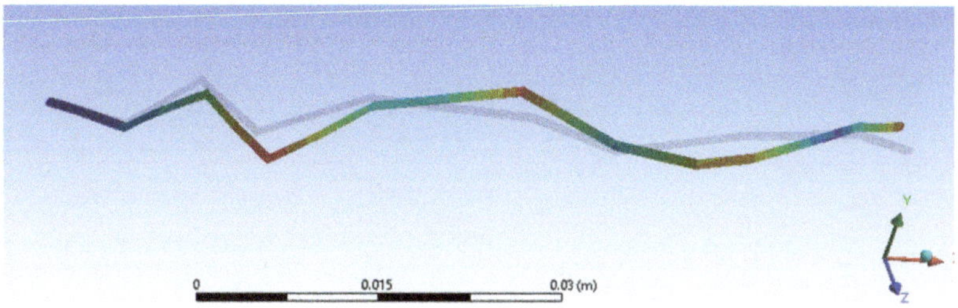

(d)

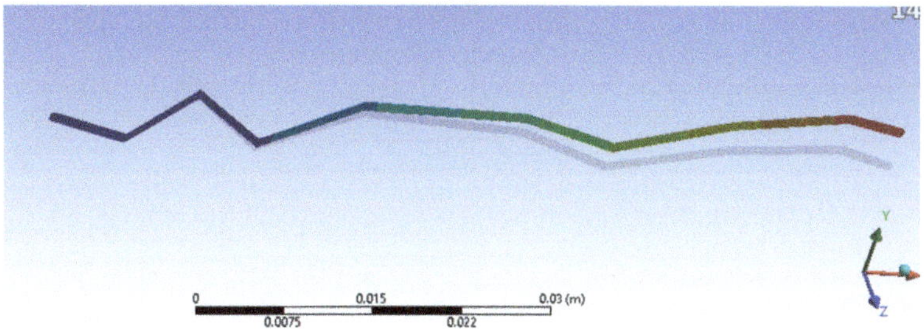

(e)

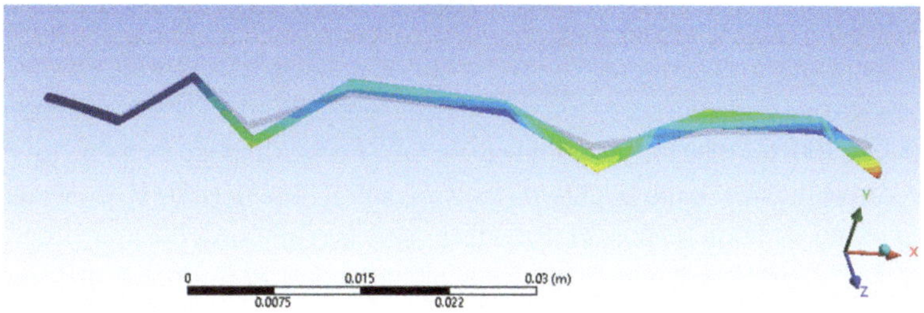

(f)

Figure 6:- (a) Mode Shape 1 at 0.69851Hz,(b) Mode Shape 2 at 4.439 Hz,(c) Mode Shape 3 at 12.523 Hz, (d) Mode Shape 4 at 21.259 Hz,(e) Mode Shape 5 at 23.561 Hz,(f) Mode Shape 6 at 34.353 Hz

Table 1:- Validation

Wing Model	Mode Shape	Natural frequency	Type	Authors
2D	1st	32.30 Hz	Bending	Wootton [12]
	5th	23.56 Hz	Bending	Present
	1st	25.40 Hz	Bending	Jongerius [20]

In this work, results of free vibration analysed. In which we neglect damping, show the same deformation for the natural modes as Wootton [12] found bending for the first natural mode and torsion for the second natural mode. In our two dimensional model the fifth natural frequency 23.561 Hz is close to the flapping frequency 32.3 Hz (Table 1).

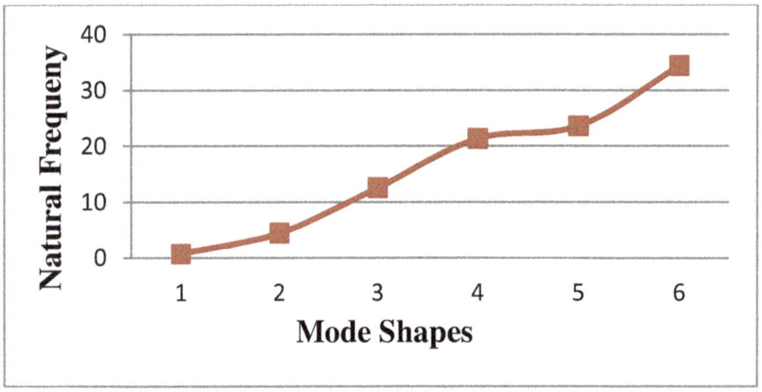

Figure 7:- Variation of Natural Frequency and Mode Shapes

7 Conclusions

In this work we find the natural frequency of dragonfly's wing in different modes. The strategy of dragonflies to flap their wings at a much lower frequency than the natural vibration frequency of their forewing is contrasted by the most advanced flapping wings designed for micro air vehicles [20]. We investigated Modal Analysis for spatio-temporal dynamics of a cut section of Aeshna Cyanea's wing. Numerical simulations were performed at ultra low Reynolds numbers (100 to 1000) at different angle of attacks ranging from 0^0 to 15^0. The results give a satisfactory measure of confidence in the fidelity of the simulation. The deflection is increased when we increased the Reynolds Number with increase their angle of attack The maximum deflection occur in Re-1000, angle of attack 15^0 i.e. 9.2597×10^{-3} mm and minimum value of deflection occurs in Re-100 with angle of attack 0^0 i.e.1.4618×10^{-4} mm..the highest natural frequency 34.353 Hz was found in mode 6 which is torsional mode, whereas minimum natural frequency 0.6951 was found in mode 1.

References:

[1] Brodsky, A K. The aerodynamics of insect flight: The Evolution of Insect flight. New York: Oxford University Press, 1994.

[2] Wakeling JM,EllingtonCP, "Dragonfly flight: III. Lift and power requirements". *The Journal of Experimental Biology* 200:583–600, 1997.

[3] Kesel, A B. "Aerodynamic characteristics of dragonfly wing sections compared with technical aerofoil." J of Exp. Biology 203 (2000): 3125-3135.

[4] Okamoto, M, K Yasuda, and A Azuma. "Aerodynamic characteristics of the wings and body of a dragonfly." J. of Experimental Biology, 1996: 281-294.

[5] Vargas, Abel, Rajat Mittal, and Haibo Dong. "A computational study of the aerodynamic performance of a dragonfly wing section in gliding flight." Bioinsp. Biomim., 2008: 1-13.

[6] Anwer SF, Ashraf I, Mehdi H, (2013) "On the Aerodynamic Performance of Dragonfly Wing Section in Gliding Mode" Advances in Aerospace Science and Applications. Vol 3, pp 227-234.

[7] Shyy W, Berg M, Ljungqvist D, " Flapping and flexible wings for biological and micro air vehicles". *Progress in Aerospace Sciences* 35(5):455–505, 1999.

[8] Mueller TJ, " Fixed and flapping wing aerodynamics for micro air vehicle applications", *Progress in Astronautics and Aeronautics* 195, 2001.

[9] Lian YS, Shyy W, Viieru D, Zhang B (2003). Membrane wing aerodynamics for micro air vehicles. *Progress in Aerospace Sciences* 39:425–465

[10] Ramasamy M, Leishman JG, Lee TE (2006). Flow field of a rotating wing MAV. *Proceedings of 62nd Annual National Forum of the American Helicopter Society*. May 9-11 2006.

[11] Combes SA, Daniel TL, "Into thin air: contributions of aerodynamic and inertial-elastic forces to wing bending in the hawkmothManducasexta". *The Journal of Experimental Biology* 206:2999–3006, 2003.

[12] Wootton RJ, Herbert RC, Young PG, Evans KE, " Approaches to the structural modelling of insect wings". *Philosophy Transaction Royal Society* 358(1437):1577-1587, 2003.

[13] Young SM, Walker RJ, Bomphrey GK, Taylor, Thomas ALR, " Details of insect wing design and deformation enhance aerodynamic function and flight efficiency", *Science* 325(5947):1549-1552, 2009.

[14] Walker J, Thomas ALR, Taylor GK, " Deformable wing kinematics in free-flying hoverflies", *Journal of Royal Society Interface* 7(42):131-142, 2010.

[15] Du G, Sun M (2010), " Effects of wing deformation on aerodynamic forces in hovering hoverflies", *The Journal of Experimental Biology* 213:2273-2283, 2010.

[16] Schlaich, J., Bergermann, R., Sobek, W, "Air-inflated roof over the Roman amphitheatre at Nimes". *Structural Engineering Review*, **6**(3-4):203-214, 1994.

[17] Bradshaw, R., Campbell, D., Gargari, M., Mirmiran, A.,Tripeny, P., 2002. Special structures: past, present, and future. *Journal of Structural Engineering ASCE*, **128**(6):691-709

[18] Wainwright SA, Biggs WD, Currey JD and Gosline JM ,"Mechanical design in organisms", Princeton University Press, 1982.

[19] Vincent JFV, Wegst UGK, "Design and mechanical properties of insect cuticle", Arthropod Stru Dev 33:187–199, 2004.

[20] Lentink D, Jongerius SR and Bradshaw NL, "The scalable design of flapping micro air vehicles inspired by insect flight, 2009.

[21] Holmes D.G. and Connell S.D., "Solution of the 2D Navier-Stokes Equations on Unstructured Adaptive Grids", Presented at the AIAA 9th Computational Fluid Dynamics Conference, June, 1989.

[22] Rauch R.D., Batira, J.T. and Yang N.T.Y., "Spatial Adaption Procedures on Unstructured Meshes for Accurate Unsteady Aerodynamic Flow Computations ",Technical Report AIAA-91-1106, AIAA, 1991.

[23] Barth T.J. and Jespersen D., "The design and application of upwind schemes on unstructured meshes", Technical Report AIAA-89-0366, AIAA 27th Aerospace Sciences Meeting, Reno, Nevada, 1989.

11

DEVELOPMENT OF HIGH ALTITUDE TEST FACILITY FOR COLD JET SIMULATION

Aldin Justin[1], Robin James, Shruti Panicker[2],A.N.Subash,Saravana Kumar

[1]Assistant Professor,Karunya University, aldinjustin@karunya.edu
[2]Student,Karunya University,shruti_panicker@yahoo.co.in
Karunya University , Karunya Nagar , Coimbatore, Email : aldinjustin@karunya.edu

Abstract

A High altitude test facility was developed in the Aerospace department, Karunya University for cold flow studies A High Altitude test facility was developed to study the jet plumes at high altitude , shock formation in a supersonic nozzle at different altitude condition above sea level and Jet interaction studies of single and multi-jets , cold and hot jets. The study basically involve the decent behavior of a nozzle. It provides altitude simulation for experimental purposes. Compressed air at high pressure is used to study the supersonic nozzle in vacuum conditions in the High Altitude Test Facility. The vacuum chamber can simulate altitude up to 32.5km. A scale model of supersonic conical nozzle of throat diameter 5mm and exit diameter 11mm was mounted inside the vacuum chamber of the high-altitude test facility. The experiment were performed at various decent conditions, static pressure along the nozzle was used to get the pressure profile using the 16 port electronic pressure scanner. CFD analysis was carried out for various inlet pressure condition, the results were found using isentropic relations are which validated with the experimental results. Air pressure decreases with increasing altitudes .A study of evacuation rate and the pressure rise with respect to time was also studied at different inlet conditions.

Keywords: CFD analysis, Isentropic relations, Electronic pressure scanner.

1. Introduction

Rocket engines and propulsion systems that are designed to operate in space often require testing under simulated altitude conditions. They often incorporate high expansion ratio nozzles for increased performance. When these nozzles are operated in an ambient pressure significantly higher than what they were designed for, there is flow separation from the nozzle wall, which results in lowering the thrust. Flow separation can cause nozzle burning and damage due to the formation of shock waves, unsymmetrical pressure distribution and excessive vibration at the separation point. Ignition characteristics of both solid and liquid propellants are significantly altered. Combustion instability triggered at engine start may be masked by ambient pressure ignitions. Convective heat transfer that readily exists at ambient pressure is nearly nonexistent in the space environment. Engines may overheat during this operation due to lack of convective cooling. All this posts a need for testing the nozzle at high altitude conditions. For studying the behavior of the nozzle for different altitude conditions a High altitude test facility was developed in Aerospace department , Karunya University. The High Altitude test facility was commissioned for studying the cold flow condition of a supersonic nozzle.

2. Experimental Setup

For developing the setup , a supersonic nozzle was to be designed and fabricated for the maximum test duration . The evacuation rate of the vacuum pump was considered to be the mass flow rate of the supersonic nozzle. The High Altitude Test facility of Karunya University is a vacuum simulation chamber that can simulate vacuum environments in the range of 0.9 to 1 mbar vacuum pressure . The High Altitude facility is a cost-efficient, secure and easy to handle configuration to perform tests and studies in the field of altitude simulation and flow separation of nozzles. The simulated altitude is approximately 32.5km in this facility. The vacuum chamber has a diameter of volume of 1.706 m^3 A vacuum pump was used for evacuating the chamber before the test. Suitable mountings were made to hold the high pressure nozzle in the test section of the vacuum chamber. The different vacuum level would give different altitudes . A positive displacement pump was used for this purpose. First it obtains a rough vacuum in the vessel before the momentum transfer pump is used to obtain a high vacuum . The specification of the vacuum pumps is of positive displacement type and the motor used of 3 phase induction type motor Pumping rate – 65 m^3/h Oil required – 2 litres. A two stage Reciprocating air Compressor was used to charge the high pressure reservoir. A High Pressure 2 stage Air cooled , Splash lubricated compressor with Displacement – 92.18 m^3/h and Free air delivery – 67.92 m^3/h was used .The working pressure of the compressor was 30kgf/cm^2 with Compressor speed of 1150 rpm. A compressed air dryer was used for removing water vapour from compressed air. This process concentrates atmospheric contaminants; including water vapour .This raises the dew point of the compressed air leading to condensation within the pipes. The air dryer is of Desiccant type with inlet flow rate of 127.42m^3/h Pressure of 6.86 to 12.3 bar and Temperature – 42^{0}C the Outlet conditions of the air dryer has Flow rate - 114.68 m^3/h Pressure – 6.67 to 12.1 bar and Temperature of 40^0 A Pressure drop of 0.2 bar Pressure was noted. A high Pressure vessel with closed container was installed to hold gases at a pressure substantially different from the ambient pressure .This Pressure Vessel was used to store the compressed dry air at high pressure for various operations to be executed in the high altitude test facility. The high pressure reservoir stored compressed air at high pressure with safe operating pressure of 20 bar. The volume of the reservoir is 6.28m^3

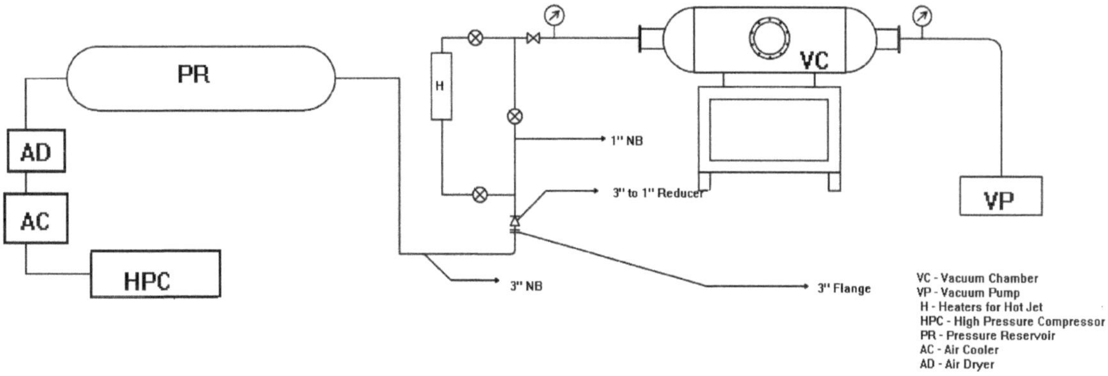

Figure 1: Experimental Setup

When High pressure is made to flow through the supersonic nozzle, the ambient pressure inside the vacuum chamber rises to higher values. A Pressure relief valve is used to maintain a controlled environment for the vacuum chamber which is calibrated for a pressure of 1.9 bar. In order to reduce the impact of vibration loads on the system due to the compressed air, we have mounted the conical nozzle with an internal rigid support structure. This will reduce the impact of vibrations. For the purpose of measuring the pressure across the different points on the contour of the nozzle, various holes of 1.5mm diameter were drilled on the surface of the nozzle. After this , the nozzle is mounted inside the vacuum chamber and silicon tubes of internal diameter 1.5mm are fixed over the steel tubes and are brought outside via the cross fittings . A 16 port electronic pressure scanner is used for measuring pressure from multiple ports simultaneously. These pressure scanners contain temperature compensated pressure sensor. The pressure gauge used is of Silicon Microstructure type with secondary storages. These ports are arranged in a manner that Port 1 can measure upto 150psi; Port 2,3,4-100psi; Port 5,6,7 -30psi; Port 8 to 16 measuring 15psi . The NI based interface connects the pressure scanner to the computer and the LAB VIEW software is used as GUI.

A supersonic nozzle was designed using conventional method. The nozzle exit angle was fixed to 15 degrees. The nozzle throat diameter was of 5 mm designed to operate at 10 bar inlet pressure. The material used to fabricate the nozzle is Stainless Steel, SS-304. The pipes used for plumbing are IBR seamless pipe made of high carbon steel. Standards flanges were used for the different joint .

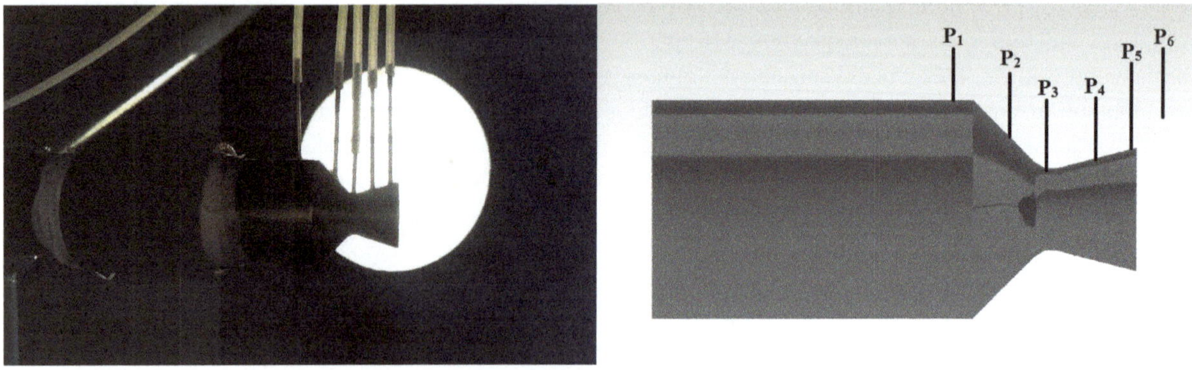

Figure 2 : P_1 - Inlet ;P_2 - Convergent section ;P_3 – Throat ;P_4- Divergent section ;P_5 – Exit ;P_6- Ambient (Vacuum chamber).

3. Results and discussion

A CFD analysis of the designed conical nozzle was carried out by generating an unstructured Quad-tri mesh using the dimensions of the nozzle in GAMBIT 2.4.6. Then the generated mesh was solved using FLUENT 6.3.26 and the results of various contours and vectors were obtained along with plots to see how the various isentropic parameters change as we go through the inlet to the throat and towards the exit of the nozzle . Then the respective boundary conditions were given for solving .

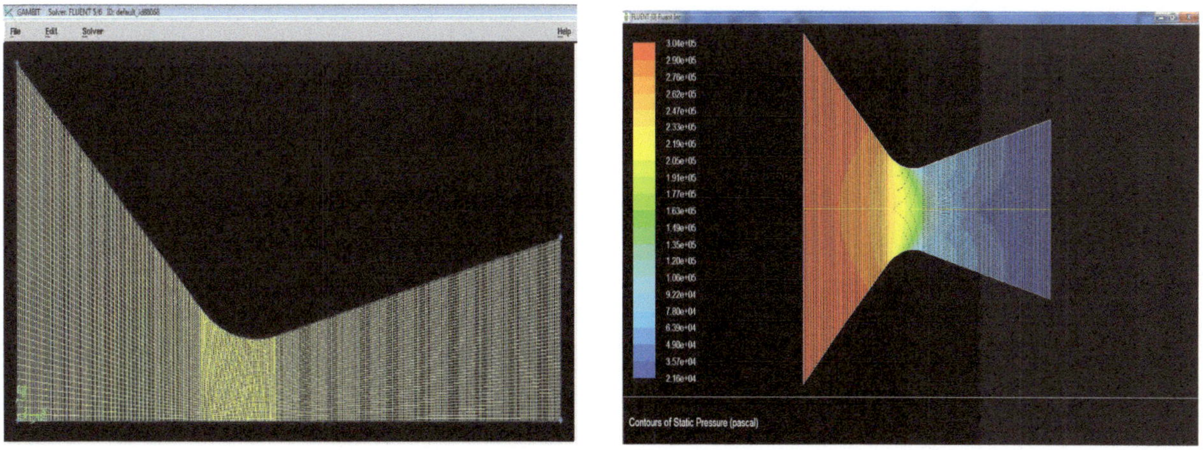

Figure 3 : (a) Nozzle mesh generated in GAMBIT 2.4.6 (b) Contour plot of Static Pressure

From the above pressure contour plot we could clearly observe that the pressure gradually started decreasing from the inlet , to medium at the throat and finally very less at the exit of the designed nozzle .[2]

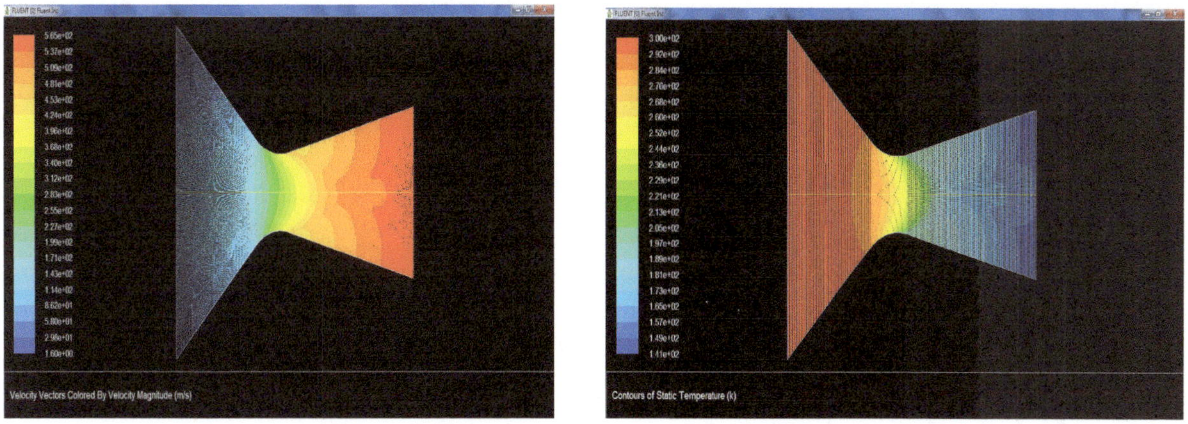

Figure 4 : (a) Vector plot of Velocity Magnitude (b) Contour plot of Static Temperature

According to the previously proved theories we have observed that from the velocity vector plot, the conical nozzle showed signs of velocity increasing from the inlet through the throat to the maximum value of the velocity at the exit . Through the CFD analysis the conical nozzle showed that the temperature kept decreasing downstream and there were freezing temperatures at the nozzle exit .

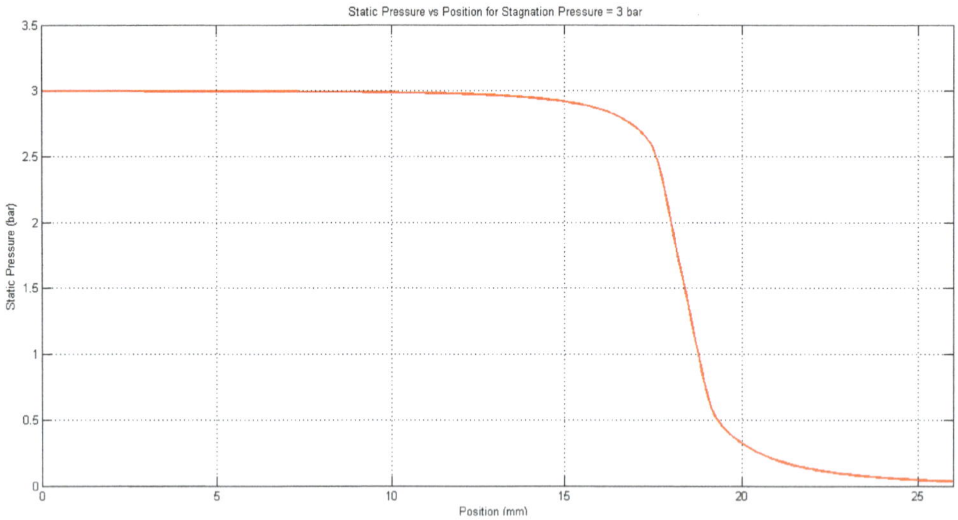

Figure 5 : Static Pressure vs Position for Stagnation Pressure of 3 bar

For the purpose of theoretically calculating the static pressure , temperature and velocity at each and every cross section of the conical nozzle all the way from the inlet , throat and the exit . We plot the curves to validate these parameters with respect to the position downstream of the nozzle. A suitable programming language was used and the results were programmed using the MATLAB R2008b .

We have plotted the various graphs for Pressure, Temperature and Velocity.

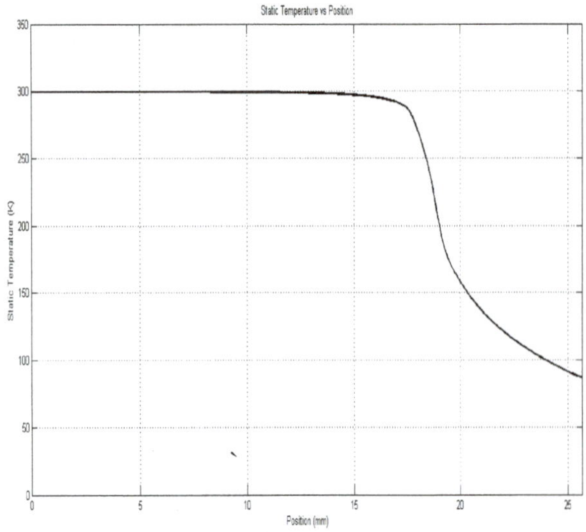

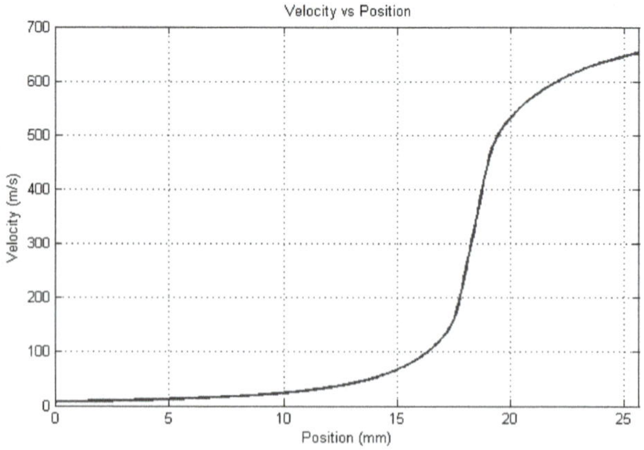

Figure6 (a) Static Temperature vs Position (b) Velocity vs Position

For conducting any experiment in vacuum chamber, it is very important to calculate the run time for the test . The run time is dependent on the pressure rise in the chamber for every inlet stagnation pressure . It is also important to note the time it takes for vacuum chamber to get filled from 0.001bar to 1 bar , which gives the run time of the experiment . Here the rise in pressure the chamber for various stagnation pressure is plotted using the MATLAB R2008b. We have got the following plots from which we know that the measured run time for 5 bar and 10 bar inlet stagnation pressure is 1minute 37 seconds and 46 seconds respectively.

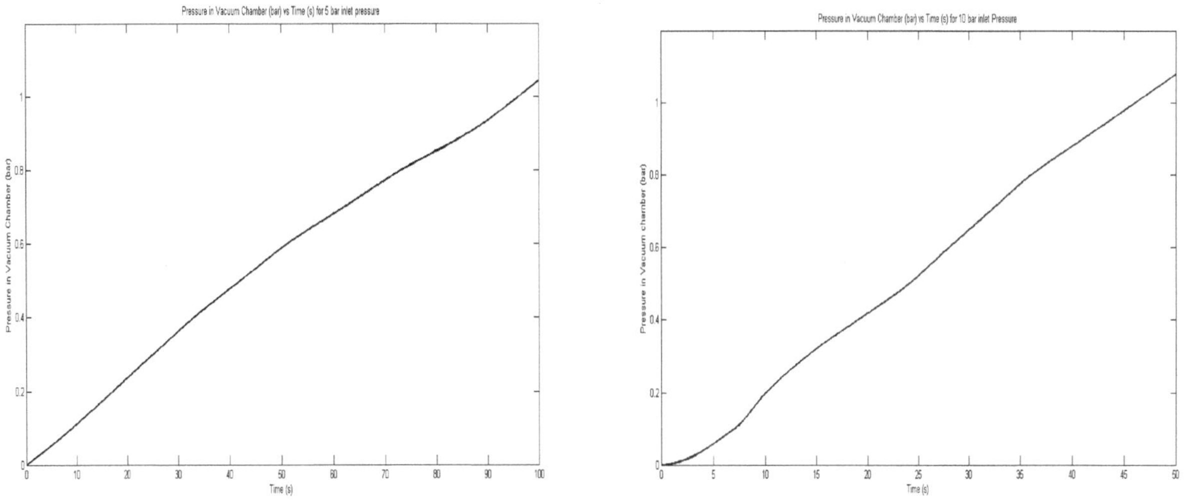

Figure 7 : (a) Pressure rise in the vacuum chamber for 5 bar inlet pressure (b) Pressure rise in the vacuum chamber for 10 bar inlet pressure

By placing the pressure port along the nozzle.An experimental pressure profile was created using a 16 port electronic pressure scanner.

4. Conclusion

At different values of stagnation pressure, the pressure at different pressure points was measured using a 16 port electronic pressure scanner along the length of nozzle from the inlet to the exit measuring the time interval of 0.1 seconds . The ratio of exit pressure to the ambient pressure (P_5/P_6) is plotted with time and the regions of under expansion and over expansion are noted along with the line where the nozzle is correctly expanded. The graphs for these values are plotted using MATLAB R2008b. The graph shows that the design value matches with the practical datas

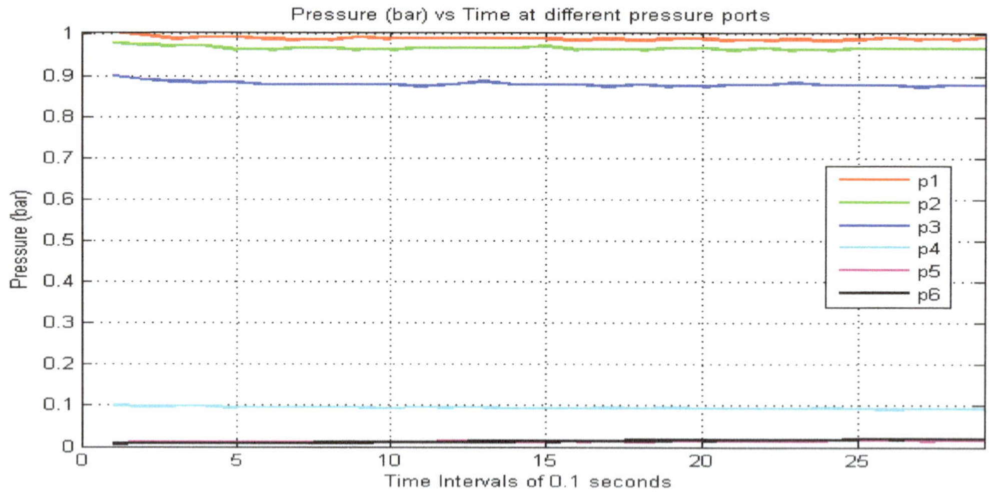

Figure 9: Pressure at different pressure port

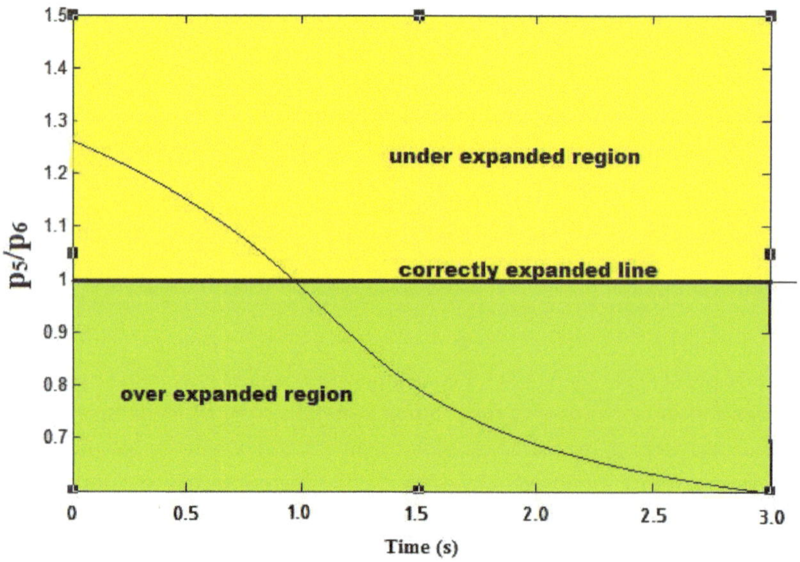

From the above graph we can interpret that the ambient pressure P6 slowly shows signs of increase , mainly due to the pressure build up in the chamber.The results of theoretical and experimental datas are found to match.

References

1. Q.Xiao & H.M.Tsai , D. Papamoschou , 2007,Numerical Investigation of Supersonic Nozzle Flow Seperation,*AIAA Journal,*Vol 45, No. 3 .

2. K.M.Pandey & A.P Singh , 2010, CFD Analysis of Conical Nozzle for Mach 3 at Various Angles of Divergence with FLUENT Software,*International Journal of Chemical Engineering & Applications,* Vol-1,No.2,ISSN:2010-0221.

3. Vladimir.V.Raibov and Fedoseyev.A.I,2013,The Analysis of Under expanded Jet Flows For Hypersonic Aerodynamic Experiments in Vacuum Chamber, *International Symposiums on Shockwaves* ,paper no:0008.

4. Md . Akhtar Khan,Sanjay Kumar Sardiwal,M.V. Sai Sharath, D.Harika Chowdary,2013,Design of a Supersonic Nozzle Using Method of Characteristics,*International Journal of Engineering Research & Technology* , Vol-2,Issue-11,ISSN: 2278-0181.

5. V.V.Volodin,T.V. Bazhenova,V.E Fortov,V.V Golub,A.A. Makeich and S.B. Shcherbak ,2005, Nozzle Start Flow Investigation in Conditions of High Altitude Test ,*24th International Symposium on Shock waves*.

6. Robert.E.Smith and Joel Ferrell , 1964,High Altitude Tests of Rocket Engines in Ground Test Facilities ,*SAE Technical Paper*.

12

Fluid Structure Interaction of Flow around a Pleated Insect 2D Airfoil at Ultra Low Reynolds Numbers

Husain Mehdi[1#], Fahad Anwer[2], Akhlaque Ahmad[2]

[1]*Assitant Professor, MED, Meerut Institute of Technology, Meerut (U.P), 250002/ INDIA*

[2]*Department of Mechanical Engineering ZHCET, AMU Aligarh, 202002/ INDIA*

[#]*Corresponding author: husainmehdi4u@gmail.com*

Abstract

In this work, Numerical study of Fluid Structure Interaction of uniform flow past a two dimensional pleated airfoil is carried out. When the wing interact with the air, it is subjected to both aerodynamic forces acting on the surface of the wing and the inertial force due to the acceleration of deceleration of the wing mass. The interaction between these inertial and aerodynamic forces resulted in wing deformation. The dynamics of a pleated insect wing subjected to aerodynamic loading is studied. The vortex induced vibration and forced vibration of a pleated flexible insect wing subjected to aerodynamic load is studied by using ANSYS-14 multi physics solver. The insect wing is of dragonfly wing cross section. In the first phase of the work, fluid flow simulation at Reynolds Number-100, 200, 500, and 1000 will be performed with angle of attack 0^0 to 15^0. The result from the CFD solver will be fed in the form of lift and drag forces are then fed into the ANSYS Workbench solver and one way Fluid Structure Interaction analysis is performed.

Keyword: Lift Coefficient, Drag Coefficient, Angle of Attack, Reynolds No, Deflection

1 Introduction

Numerous experimental and simulation studies on flapping wings have been conducted, and significant understanding of the aerodynamic-force generation mechanism has been achieved (Ellington et al., 1996; Wakeling and Ellington, 1997; Dickinson et al., 1999; Liu et al., 1998; Sane and Dickinson, 2001; Usherwood and Ellington, 2002; Sun and Tang, 2002a and b; Ramamurti and Sandberg, 2002 and 2007). However, in most of these studies, rigid-wing models are employed although wing flexibility has been recognized as an important factor in flapping-wing aerodynamics (Young et al., 2009; Walker et al., 2010; Du and Sun, 2010; Wootton et al., 2003; Combes and Daniel, 2003).

Micro air vehicles (MAVs offer great potentials for various civil and military applications in revolutionizing sensing and information gathering, such as surveillance, inspection, targeting, or biochemical sensing tasks at remote and hazardous locations. Numerous vehicle concepts, including fixed, rotary, and flapping wings, have been proposed (Shyy et al., 1999; Mueller, 2001; Lian et al., 2003; Stanford et al., 2004; Bohorquez et al., 2003; Ramasamy et al., 2006). However, the advantage of a flapping wing in MAV compared with fixed or rotary-wing aircraft of the same scale is emphasized by its maneuverability and smooth transition from hovering to forward flight signature.

The structural deformation during flapping can significantly influence the flow behavior around the wing and, consequently, brings an important effect on its aerodynamic performance (Combes and Daniel, 2003. Physiological studies have shown that insect wing deformation is passive in which the wing is deformed by both inertial acceleration and deceleration of the wing mass, by the aerodynamic force from surrounding air, or by the combination of both (Wootton et al., 2003). Combes and Daniel, (2003b) discovered that the overall bending patterns of a hawkmoth wing are quite similar when flapped (single degree-of-freedom flap rotation) in air and in helium, despite an 85% reduction in fluid density.

Young et al., (2009) found out that wing deformation leads to substantial power economy for lift production. Moreover, they noticed that the leading-edge vortex (LEV) in the full- kinematic model remains attached to the wing during the entire flapping cycle, which may have contributed to the aerodynamic power efficiency of the wing. Similarly, Walker et al. (2010) obtained quantitative data on the time-varying camber and twist of the wing in free-flying drone flies. A further study was conducted by Du and Sun (2010) where they removed the camber and spanwise twist one by one. They observed that the lift in the free-flying drone flies is principally caused by camber deformation. Mueller (2001) reported that a fundamental difference exists between the interest of an aerospace engineer in flapping flights and that of a biologist or zoologist.In recent years, numerous experimental and numerical studies have been carried out to investigate the effects of flexibility on flapping airfoil/wings. Heathcoat et al. (2004 and 2008). Hu et al. (2010) studied the aerodynamic benefits of flexible membrane wings in the development of flapping-wing MAVs. The time-averaged lift and drag generation of two flexible membrane wings with different skin flexibilities (flexible nylon and flexible latex wings) are compared with those of a rigid wing.The effect of the Reynolds number on the gliding ratio is that at Re 1000 and at angle of attack (here after, AOA) 15^0, the largest gliding ratios are obtained. Flow invariably for all Reynolds number, minimum Drag coefficient is obtained at AOA 15^0 Anwer et al (2013). Aeshna genus dragon fly can glide up to 30s without any significant loss of altitude (Brodsky 1994). Wakeling and Ellington (1997) filmed a smaller dragon flies and found there gliding periods lasting 0.5s, covering a distance of approximately 1m and thus achieving gliding speed of $2.5ms^{-1}$. Thus the typical range of Reynolds number in gliding flight can vary from 10^2 to 10^4 as suggested by Wakeling and Ellington (1997). This regime of flow can be thus termed as an ultra low Reynolds number regime.Kesel (2000) compared aerodynamic performance of cross-sections at different positions along the span of a wing of an Aeschna cyanea to develop the pleated models and its corresponding profiled airfoil at a chord Reynolds number of 10 000, and the results showed that the pleated airfoils generated higher lift than the profiled airfoils. However, Okamoto et al (1996), Kesel (2000) revealed that the orientation of the leading edge does not have an effect in enhancing the lift production. As with the early flow visualization experiments, Kesel (2000) noticed trapped vortices present in the folds that serve to change the effective profile of the airfoil. Dragon flies have corrugated wings where the pleated configuration varies along the span-wise and chord-wise directions (Vargas and Mittal 2006). Vergas and Mittal (2006) also studied the aerodynamic performance of a pleated wing section based on the wing of Aeshna cyanea has been performed at ultra-low Reynolds numbers (5000-10000) corresponding to the gliding flight of these dragonflies in two dimension using computational fluid dynamics. The simulations demonstrate that the pleated airfoil produces comparable and at times higher lift than the profiled airfoil, with a drag comparable to that of its profiled counterpart.

Mehdi. H (2014) The deflection is increased when we increased the Reynolds Number with increase their angle of attack The maximum deflection occur in Re-1000, angle of attack 15^0i.e. $9.2597x10^{-3}$mm and minimum value of

deflection occurs in Re-100 with angle of attack 0^0 i.e.1.4618 x 10^{-4} mm..the highest natural frequency 34.353 Hz was found in mode 6 which is torsional mode, whereas minimum natural frequency 0.6951 was found in mode .
Mehdi. H (2014)It was found that for all the simulations performed flow always remained steady at Re 100 and 200 at all angle of attack (0^0 to 15^0). First unsteady flow was obtained at Re 500 and AOA 10^0. But flow always remained steady at AOA 0^0 and 5^0 for all the Reynolds numbers
In this work, we investigate Fluid structure interaction of flow around a pleated insect 2D airfoil for spatio-temporal dynamics of a cut section of Aeshna Cyanea's wing at ultra low Reynolds numbers (100 to 1000) at different angle of attacks ranging from 0° to 15°. To the best of author's knowledge, The Fluid structure interaction of 2D pleated airfoil is not studied on these reynolds Number and their Angle of attack. These parameter ranges are relevant for both dragon flies and micro-aerial vehicles. Past experimental studies (Buckholz, 1986, Okamoto et al 1996, Kesel, 2000) have found no intrinsic three-dimensional effects at these low Reynolds numbers. Thus, 2D simulations are implemented in this study to encompass a relatively wide range of the parameter space necessary to draw some general conclusions regarding pleated airfoils. The pleated airfoil implemented in the numerical simulation corresponds to a cross-section located at the mid-section of the forewing of a dragonfly (Aeshna cyanea). The specific profile chosen for the numerical simulations corresponds to 'Profile 2', which was digitally extracted from the paper of Vargas et al (2008) (which is the same as Profile 2 from Kesel (2000)). From the three pleated geometries to select from the paper of Kesel (2000), 'Profile 2' was chosen due to its horizontal leading edge, thus eliminating the issue that the orientation of the leading edge has an influence on the aerodynamic performance (Okamoto et al 1996).

(a)Profile 1 (b) Profile 2 (c) Profile 3

Figure 1: An Illustration of different cut section profiles used by Kessel (2006) and Vargas et al (2008)

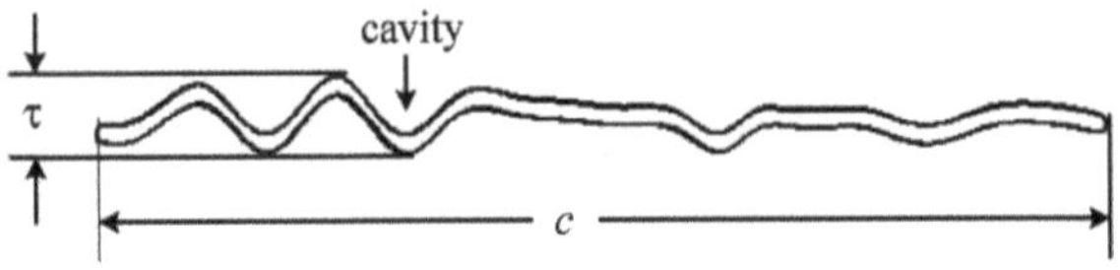

Figure 2: The two-dimensional airfoils used in the numerical simulation. Pleated airfoil representing a cross-section of the forewing of a dragonfly (Aeshna cyanea) having τ /c = 7.531%

2 Governing Equation

2.1 Fluid Flow

The equations governing the flow in the numerical solver are the time-dependent, viscous incompressible Navier–Stokes equations. The non-dimensional momentum and continuity equations are as follows:

$$\frac{\partial u_i}{\partial x_i} = 0 \tag{1}$$

$$\frac{\partial u_i}{\partial t} + \frac{\partial u_i u_j}{\partial x_j} = -\frac{\partial p}{\partial x_i} + \frac{1}{Re}\frac{\partial^2 u_i}{\partial x_j \partial x_j} \tag{2}$$

The equations are non-dimensionalized with the appropriate length and velocity scales, in this case the airfoil chord and free stream velocity. Here Re corresponds to the Reynolds number which is defined as below:

$$Re = \frac{\rho u_\infty c}{\mu} \tag{3}$$

Using the same flow parameters and geometrical dimensions as Kesel (2000), this allowed for validation and a critical analysis of the numerical results. The key quantities examined are the lift and drag coefficients which are defined as

$$C_L = \frac{F_L}{\frac{1}{2}\rho u_\infty^2 c} \tag{4}$$

$$C_D = \frac{F_D}{\frac{1}{2}\rho u_\infty^2 c} \tag{5}$$

$$\text{Gliding Ratio} = \frac{C_L}{C_D} \tag{6}$$

2.2 *Structure Analysis:*

The interaction of the fluid and the structure at a mesh interface causes the acoustic pressure to exert a force applied to the structure and the structural motions produce an effective "fluid load." The governing finite element matrix equations then become

$$[M_S]\{\ddot{U}\} + [K_S]\{U\} = \{F_S\} + [R]\{P\} \tag{7}$$
$$[M_f]\{\ddot{P}\} + [K_f]\{P\} = \{F_f\} - \rho o\,[R]^T\{\ddot{U}\} \tag{8}$$

[R] is a "coupling" matrix that represents the effective surface area associated with each node on the fluid-structure interface (FSI). The coupling matrix [R] also takes into account the direction of the normal vector defined for each pair of coincident fluid and structural element faces that comprises the interface surface. The positive direction of the normal vector, as the program uses it, is defined to be outward from the fluid mesh and in towards the structure. Both the structural and fluid load quantities that are produced at the fluid-structure interface are functions of unknown nodal degrees of freedom. Placing these unknown "load" quantities on the left hand side of the equations and combining the two equations into a single equation produces the following.

2.3 *Aerodynamic Performance*

The aerodynamic performance of airfoil can be determined by various lift and drag ratio (Kesel, 2000). The gliding ratio ($\in_r$) is defined as

$$\text{Gliding Ratio } (\in_r) = \frac{C_L}{C_D} \tag{9}$$

It gives the maximum gliding distance per unit height. The corresponding gliding angle can be calculated as

$$\text{Gliding Angle } (\gamma) = \tan^{-1}\frac{C_D}{C_L} \tag{10}$$

The maximum sinking rate ($\in_s$) is calculated from

Sinking Rate $(\in_s) = \frac{C_L^3}{C_D^2}$ (11)

3 Numerical Methodology

Ansys Fluent 14.0 is used to solve the above equations. The solver used is based on collocated methodology and finite volume discretization technique is used.

$$\iiint \frac{\partial \rho \emptyset}{\partial t} dV + \oint \rho \emptyset \vec{v}.d\vec{A} = \oint \Gamma_\phi \nabla \emptyset.d\vec{A} + \iiint S_\emptyset \, dV \quad (12)$$

$$\frac{\partial \rho \emptyset}{\partial t} V + \sum_f^{N_{faces}} \rho_f \vec{v}_f \, \phi_f.\vec{A}_f = \sum_f^{N_{faces}} \Gamma_\phi \, \nabla \phi_f.\vec{A}_f + S_\phi V$$

The gradients are calculated using Green-Gauss Cell-Based methodology proposed by Holmes and Connel (1989) and Rauch *et al.*(1991).

$$\overline{\emptyset_f} = \frac{1}{N_f} \sum_n^{N_f} \overline{\emptyset}_n \quad (13)$$

where N_f is the number of nodes on the face Convection terms in the momentum equation are discretized using the second order upwind methodology given by in the second order up wind quantities at cell faces are computed using a multidimensional linear reconstruction approach (Barth and Jespersen, 1989). The result from the CFD solver will be fed in the form of lift and drag forces are then fed into the ANSYS Workbench solver and one way Fluid Structure Interaction analysis is performed.

4 Grid Design:

The pleated cross section of modeled dragon fly is taken from the paper [5]. The length of pleated plate is assumed to be of unit length by giving X coordinate=1, Height is also given to pleated section in the Y direction and it is 0.05. The pleated section is covered with a circular region of diameter 20C where C is set to 1.This circular region is then divided into 2 parts along semi-circle parallel to Y axis.Then 150 nodes are taken on each semi circle.60 nodes are also taken on airfoil's upper and lower part.Then command of triangular meshing is giving and the required meshed profile is obtained.

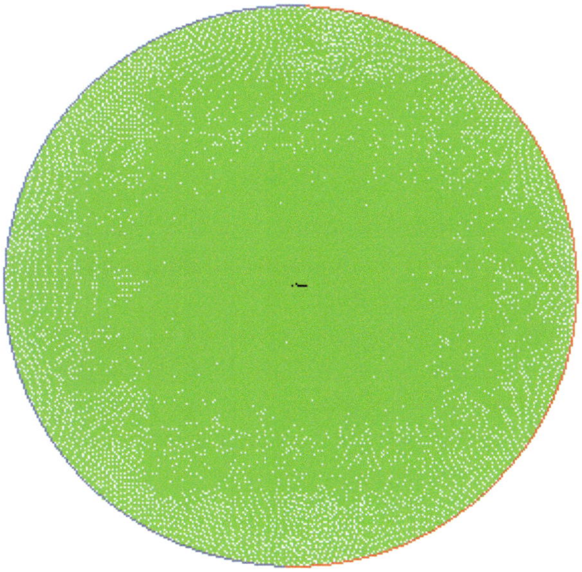

Figure 3:- Mesh of the problem

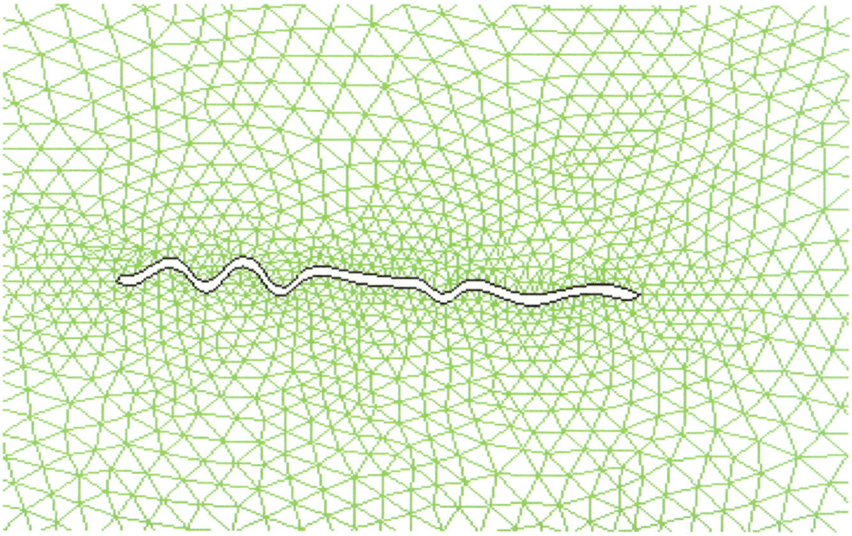

Figure 4:- The zoomed view of the mesh

Grid information:

- 195834 triangular cells
- 196878 nodes,
- Inflow and out flow boundaries are pressure far field
- Inner meshed region is fluid.

Face area statistics:

- Minimum face area (m^2) : 4.428917 e-003
- Maximum face area (m^2) : 2.160249 e-001

5 Boundary Condition

A constant velocity u= 0.14607 m/sec is imposed on the left side of grid, and the right side set as an outflow region where the gradient values are set to zero. The components are taken in accordance with the angle of attack. Pressure on the both sides was taken as atmospheric i.e. $P=P_{atm}$

Velocity inlet boundary conditions are used to define the velocity and scalar properties of the flow at inlet boundaries. Out flow boundary condition are used to model flow exits where the details of the flow velocity and pressure are not known prior to solution of the flow problem. They are appropriate where the exit flow is close to a fully developed condition, as the out flow boundary condition assumes a zero normal gradient for all flow variables except pressure.

6 Results and Discussions

Validation study was carried out for, Profile 2 as shown in table 1. Simulations were carried out for Reynolds number of Re =100, 1000 and zero angle of attack for profile 2. These results are compared with Anwer et al (2013). The global parameters, C_D, C_L were calculated and were well in agreement with results of Anwer (2013).

Table 1:- Comparison of force coefficients for Profile 2, Anwer (2013)

Re	α (AOA)	Lift Coefficient C_L **Present**	Drag Coefficient C_D **Present**	Lift Coefficient C_L Anwer **(2013)**	Drag Coefficient C_D Anwer **(2013)**
100	0^0	0.174	0.41	0.172	0.399
1000	0^0	0.221	0.121	0.216	0.119

In all twenty simulations were performed to determine the aerodynamic characteristics of spatio temporal dynamics of a cut section of Aeshna Cyanea's wing has been performed at ultra low Reynolds numbers (100 to 1000) at different angle of attacks ranging from 0^0 to 15^0. The effect of the Reynolds number on the gliding ratio is shown in figure 5a, at Re 1000 and angle of attack 15^0, the largest gliding ratios are obtained. Figure 5b shows that invariably at all Reynolds number, minimum Drag coefficient is obtained at AOA 15^0. The drag production leads to some interesting observations. As expected, the overall drag coefficient increases as Re is decreased .because the viscous effects are more dominant at lower Reynolds numbers which cause the skin friction to be the major contributor to the overall drag. As the angle of attack is increased, drag coefficient further decreases.

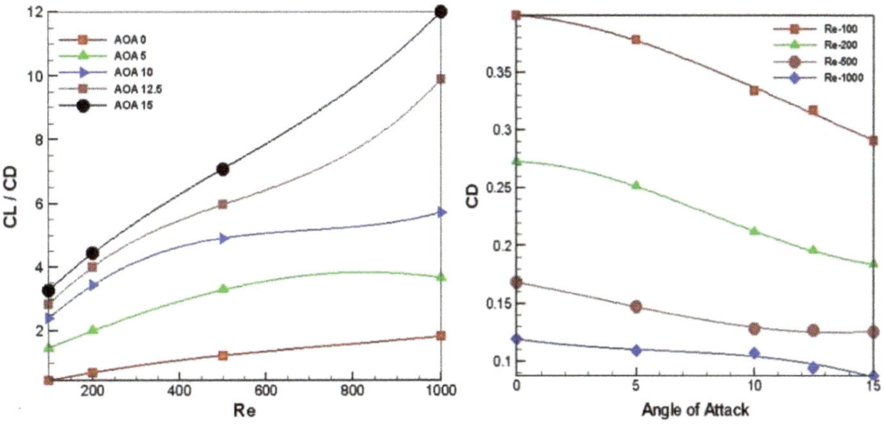

(a) (b)

Figure 5:- (a) Comparison of Gliding Ratios for different Angle of Attacks (AOA), (b) Comparison of Average Drag Coefficient for different Angle of Attacks (AOA)

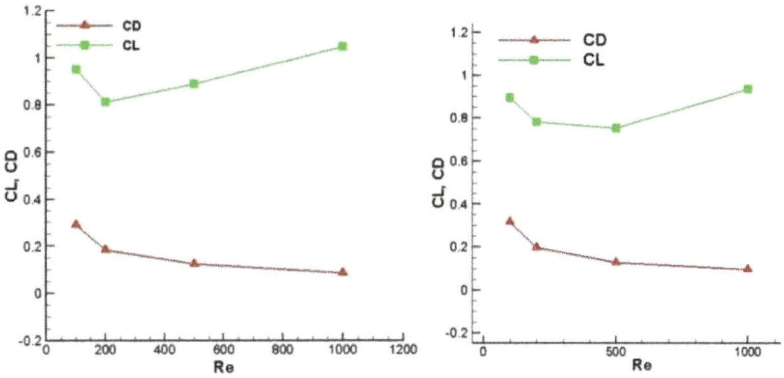

(a) (b)

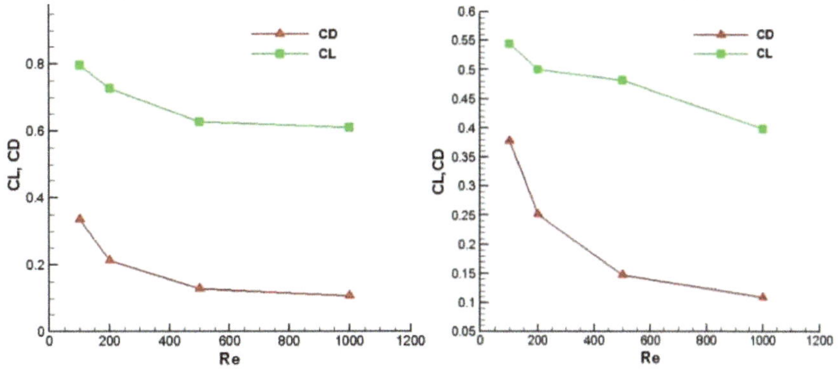

(c) (d)

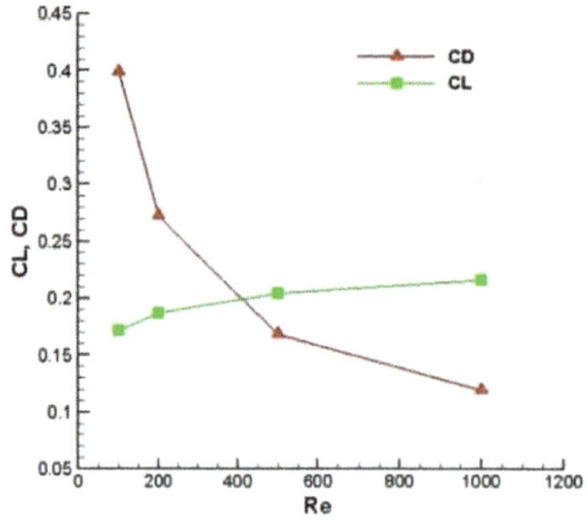

(e)

Figure 6:- Variation of mean Lift and Drag coefficient with Reynolds number for angle of attack (a) $\alpha = 0^0$, (b) $\alpha = 5^0$, (c) $\alpha = 10^0$, (d) $\alpha = 12.5^0$ (e) $\alpha = 15^0$

It can be seen that as Reynolds no increased, lift coefficient increases for angle of attack 0^0 while C_D decreases. For angle of attack 5^0 and 10^0 both the coefficient decreases monotonically.While at 12.5^0 and 15^0, C_L first decreases to a minimum but then increases to a maximum value, while C_D always decreases. It can be seen that as Reynolds no increased, lift coefficient increases for angle of attack 0^0 while C_D decreases. For angle of attack 5^0 and 10^0 both the coefficient decreases monotonically.While at 12.5^0 and 15^0, C_L first decreases to a minimum but then increases to a maximum value, while C_D always decreases as shown in figure 6.

6.1 Steady Flow:

The time histories of lift and drag coeffecient shows that for all angle of attacks at Re 100 and 200 shows that flow remains steady. Same is the case for Re 500 and 1000 at 0° and 5° angle of attacks. First unsteady flow was encountered at Re 500 at 10^0 and similarly for Re 1000. The results shown here were obtained after running simulations for around 5000 non-dimensional time units

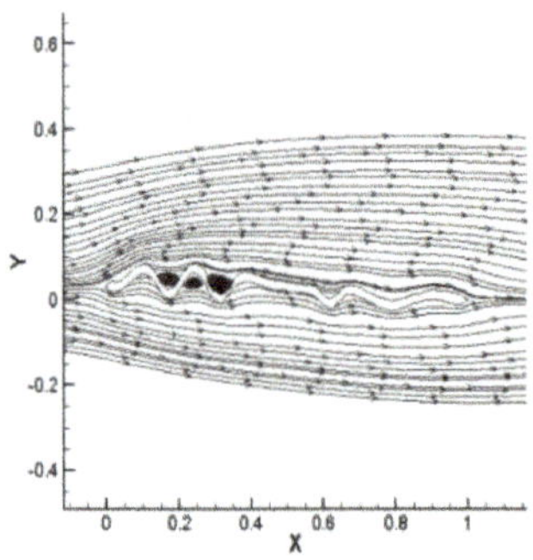

(a)

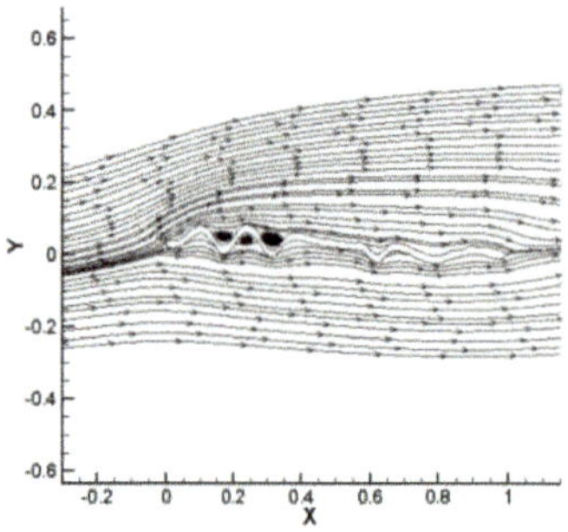

(b)

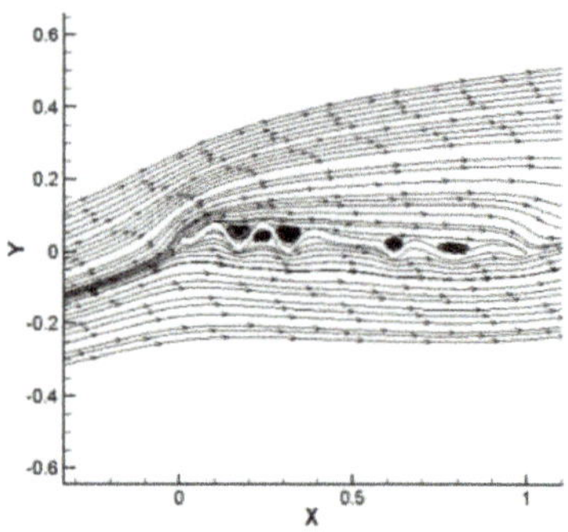

(c)

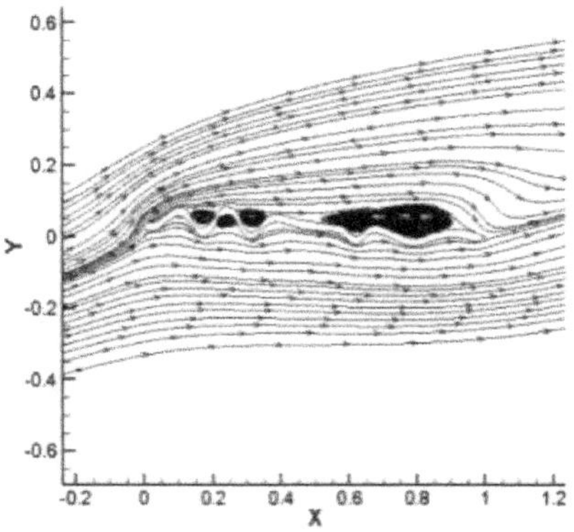

(d)

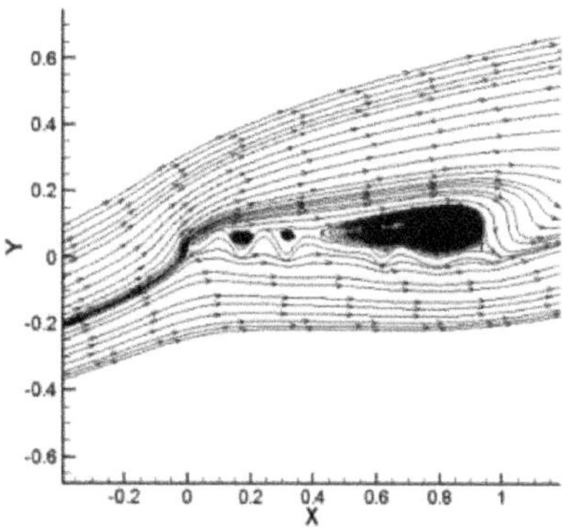

(e)

Figure 7:- Stream-traces at Re = 100 for (a) α= 0^0, (b) α= 5^0, (c) α= 10^0, (d) α=12. 5^0 , (e) α=15^0

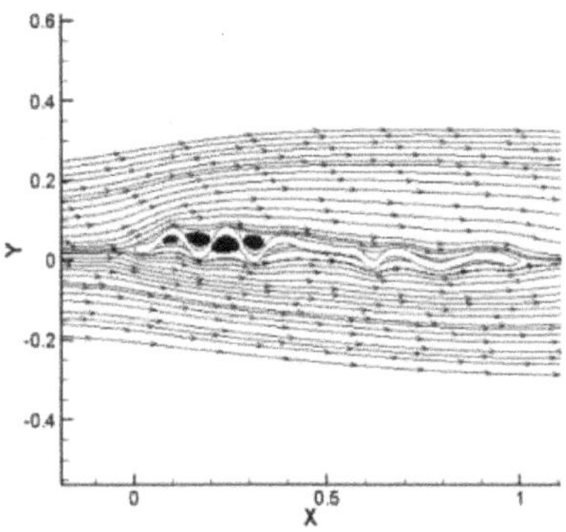

(a)

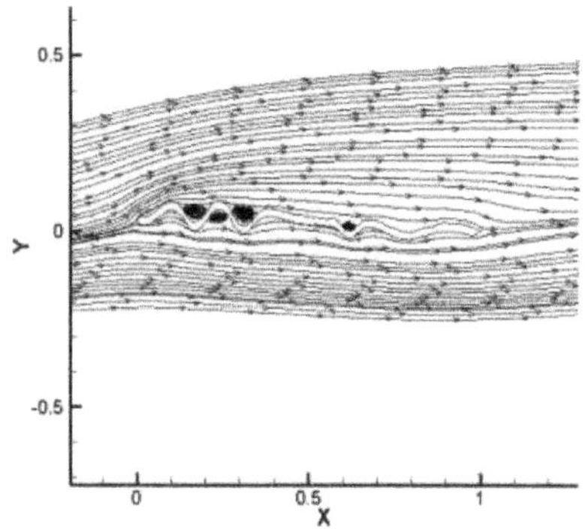

(b)

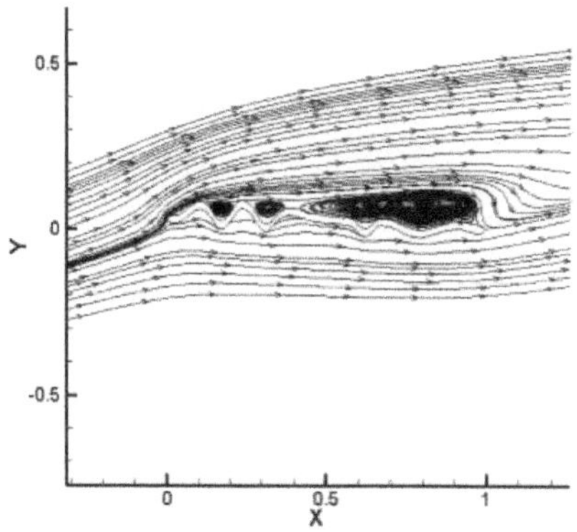

(c)

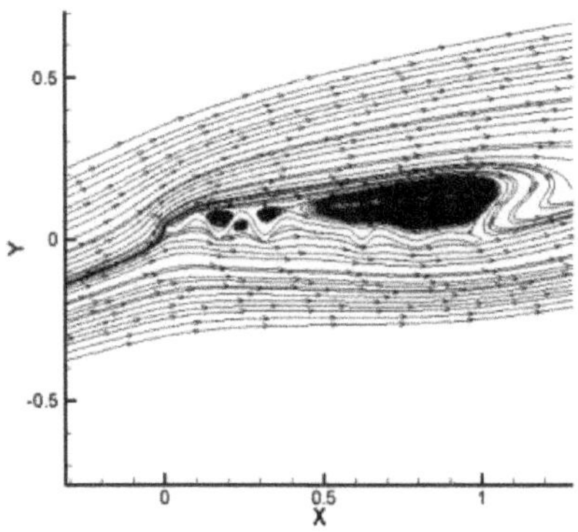

(d)

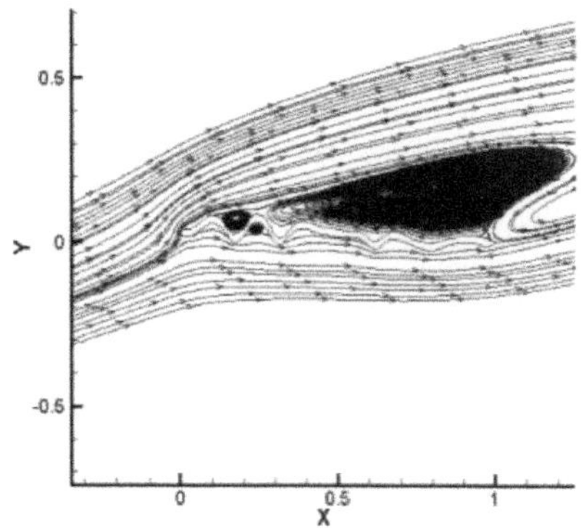

(e)

Figure 8:- Stream-traces at Re = 200 for (a) α= 0^0, (b) α= 5^0, (c) α= 10^0, (d) α=12. 5^0 and , (e) α=15^0

6.1.1 Effect of Angle of Attack:

It is seen in figure 7-8 the stream-traces on the pleated airfoil show that there is a trapped vortex in each cavity which causes the overall flow), who hypothesized that flow would behave in such a way due to the trapped vortex inside each cavity. Furthermore, the flow separates just downstream of the tip of each corrugation and reattaches just upstream of the tip of the next corrugation. The region between the separation and reattachment points is a region with a re-circulating flow and there is negative shear drag generated in this region, as shown in figure 7-8, since these are the regions where the surface normal gradient in the tangential velocity is large. The pleated airfoil is therefore able to produce total drag comparable to that of its smooth counterpart by reducing the shear drag as a direct result of flow reversal occurring in each cavity

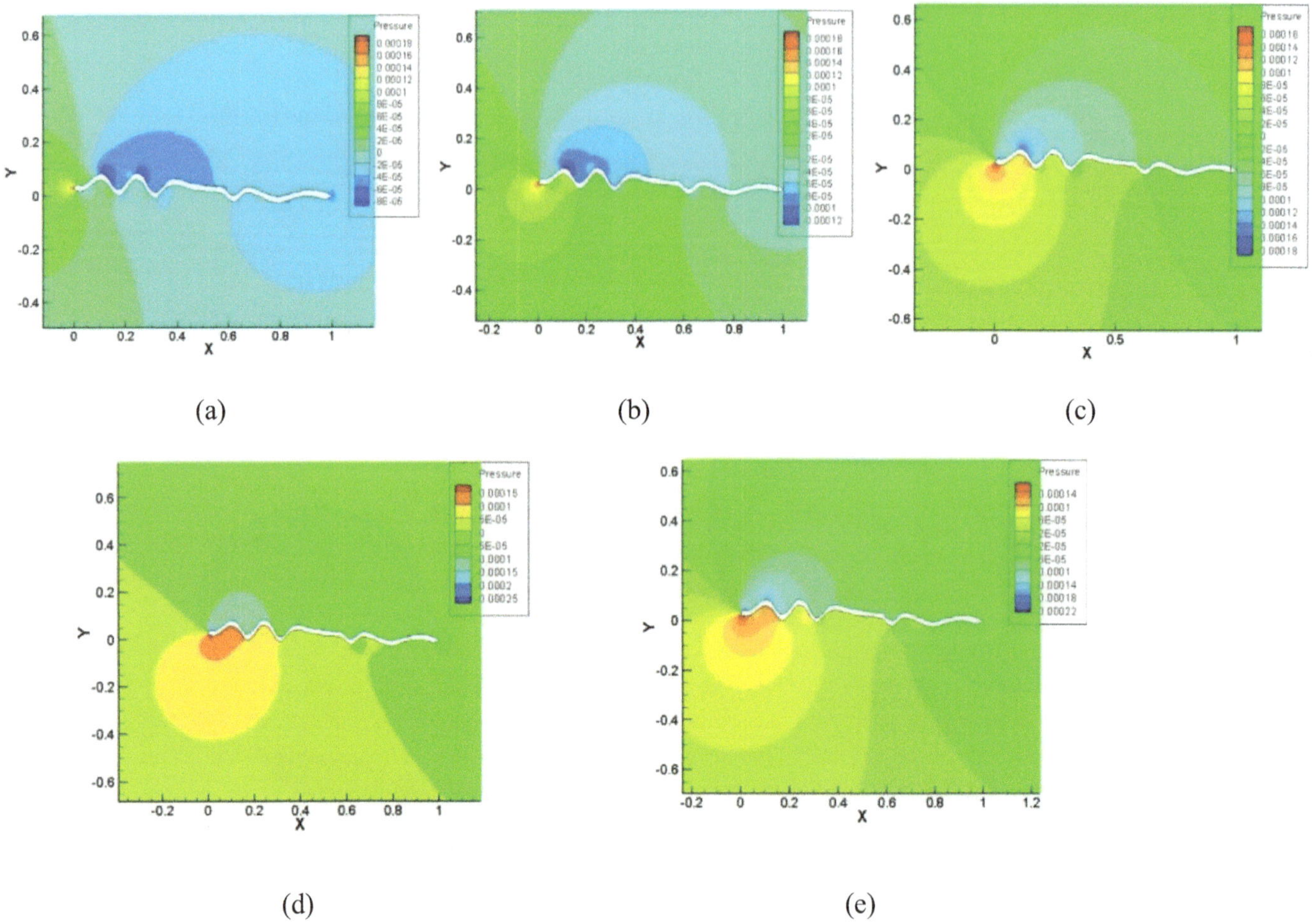

Figure 9:- Pressure contours at Re = 100 for (a) $\alpha = 0^0$, (b) $\alpha = 5^0$, (c) $\alpha = 10^0$, (d) $\alpha = 12.5^0$ and , (e) $\alpha = 15^0$

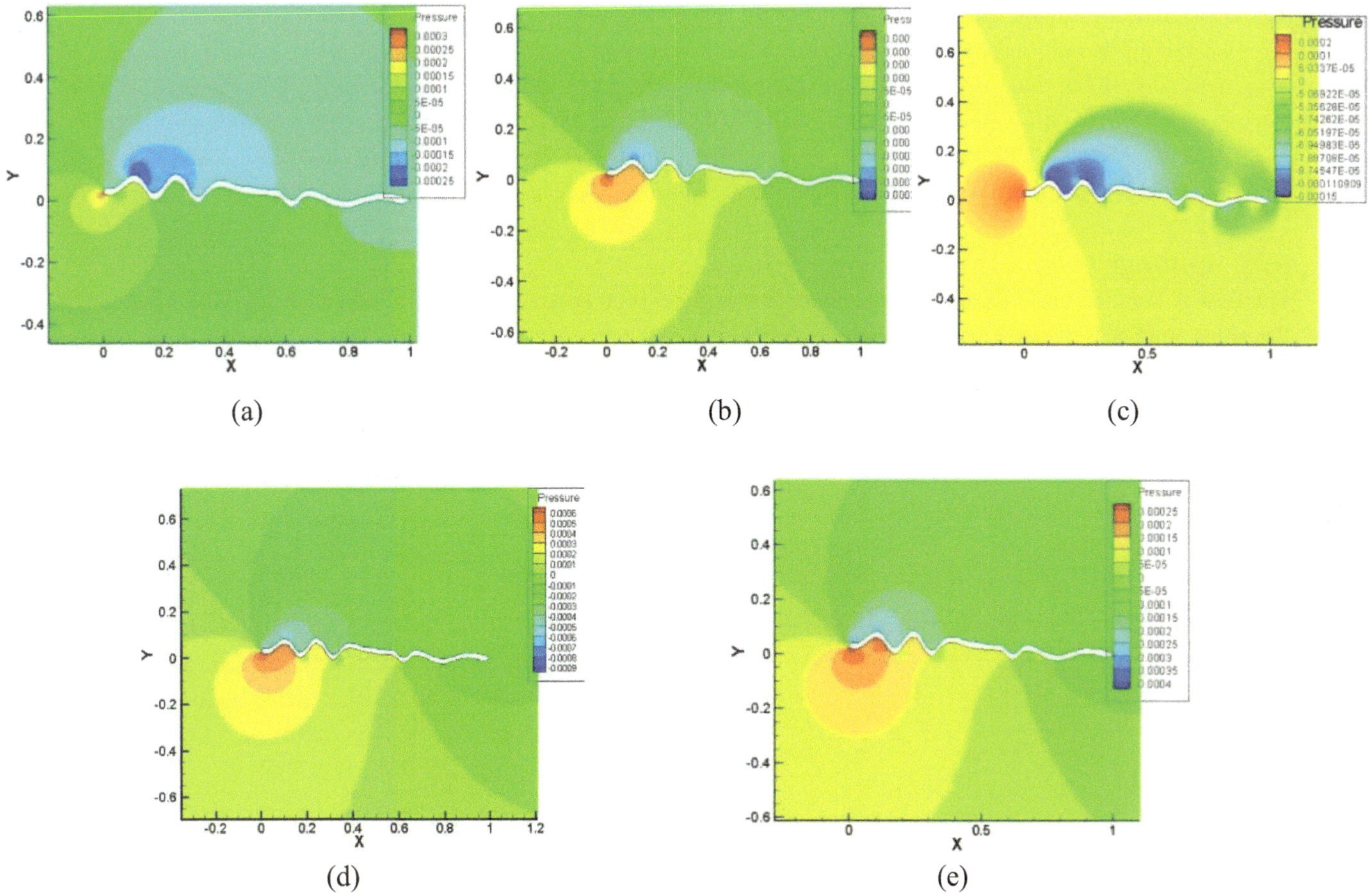

Figure 10:- Pressure contours at Re = 200 for (a) $\alpha = 0^0$, (b) $\alpha = 5^0$, (c) $\alpha = 10^0$, (d) $\alpha = 12.5^0$ and , (e) $\alpha = 15^0$

The observations of demonstrated that pleats cause an increase in negative pressure on the upper surface of the profile and, thus, an increase in lift production Vargas (2008). This can be clearly seen in figure 9-10 which shows pressure contours as a function of angle of attack. The negative pressure is found in all profile valleys, regardless of the profile side. This is contradictory to our observations. We found that, in all the cases pressure on the upper side of the pleated airfoil remains negative while at the lower side, it always remains positive Kesel (2000). As the angle of attack increases, drag force continue to decrease while lift force continuously increases. This is due to a larger attached vortex on the upper surface of the airfoil. The decrease in drag is due to lower shear drag as the strength of the trapped vortex in the pleats of the airfoil.

It is important to examine the effect of the Reynolds number on the airfoil performance since this has implications of how well these wings would work for dragonflies (and MAVs) of different sizes. This issue is also important from the point of view of use of such wings in micro-aerial vehicles. Therefore, a numerical study which focuses on the effect of the Reynolds number on the foil performance was conducted . A comparison of lift, drag coefficients and glide ratio was shown in figure 5. at Re =100, $C_L = 0.545$, $C_D = 0.3783$ and ϵ_r is 1.44. As the

Reynolds number is doubled to 200, the lift decreases marginally to 0.5 but drag decreases by 0.1 approximately to 0.2514 while glide ratio increases to 2.0. Now, as Re is increased by a factor of 5 to 500, the lift still decreases marginally to 0.48 but drags decreases by 0.1 approximately to 0.12 and glide ratio reaches 3.27. Further, as Re is increased to 1000, the lifts still decreases marginally to 0.4 and drag also decreases marginally to 0.1 and glide ratio reaches 3.65. Vargas (2008) also report similar observations. They also confirmed that glide ratio has a monotonous behaviour for pleated airfoil in this ultra Reynolds number regime

6.2 Unsteady Flow:

The flow is found to unsteady and periodic at Re-500 and Re-1000 at AOA 10^0, 12.5^0, and 15^0. It was found that as Reynolds number is increased then strouhal number also increased

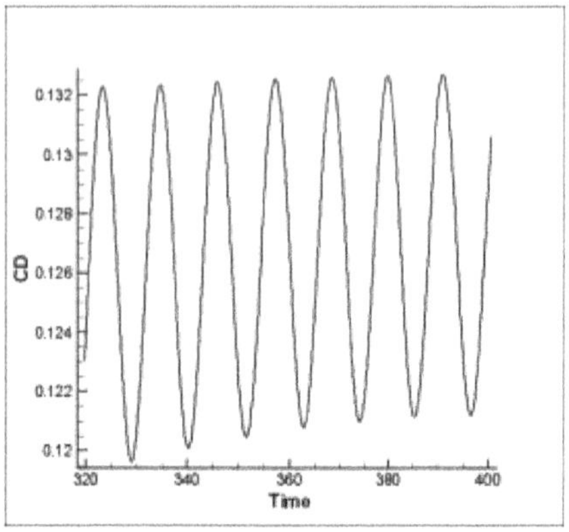

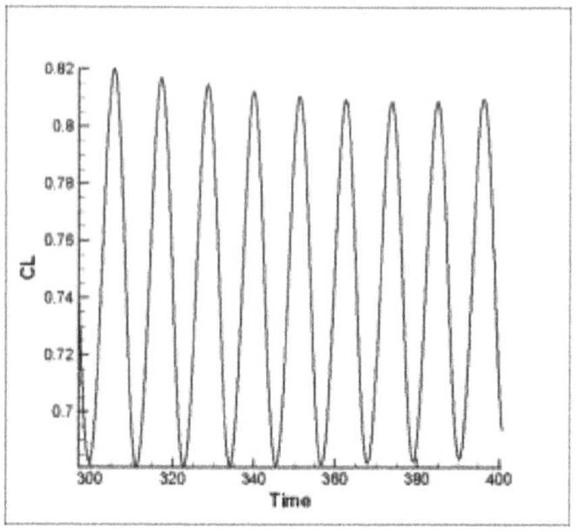

(a) (b)

Figure 11:- Time variation coefficient of (a) Drag (b) Lift at Re-500 AOA12.5^0

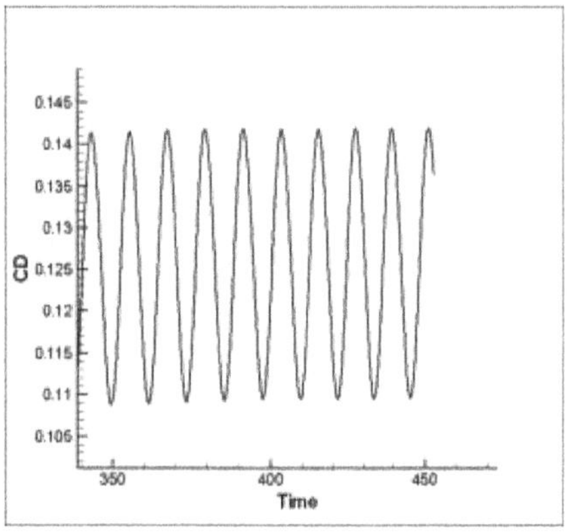

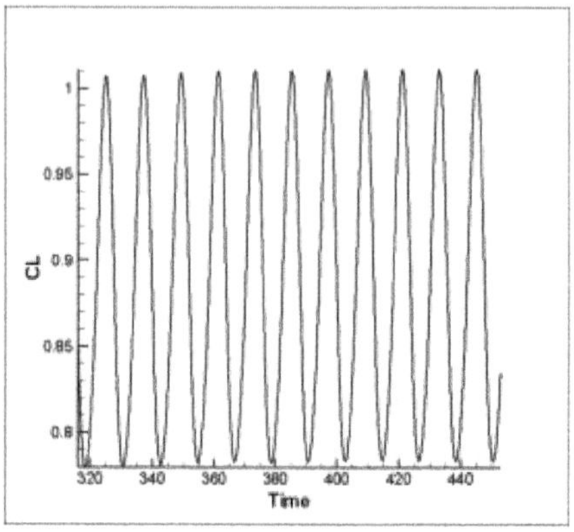

(a) (b)

Figure 12: Time variation coefficient of (a) Drag (b) Lift at Re-500 AOA15^0

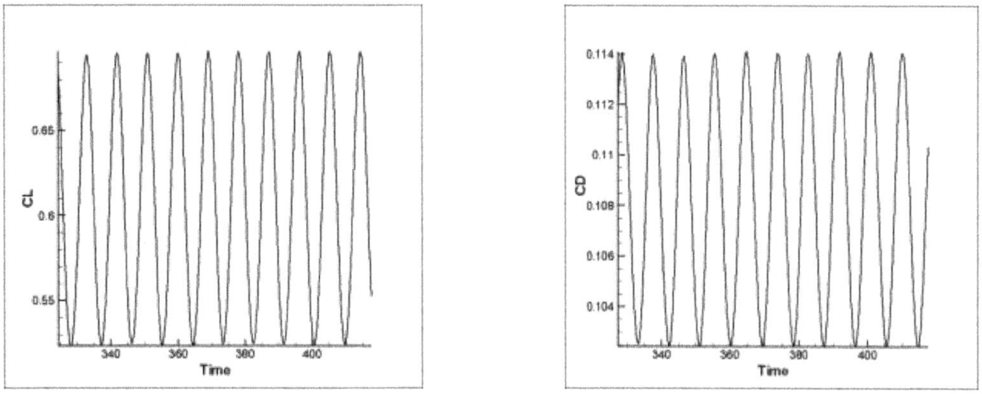

(a) (b)

Figure 13:- Time variation coefficient of (a) Drag (b) Lift at Re-1000 AOA10^0

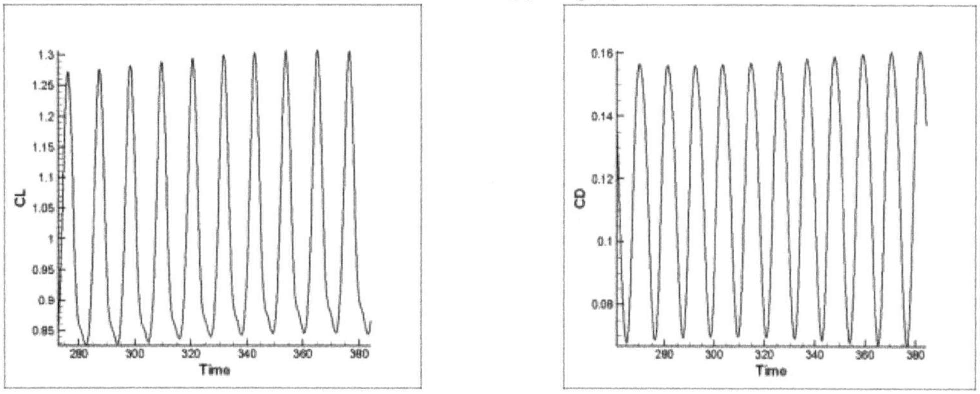

(a) (b)

Figure 14:- Time variation coefficient of (a) Drag (b) Lift at Re-1000 AOA15^0

When we analyze the Reynolds no of 500 and 1000 with angle of attack (10^0, 12.5^0, 15^0) it shows the unsteadiness because of variation of C_D and C_L values. The time history of C_L and C_D for these cases is shown in Figure 11-14. The pleated airfoil reaches a maximum C_L= 1.048 at Re-1000 with angle of attack 15^0 and minimum C_L=0.1715 at Re-100 with angle of attack 0^0. Whereas the maximum value of C_D =0.4 at Re-100 with angle of attack 0^0 and the minimum value of C_D= 0.0873 at Re-1000 with angle of attack 15^0

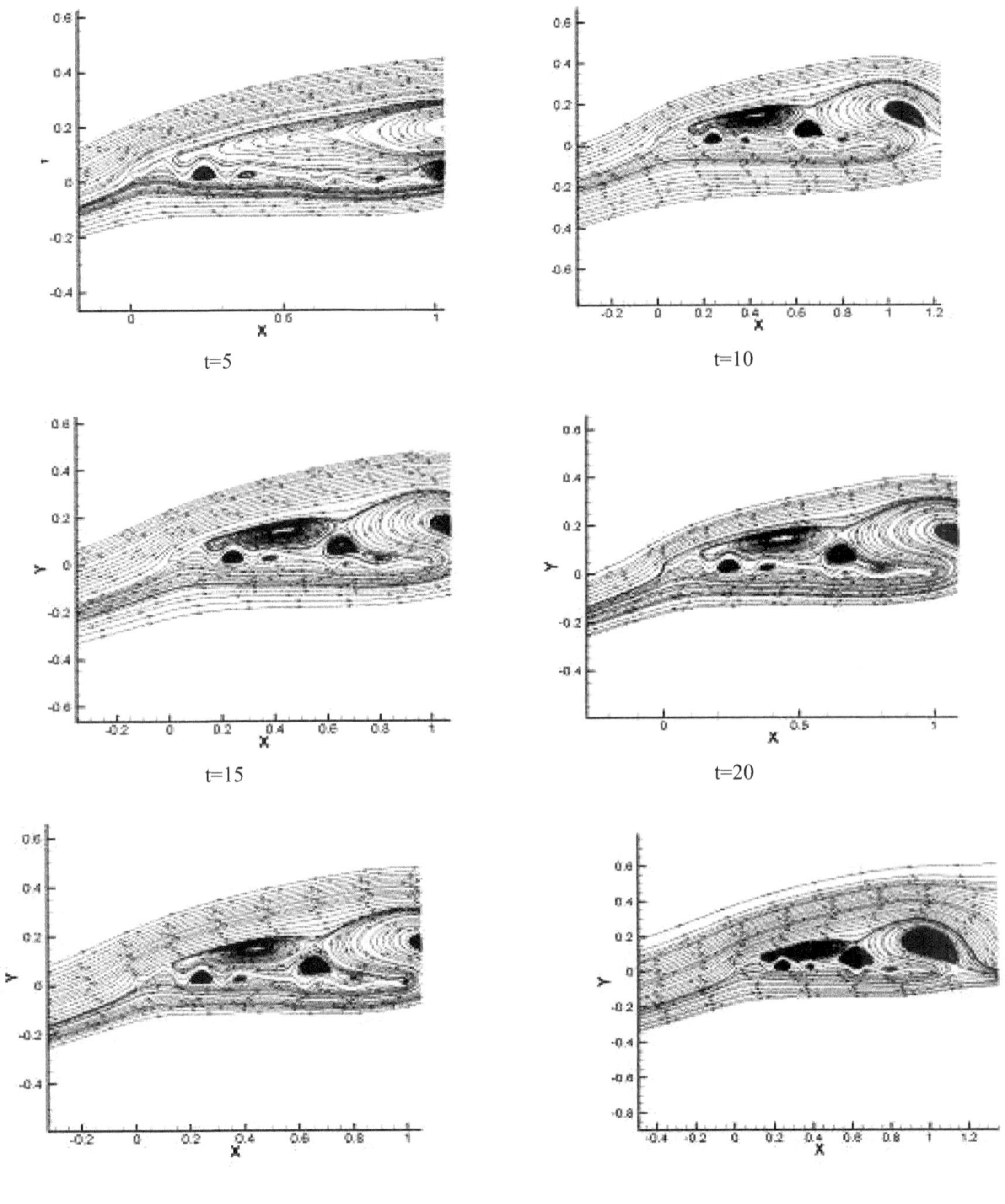

Figure 15:- Streams Lines at dimensional less time units illustrating of the pleated airfoil 1000 at (AOA 15^0)

The stream lines of Re-1000 at Angle of attack 15^0 as shown in figure-15. In this figure when we increase the dimensional less time unit at every t=5. Then attached vortex increased in leading edge, while there is no change in vortex in trailing edge, so according to figure 15 increased in attached vortex, increased in lift coefficient. So the maximum attached vortex is found in Re-1000 at angle of attack 15^0.

7 Structure Analysis:

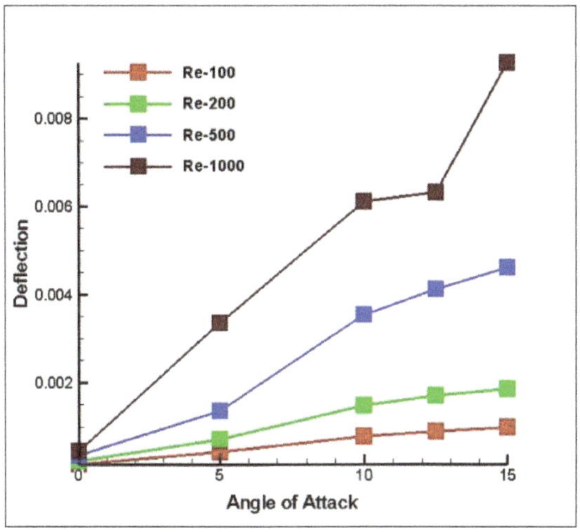

(a)

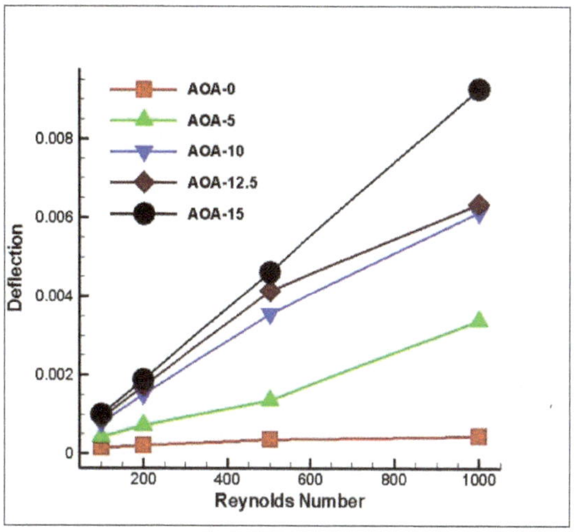

(b)

Figure 16:- (a) Variation of deflection and AOA, (b) Variation of deflection and Re

The result in the form of lift and drag forces are then fed into the ANSYS Workbench solver and one way Fluid Structure Interaction analysis is performed. Wing deformation leads to substantial power economy for lift production Young et al., (2009). Figure 16 (a) shows the variation of deflection and angle of attack. It is seen to be that deflection is increased with increasing the Reynolds Number with their angle of attack. The maximum deflection occur in Re-1000, angle of attack 15^0 i.e. 9.2597×10^{-3} mm and minimum value of deflection occurs in Re-100 with angle of attack 0^0 i.e.1.4618×10^{-4} mm.

8 Conclusions

In this work, we investigated fluid structure Interaction for spatio-temporal dynamics of a cut section of Aeshna Cyanea's wing. Numerical simulations were performed at ultra low Reynolds numbers (100 to 1000) at different angle of attacks ranging from 0^0 to 15^0. The results give a satisfactory measure of confidence in the fidelity of the simulation. It was found that for all the simulations performed flow always remained steady at Re 100 and 200. First unsteady flow was obtained at Re 500 and AOA 10°. But flow always remained steady at AOA 0° and 5° for all the Reynolds numbers. The deformation of the dragonfly wing airfoil was measured to the effect of wing deformation on the aerodynamic performance of the dragonfly wing, the deflection is increased when we increased the Reynolds Number with increase their angle of attack and the maximum deflection occur at Re-1000, angle of attack 15^0 i.e. $9.2597x1\,0^{-3}$ mm.

References:

Anwer SF, Ashraf I, Mehdi H, (2013) "On the Aerodynamic Performance of Dragonfly Wing Section in Gliding Mode" Advances in Aerospace Science and Applications. Vol 3, pp 227-234.

Bohorquez F, Samuel P, Sirohi J, Pines D, Rudd L, Perel R (2003). Design analysis and hover performance of a rotary wing micro air vehicle. *Journal of the American Helicopter Society* 48(2):80 – 90.

Combes SA, Daniel TL (2003a). Flexural stiffness in insect wings II. Spatial distribution and dynamic wing bending. *The Journal of Experimental Biology* 206:2989-2997.

Combes SA, Daniel TL (2003b). Into thin air: contributions of aerodynamic and inertial-elastic forces to wing bending in the hawkmoth Manducasexta. *The Journal of Experimental Biology* 206:2999–3006.

Dickinson MH, Lehman FO, Sane SP (1999). Wing rotation and the aerodynamic basis of insect flight. *Science* 284:1954-1960.

Du G, Sun M (2010). Effects of wing deformation on aerodynamic forces in hovering hoverflies. *The Journal of Experimental Biology* 213:2273-2283.

Ellington CP, Van Den Berg C, Willmott AP, Thomas ALR (1996). Leading-edge vortices in insect flight. *Nature* 384:626-630.

Heathcote S, Martin D, Gursul I (2004). Flexible flapping airfoil propulsion at zero freestream velocity, *AIAA Journal* 42(11):2196–2204.

Heathcote S, Wang Z, Gursul I (2008). Effect of spanwise flexibility on flapping wing propulsion. *Journal of Fluids and Structures* 24(2):183–199.

Ho S, Nassef H, Pornsinsirirak N, Tai Y, Ho C (2003). Unsteady aerodynamics and flow control for flapping wing flyers. *Progress in Aerospace Sciences* 39(8):635-681.

Hu H, Kumar A G, Abate G, Albertani R (2010). An experimental investigation on the aerodynamic performances of flexible membrane wing inn flapping flight. *Aerospace Science and Technology* 575– 586.

Kesel, A B. (2000) "Aerodynamic characteristics of dragonfly wing sections compared with technical aerofoil." J of Exp. Biology 203 3125-3135.

Kwok, M.and Mittal, R., "Experiment investigation of the aerodynamics of a modeled Dragon fly Wing Section".AIAA Journal 2005, pp 1-7.

Lian YS, Shyy W, Viieru D, Zhang B (2003). Membrane wing aerodynamics for micro air vehicles. *Progress in Aerospace Sciences* 39:425–465.

Liu H, Ellington CP, Kawachi K, Van Den Berg C, Willmott APA (1998). Computational fluid dynamic study of hawkmoth hovering. *The Journal of Experimental Biology* 201:461-477.

Mehdi H, Anwer SF, Ahmad A, (2014) "Vibration Analysis of Dragonfly wing section in Gliding Mode at Low Reynolds Numbers" International Journal of Research in Aeronautical and Mechanical Engineering. Vol 2, issue 12, pp 11-23.

Mehdi H, Lakhera B K, ,Kumar A, (2014) "Numerical Analysis of Steady and Unsteady Flow for Dragonfly Wing Section in Gliding Mode" International Journal of Advanced Mechanical Engineering. Vol 4, pp 365-370

Mittal, S., and Tezduyar, T. E., (1992)"A Finite Element Study of Incompressible Flows Past Oscillating Cylinders and Airfoils," International Journal for Numerical Methods in Fluids, Vol. 15, pp. 1073–1118.

Mueller TJ (2001). Fixed and flapping wing aerodynamics for micro air vehicle applications. *Progress in Astronautics and Aeronautics* 195.

Okamoto, M, K Yasuda, and A Azuma (1996). "Aerodynamic characteristics of the wings and body of a dragonfly." J. of Experimental Biology, : 281-294.

Ramamurti R, Sandberg WC (2002). A three-dimensional computational study of the aerodynamic mechanisms of insect flight. *The Journal of Experimental Biology* 205:1507– 1518.

Ramamurti R, Sandberg WC (2007). A computational investigation of the three-dimensional unsteady aerodynamics of Drosophila hovering and maneuvering. *The Journal of Experimental Biology*, 881–896.

Ramasamy M, Leishman JG, Lee TE (2006). Flow field of a rotating wing MAV. *Proceedings of 62nd Annual National Forum of the American Helicopter Society*. May 9-11 2006.

Sane SP, Dickinson MH (2001). The control of flight force a flapping wing: lift and drag production. *The Journal of Experimental Biology* 204:2607-2626.

Shyy W, Berg M, Ljungqvist D (1999). Flapping and flexible wings for biological and micro air vehicles. *Progress in Aerospace Sciences* 35(5):455–505.

Stanford BK, Ifju P, Albertani R, Shyy W (2004). Fixed membrane wings for micro air vehicles:experimental characterization, numerical modeling, and tailoring. *Progress in Aerospace Sciences* 44(4):258-94.

Sun M, Tang J (2002a). Lift and power requirements of hovering flight in Drosophila virilis. *The Journal of Experimental Biology* 205:2413–27.

Sun M, Tang J (2002b). Unsteady aerodynamic force generation by a model fruit fly wing in flapping motion. *The Journal of Experimental Biology* 205:55-70.

Usherwood JR, Ellington CP (2002). The aerodynamics of revolving wings I. Model hawkmoth wings. *The Journal of Experimental Biology* 205:1547-1564.

Vargas, Abel, Rajat Mittal, (2008). "A computational study of the aerodynamic performance of a dragonfly wing section in gliding flight." Bioinsp. Biomim. 1-13.

Wakeling JM, Ellington CP (1997). Dragonfly flight: III Lift and power requirements. *The Journal of Experimental Biology* 200:583–600.

Walker J, Thomas ALR, Taylor GK (2010). Deformable wing kinematics in free-flying hoverflies. *Journal of Royal Society Interface* 7(42):131-142.

Wootton RJ, Herbert RC, Young PG, Evans KE (2003). Approaches to the structural modelling of insect wings. *Philosophy Transaction Royal Society* 358(1437):1577-1587.

Young SM, Walker RJ, Bomphrey GK, Taylor, Thomas ALR (2009). Details of insect wing design and deformation enhance aerodynamic function and flight efficiency *Science* 325(5947):1549-1552.

Yusoff H, Abdullah MZ, Mujeebu MA, Ahmad KA (2011). Development of flexible wings and flapping mechanism with integrated electronic control system for micro air vehicle research. *Experimental Techniques* 33(6):53-58.

13

Drogue Deflections in Low Speed Unmanned Aerial Refuelling

Dr Ian R. McAndrew[1], Elena Navarro[2] & Dr Kenneth Witcher[3]

[1]*Embry Riddle Aeronautical University, College of Aeronautics, McAnd4f1@erau.edu*
[2] *Embry Riddle Aeronautical University, College of Arts &Science, Navarrj1@erau.edu*
[3]*Embry Riddle Aeronautical University, College of Aeronautics, witchea8@erau.edu*

Earnley Place, Clappers Lane, Chichester, Hants PO20 7JL UK

Abstract

In this research paper the implications of Unmanned Aerial Vehicles (UAV) being refuelled in flight by another UAV is addressed for specific requirements relating to aerodynamics and flight stability. Low speed flight, less than 150KTS, introduces unique flight and stability problems that can be exacerbated by weather conditions. The methods suitable for refuelling have been researched and presented in previous papers by the authors and drogue systems are justified as being operationally feasible with significant aerodynamic characteristics. Nevertheless, these drogue systems do have flight stability problems needing addressed. Not least of these stability problems is docking when weather conditions are less than optimal. In this research the data for Wind tunnels and Computational Fluid Dynamics are used in an Experimental Design system based on a full factorial orthogonal array. The data generates statistical values offering designers and users of UAV that require refuelling a unique insight for allowing operational usage in weather conditions previously problematic. This increase operational possibility extents not only the range but regions where weather might influence deployment of UAVs

Keywords: UAVs, refuelling, Low speed aerodynamics Experimental Design.

1. Introduction

In the last 10 years the use of Unmanned Aerial Vehicles (UAV) has expanded exponentially, furthermore, they have become accepted as a technology that offers boundless opportunities at high technical levels. They have progressed from those that fly within visual sight to those that are operated from the opposite side of the world. Of course, this does not address the practical implications of flying in controlled air space and near misses have been recorded, McAndrew & Moran (2013).

One of the very unique aspects of UAVs, small, light and cheaper than conventional aircraft is also one of its principal weaknesses. The lack of space limits the range they can fly and as these UAVs have been limited by range due to their low speed, McAndrew & Witcher (2013). One way to increase the range is by refueling in operational use. Previous papers have addressed the implications and applications of refueling. Here the focus is towards the design and considerations of drogues and how best to minimise drogue deflection (docking) when refueling.

2. Drogue refueling

Drogue refueling is where the receiver docks with the deliverer of the fuel. The drogue is the interface device deployed. In Figure 1 below it shows the deliverer and receiver and the drogue. This drogue is deployed from the deliverer and extended rewards in the free air flow to a position sufficiently behind that does not cause aerodynamic considerations and the drogue remains relatively static for the receiver to dock. As with any flow in air there is a tendency for movement and this movement can inhibit if not prevent docking in severe cases.

Figure 1: Receiver docked to drogue

Aerodynamics and atmospheric pressures will influence the free movement of the drogue. These parameters are beyond the direct control of the process, McAndrew et al. (2014). To establish the direct influences of the design the thre principal parameters selected for consideration were: speed, lenth of the droge from the refueler (extended position) and mass of the drogue itself, Anderson (2005). To investigate these parameters an three factor, two positional full factorial array was proposed. This would not only investiaget the influence of each factor, it would also establish if any interactions were present. With a two level design of experiment it is critical to determine the input values of each. Too close together and no insight to full use, too far apart and the extrapolation is superflouous. In table 1 below the parameters selected are shown along with their values.

Table 1: Factor values and units

	Low	**High**	
A	100	130	Speed, in knts
B	10	15	Length, in m
C	12	16	Weight, in kg

In table 2, below, it shows the full array for this full factor where it balanced to produce an orthogonal array of values where there is an equal probability within each column as there is within each row. Thus a non-parametric set of results for analysis of data. Each value in the main array is assigned a negative (-) or positive (+) value corresponding to the data shown in table 1.

In the left hand column each experimental run from y_1 to y_8 and represents the eight experimental settings. The 5th, 6th 7th and 8th columns are the signage for any interactions between each input factor and another (two level interactions) and the 8th for any three level interactions. Each value is divided by four in order to average the values in the positive and also negative. In effect these column numbers show the difference between the average of the four positive numbers and the average of the four negative numbers. The larger they are the most probability of a statically significant effect. Likewise, the closer to zero it suggests they have negligible influence on the outcome. Generally (ref) three level interactions are very unlikely to have an influence and this is kept in the experiment as a benchmark comparison. Some experiments use this column for a forth factor and thus a half full factorial array is developed and increase error but widens scope of comparisons, Bertin & Cummings (2008).

In the final four columns are the recorded values for each experimental run (rows 2 to 9) of deflections in the x-axis (repeated 5 times) and its corresponding standard deviation, σ_x and y-axis deflection with its standard deviation, σ_y. The bottom row, labelled L is calculated by determining the difference between the average low level and the average of the higher level. Equation 1 displays such for Column A. The combination of – and + signs follows that in the column from top to bottom, Box & Norman (2007). Each Factor and Interaction column has its unique formulae. These are subsequently used to calculate the values for the last four column results that are analysed and presented. Having repeated runs allows for comparisons between experiments and more importantly within experiments. This is addressed and compared later with the N Score plots. For now, each sample is five and the standard deviation is σ_{N-1} to represent the discrete statistical sample used.

Table 2: Formatting guidance

	A	B	C	AxB	BxC	AxC	AxBxC	$\bar{x}$	σ	$\bar{y}$	σ
y_1	-	-	-	+	+	+	-	1200	80	800	60
y_2	+	-	-	-	+	-	+	1400	100	950	85
y_3	-	+	-	-	-	+	+	1300	90	900	80
y_4	+	+	-	+	-	-	-	1500	110	1000	90
y_5	-	-	+	+	-	-	+	800	60	650	40
y_6	+	-	+	-	-	+	-	900	70	700	50
y_7	-	+	+	-	+	-	-	850	65	675	40
y_8	+	+	+	+	+	+	+	950	75	750	50
L	L_A	L_B	L_C	L_{AB}	L_{BC}	L_{AC}	L_{ABC}				

$$L_A = \frac{-y_1+y_2-y_3+y_4-y_5+y_6-y_7+y_8}{4} \tag{1}$$

2.1 Results from the full factorial array

It is shown below in Table 3 it is divided into sections, x deflections, y deflections, σ_x and σ_y values for the data. The left hand column is listed chronologically from L_a to L_{abc}. The second column lists in this order the results obtained and these results correspond to the first column of each four sections. The next column sorts them from lowest to highest. This is on the assumption of the Central Limit Theorem, CLT, where we are reviewing to separate the random variations to spurious ones that explain the vast variation. An ANOVA was calculated to determine the expected outcomes. The i column iterated the sorted outcome and f_i the median probability score for individual probability, not division of probability in equal measure to accommodate the first value can happen at t=0 and not 14.28 (100 divided by 7) O'Conner (1992).

Table 3: Individual results for each column and output

	$\bar{x}$	Sorted	Expected	i	f_i	$\bar{y}$	Sorted	Expected	i	f_i	σ_x	Sorted	Expected	σ_y	Sorted	Expected
L_A	150	-475	-1.36449	1	0.086207	93.75	-218.75	-1.36449	1	0.086207	15	-27.5	-1.36449	13.75	-33.75	-1.36449
L_B	75	-50	-0.75829	2	0.224138	56.25	-31.25	-0.75829	2	0.224138	7.5	-5	-0.75829	6.25	-6.25	-0.75829
L_C	-475	-25	-0.35293	3	0.362069	-218.75	-18.75	-0.35293	3	0.362069	-27.5	-2.5	-0.35293	-33.75	-3.75	-0.35293
L_{AB}	0	0	0	4	0.5	-6.25	-6.25	0	4	0.5	0	0	0	-3.75	-3.75	0
L_{BC}	-25	0	0.352934	5	0.637931	-18.75	18.75	0.352934	5	0.637931	-2.5	0	0.352934	-6.25	3.75	0.352934
L_{AC}	-50	75	0.758293	6	0.775862	-31.25	56.25	0.758293	6	0.775862	-5	7.5	0.758293	-3.75	6.25	0.758293
L_{ABC}	0	150	1.364489	7	0.913793	18.75	93.75	1.364489	7	0.913793	0	15	1.364489	3.75	13.75	1.364489

This data is analysed in the following sections with N Score plots as it highlights the variations that are normally distributed or not. A Normal distribution will have an average of zero and standard deviation of 1. If this is plotted then those influenced more by the CLT will lay closer to a straight line passing through zero (average) and 0 (standard deviation). Alternatively, the further they are from a straight line they are more spurious results that cannot be explained by randomness either within the samples or between the samples. Thus it can be concluded that they are the significant values influencing the output variations that are shown.

3. Normal Probability plots of data outputs

Normal Probability plots demonstrate visually and numerically the separation of normally distributed and spurious outputs. Figure 2 is for the data set allied to the x axis movement of the drogue. There are seven plots that correspond to the seven columns in the full factorial array. Basing the origin of (0,0) a projected line can be seen that would encompass the majority of the points. This line clearly demonstrates C as an outlier to the others and it significantly accommodates for the majority of the variation.

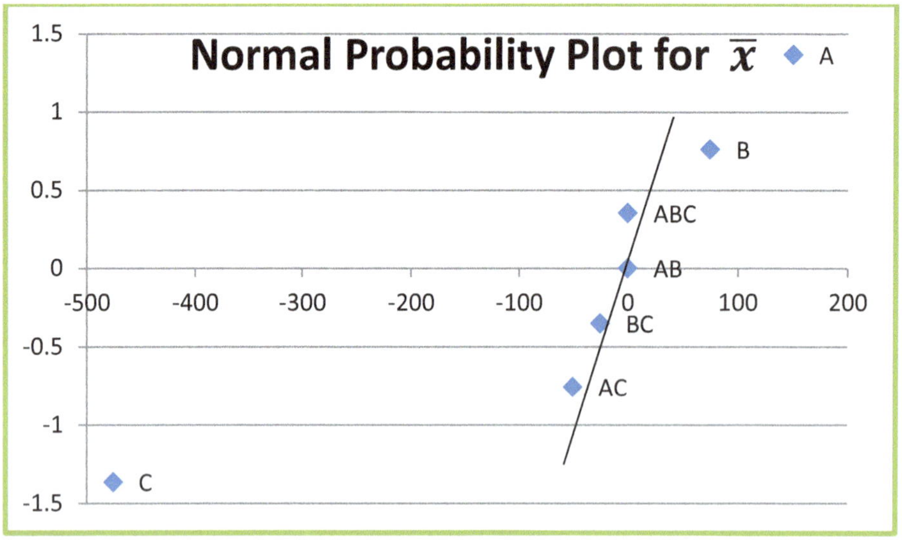

Figure 2: N Score plots average in X axis

It is worth nothing that the (0,0) line also has A relative far away, not as significantly, sufficiently far enough to be able to say that it too has an input to the total variation. Starting with this it can be hypothesised that the drogue movement weight is significant and the heavier the weight the more reduction in free movement. In addition, as the speed increases this too assists in reducing. Inversely, the length of the drogue from the aircraft is not influential to a significant level. None of the interaction plays any significant role, which is surprising given that A and C are both. However, given it is when they are closer to maximum individual values they have their greatest influence. Noting ABC is close to the centre supports the validation of not significantly active.

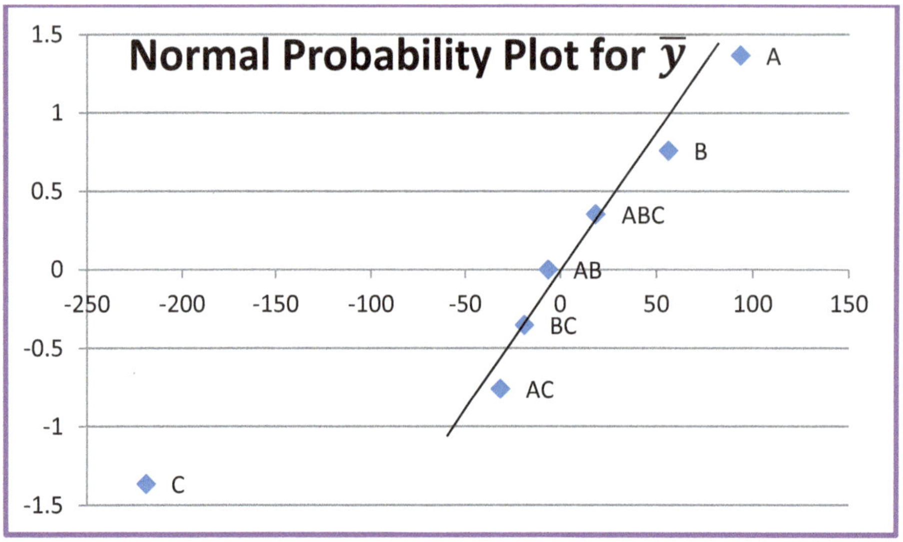

Figure 3: N Score plots average in Y axis

The variations for the Y axis deflections show a similar trend tho those for the X axis deflection. This may seem a natural response; it must be noted that Dutch Role is more influential in the X axis and there could be other factors that have influenced, McAndrew & Navarro (2014). In figure 3 factor C is still the largest influence, factor A also and this time factor B has a greater significance. The inter-relationship will be covered more in the discussion of figures 4 and 5 where the variations within samples are addressed in greater detaial. This data also shows that there are no interactions that are influencial.

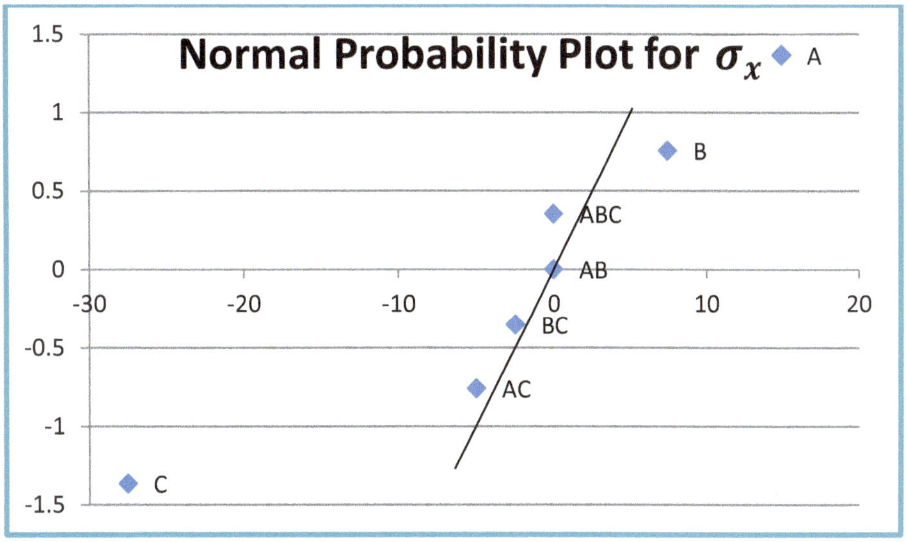

Figure 4: N Score plots standard deviations in X axis

Standard deviations of each experiment considering the X axis are based on a sample of five. Unlike the N Score plots above for the X and Y axis deflections that have various reasons and factors the there is only one factor, C, that is influential in explaining the variation of the drogues movement, Nair (1992). This is by far the greatest influence in variations within each sample and no other factor or interaction has any principal affect. All the other variation can be explained by the randomness within the sample, Leon et al (1987). Below, in figure 5 shows the corresponding values for the standard deviation of the Y variations. This too follows a pattern in logic and results for the data expressed within the individual samples, and the individual implications and results are discussed further on the following page. These results need to be reviewed in isolation and also in comparison of their total implications from the total array of all experiments.

other factors that have influenced, McAndrew & Navarro (2014). In figure 3 factor C is still the largest influence, factor A also and this time factor B has a greater significance. The inter-relationship will be covered more in the discussion of figures 4 and 5 where the variations within samples are addressed in greater detaial. This data also shows that there are no interactions that are influencial.

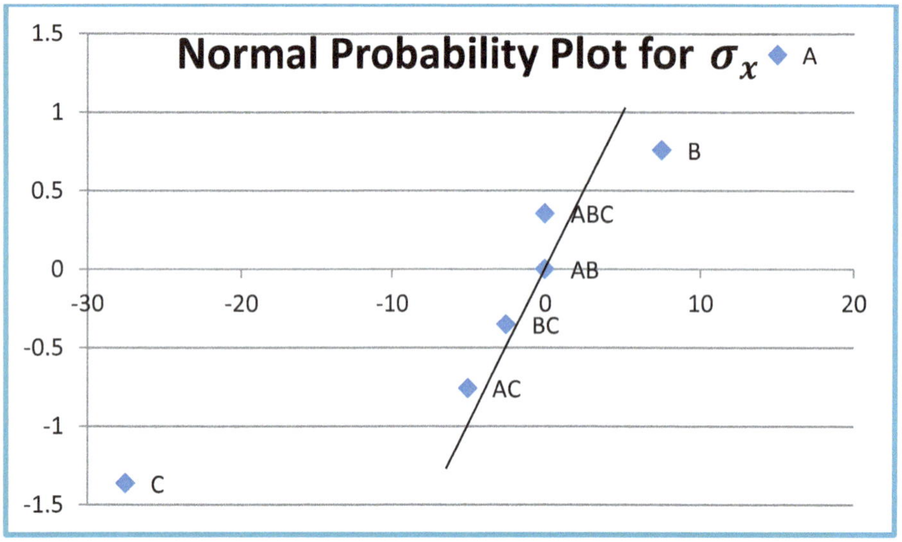

Figure 4: N Score plots standard deviations in X axis

Standard deviations of each experiment considering the X axis are based on a sample of five. Unlike the N Score plots above for the X and Y axis deflections that have various reasons and factors the there is only one factor, C, that is influential in explaining the variation of the drogues movement, Nair (1992). This is by far the greatest influence in variations within each sample and no other factor or interaction has any principal affect. All the other variation can be explained by the randomness within the sample, Leon et al (1987). Below, in figure 5 shows the corresponding values for the standard deviation of the Y variations. This too follows a pattern in logic and results for the data expressed within the individual samples, and the individual implications and results are discussed further on the following page. These results need to be reviewed in isolation and also in comparison of their total implications from the total array of all experiments.

4.2 Combined Affect

Experimental arrays combine all inputs to determine their significance of the outputs. Minimum deflections for the X and Y axis do not conflict and mirror the requirements needed. The greatest influences are the firstly the factor C, the weight of the drogue. Next follows the speed and finally the distance trailing from the deliverer. Optimum setting is factor A negative, factor B negative and factor C positive. That means the setting should be as close to the settings used in the experiment, see table 1. Thus, a lower speed, shorter length, and heaviest weight is the focus for flight operations with minimal external influences on the deflections. With this array it is not significantly robust to extrapolate the results as they are based on only two levels that have no data from the points plotted. An array with multi-level is needed to fully evaluate.

5. Conclusion

The results from this study have shown that there is one significant input factor that influences the movement in the X and Y axis, the weight of the drogue. The greater the weight of the drogue the lower is the deflections in the movement for docking. Inversely, the distance the drogue is left trailing behind has negligible affect and only accounts for those by randomness in values. Factor A, in the Y axis had an influence in reducing the deflection; although, the aerodynamic reasons why are not clear from this study and will be reviewed in further studies.

A unique and not predicated at the onset was that there was no interaction at two or three levels. This does allow for the next set of experiments to focus on multi-level and Factor input to determine if the responses between settings are linear or curved. This will enable predicting setting levels. Results have suggested optimum settings for the parameters used and application are based on two level linear responses.

References

Anderson, J.D., (2005) Introduction to Flight. McGraw-Hill Higher Education, USA.

Bertin, J.J., Cummings, R.M., (2008) Aerodynamics for Engineers. 5th Edn. Prentice Hall, Inc. USA.

Box, G. E. P. and Draper, Norman. (2007). *Response Surfaces, Mixtures, and Ridge Analyses*, Second Edition Wiley.

León, R V; Shoemaker, A C & Kacker, R N (1987) Performance measures independent of adjustment: an explanation and extension of Taguchi's signal-to-noise ratios (with discussion), Technometrics vol 29, pp. 253–285.

McAndrew, I. R., & Witcher, K. (2013). Design consideration and requirements for in-flight refuelling of unmanned vehicles. Journal of Aeronautics & Aerospace Engineering, 2(2), 1-4. doi 10.4172/2168-9792.1000108

McAndrew, I. R., Godsey, O., & Navarro, E. (2014). Proceedings from the 5th International Conference on Mechanical and Aerospace Engineering: Low Speed Aerodynamics of Drogue System Refuelling of Unmanned Aerial Vehicles. Madrid, Spain: Applied Mechanics and Materials Journal.

McAndrew, I, R. & Navarro, E. (2014). Low speed refuelling of unmanned aerial vehicles using the Drogue system. European Journal of Engineering and Technology, 2 (1), 1-8.

McAndrew, I. R. & Moran, K. (2013), Aerodynamic Considerations for in-flight Refuelling of Unmanned, International Conference of Aeronautical and Mechanical Engineering, July 20-21st. Moscow, Russia

Nair, V N (ed.) (1992) Taguchi's parameter design: a panel discussion, Technometrics vol34, pp. 127–161

O'Conner, P (1992) Practical reliability, Prentice Hall, Chichester, UK.

A Brief Author Biography

1st Dr Ian R. McAndrew – has been in academia for the past 25 years, building on his industrial experience and used this to develop his extensive knowledge of aerodynamics in subsonic and sonic flight. Dr McAndrew has published widely and been involved with many PhDs in both Europe and USA. He is the author of two books and in excess of 20 international conferences and journals. His current research is with a focus of low speed flight requirements in refuelling of Unmanned Aerial Vehicles in flight. He is currently the Department Chair of Graduate Studies for Worldwide at Embry Riddle Aeronautical University (ERAU).

2nd Elena Navarro – has been working with ERAU since 2008, first as an adjunct instructor. Between 2010 and 2013 she served as a Director of Academics at several USAF bases in Germany. Currently, she is a Full Time Faculty within the College of Arts and Sciences. Elena Navarro has both her undergraduate and graduate degrees in Computer Science from Riga Technological University of Civil Aviation. She has taught various undergraduate Computer Science courses at the New Mexico State University, where she also worked as a tutor for all Math courses. Elena has hands on experience with substantial programming and application development at Latvian Environment, Geology, and Meteorology Agency. She has co-authored several publications for conferences and journals. Her current research focuses on the construction and analysis of data from factorial experiments. Elena is also assisting on the numerical side of several aerodynamic research projects and works supporting research in the University where computing and numerical analysis is needed.

3rd Dr Ken Witcher – aviation experience includes 20 years of service in the United States Air Force. During this time, he served as superintendent of an operational test and evaluation squadron and field training detachment chief for F-15, F-16, F-22, H-60, A-10, MQ-1, and MQ-9 aircraft. He also served as a team member of the United States Air Force Air Demonstration Squadron, *Thunderbirds*. Subsequently, he joined Embry Riddle Aeronautical University (ERAU) as Faculty, Director of Academics, Program Chair, and now Dean of the College of Aeronautics, which is responsible for the largest Graduate and Undergraduate aviation degrees from the world's only dedicated aviation university. Dr Witcher's areas of research interest are in effective use of airport capacity and integrating Unmanned Aerial Vehicles (UAV) into commercial airspace.

14

UAV- A MANNED VEHICLE IN ADVERSE ENVIRONMENTS

Sriram.R[1], Nandhini.S[2]

[1]PG- SVCE, Sriperumbudur, sriram.aviator12@gmail.com
[2] PG- SVCE, Sriperumbudur, nandhukohi@gmail.com
8/42 Salem mail road, Elavanasur Kottai, Ulundurpet TK, Villupuram

Abstract

Now a day plenty of amounts are being debilitated in defence sector. Gross Domestic Product (GDP) is an indicator which reveals the prosperity of a nation. Nation with rising riches needs adequate physique in protecting the country's hard earned assets along with its dwellers. A Precinct which plays dominant role for the triumph of these above stated devoir is the defence sector. Aviation industry earned a blossoming reputation right from the opus of Unmanned Aerial Vehicle (UAV) which penetrates the areas where personage cannot. The Aerosonde UAV is a small unmanned aerial vehicle (UAV) designed to collect weather data including Temperature, atmospheric pressure, humidity and wind measurement over oceans and remote areas. This research work address the disasters occurred when the UAV had maneuvered in adverse conditions. An alternate approach to evade this failure is by employing control horns and positioning the on board for acquiring values. Design of the proposed system is sculpted by means of CATIA V5 software which proves to be best suited for reverse engineering. Fabrication of proposed model is accomplished and tested in adverse environments. Proposed model proved to be a new resolution in the field of RC modeling during test flight.

***Keywords**:* Aerosonde UAV, E-logger, Control horns

1. Introduction

UAV's commonly find its application in defence and patrol in situations where the risk of sending a human piloted aircraft found to be threatening, or the situation of employing piloted aircrafts proves to be impractical. One of the predecessors of today's completely automated Unmanned Aerial Vehicles was the "aerial torpedoes", designed and built during First World War.

Advanced UAV's used radio waves for guidance, allowing them to fly missions and return. They were constantly controlled by a human pilot, without deserving a place in cockpit. This made them much like today's RC model airplanes which many people fly as a hobby. It is interesting to note that the government consider all aircraft UAV's, if they are unmanned and used by a government or business. After the invention of the integrated circuit, engineers were able to build sophisticated UAV's, using electronic autopilots which replicates the work of pilot.

It was at this stage of development that UAV's became widely used in defence applications. UAV's could be deployed, fly themselves to a target location, to target the enemy with weapons or survey it with cameras and other sensor equipment. Modern UAV's are controlled with both autopilots, and human controllers in ground stations. This allows them to have long flights, uneventfully flights under their own control, and fly under the command of a human pilot during complicated phases of the mission.

2. Methodology

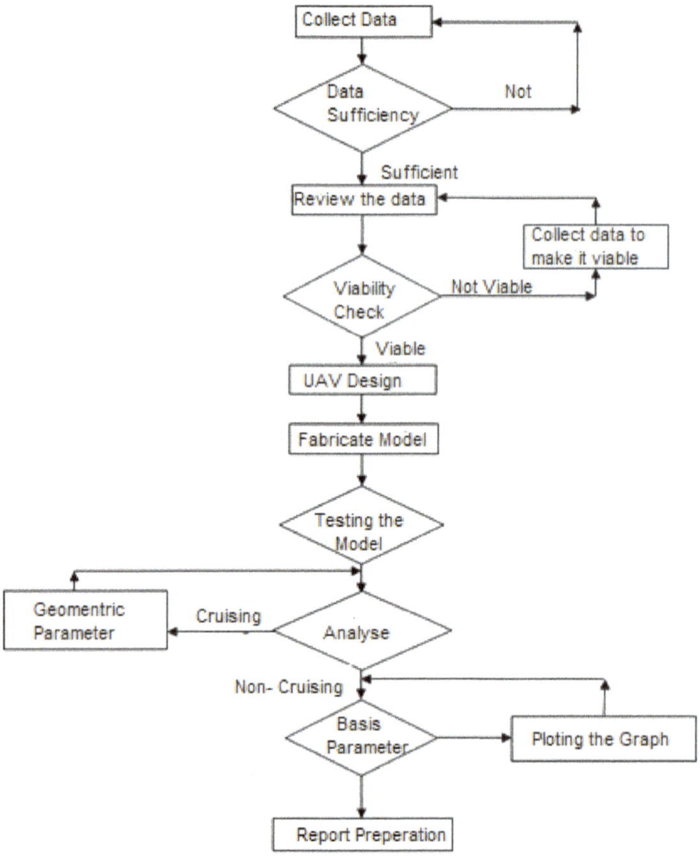

Figure 1: Methodology of our paper

3. Fabrication of Modified UAV Requirements

3.1 Materials used For Fabrication of RC model

- Expanded polystyrene
- Cyno glue

3.2 Equipment details of RC model of Aerosonde UAV

There are 3 main types of foams used for RC airplanes they are:

- EPP (Expanded poly propylene).
- EPO (Expanded poly olefin).
- EPS (Expanded poly styrene).

EPP foams are preferred because of its toughness and flexibility. It is light and used for Combat wings, durable by demanding carbon rods for reinforcement. Density of the material varies from surface to core. It has a waxy feel which requires spraying of 3M adhesive.

Table 1: Equipment required and its specification

Battery (B3 Pro Compact Charger)	
Balance charge current	850mA
Bush Less Motor (2840kw) -20AMPS	
Cable length	28cm
Case	Plastic
Compliant with Lithium battery	**1**
Continuous current	3A
Control Horns	Nylon control horns
Data Recorder	Eagle Tree V4 eLogger
DC	5.6V-20V
Dimensions	90mm*55mm*35mm
Dimensions (L*W*H)	13*11*4mm
Display	Green, Red.
Electronic Speed Controller-3AMPS	
Global Positioning System (GPS)	
IBM	2205-14
Input voltage	110V-240V.
Max charge current	3*700mA
Pepersistent current	5A

Propeller- Double blade (6*4E)	
Receiver- 2.4HZ	Channel 2.4 GHz, Battery needed of 1.5V, Model no: FS-R6A
Servos - 9gms.	4 Channels
Transmitter	FS T4A, Channel 4, 100% Digital proportional radio control system. Model no: H700183
Weight	180g
Y Connector	3 Phase Y Connection

3.3 Software Used For Data Collection

Figure 2: Eagle Tree V3 eLogger.

It is an extremely portable unit that can be used in-flight or on the workbench for both gathering and analyzing data via both Live mode in conjunction with a laptop or the Eagle Tree Power Panel. Data storage is accomplished by employing eLogger's internal memory and in turn could be downloaded to your computer via USB.

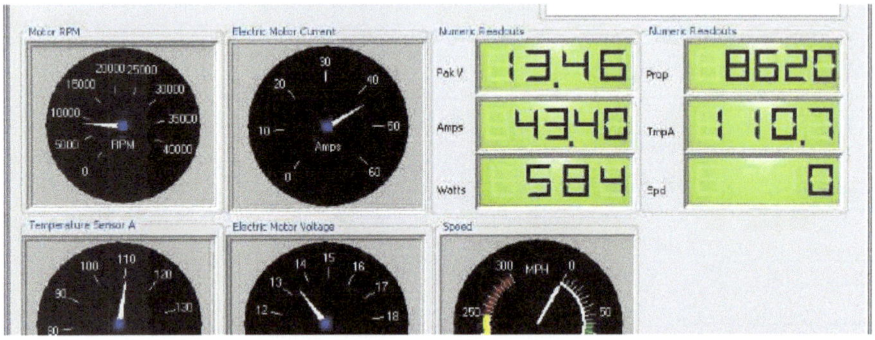

Figure 3: Parameter measured using Eagle Tree eLogger V3.

Once the data is collected, it is extremely easy to extract. Plug the eLogger into the computer, bring up the Data Recorder software and click download. Once the data is downloaded the maximum readings will all be displayed. Next, it is either possible to watch the flight via the gauges displayed or go directly to the graphing portion of the software. In order to watch the data you have just recorded via the gauges, hit Play. To begin graphing, a quick click on Graph Data is all that is needed.

Customizing the graphs is an easiest task. Parameters to graph, color, line thickness, 2D, 3D, axis placement, etc. could be easily accessed. The length to which you can customize the graphs is limitless. It all happens in the Edit menu, which could be accessed through the Graph page.

3.4 Fabrication Procedure

Equipments needed includes such as 4 Servos, ESC, GPS, Temperature sensor, Data recorder, Battery, Receiver and other materials such as Y Connector, Control horns, Transmitter, Carbon rods, Foams and Adhesives.

- Place two servo in each wing by using adhesives (Cyno glue) and place the push pull rods at the wing portion in the respective space provided for yielding the strength characteristics.
- Place other two servo inside the fuselage its gets connected with elevator and rudder for deflections by using control rods.
- Each Servo is attached with control horns using screws and control rods.
- By using Y Connector and push pull rods connect the port and starboard wing to the fuselage.
- Use the Bushless motor for mounting the propeller (2 bladed) it is connected with ESC which is mounted on center of the fuselage portion.
- At nose portion of UAV, GPS and temperature sensor is to be placed and inside of rear fuselage portion battery, Eagle tree V4 and 6 channel receiver is mounted by using adhesives and tapes.
- Receiver has 6 channels in that 4 servos and ESC gets connected with it.
- For starting the operation Battery is used, it is connected to ESC and Eagle tree V4 data recorder system by using cables.
- Eagle tree v4 is connected with GPS, Temperature sensor, Battery and ESC.

4. Modelling and Fabrications of Modified Aerosonde UAV

4.1 Theoretical Background

Table 2: Formula used for designing Modified UAV model

$AR = b^2/s$	AR - Aspect ratio b - Wing span (cm) s - Wing planform area (cm^2)
(w/s) = weight/wing area	(w/s) - Wing loading (kg/cm^2)
$s = b^2/AR$	AR - Aspect ratio b - Wing span (cm)
b = sqrt (AR*s)	AR - Aspect ratio s - Wing planform area (cm^2)
$w_{crew} + w_{payload} + w_{fuel} + w_{empty}$	W_o - Overall weight (gm)

4.2 Design Calculation

Table 3: Design dimensions of UAV model

General Characteristics	Weight Configuration	Performance Calculation

Length - 77cm Wing span - 135cm Wing area - $1764cm^2$	Overall weight - 600gm	Wing loading - 0.340 kg/cm^2 Range - up to 1km Aspect ratio - AR = 10.3 Wing loading - 0.340 kg/cm^2 Wing area - $1769cm^2$ Wing span - 134.98cm

4.3 Aerosonde UAV model in CATIA V5

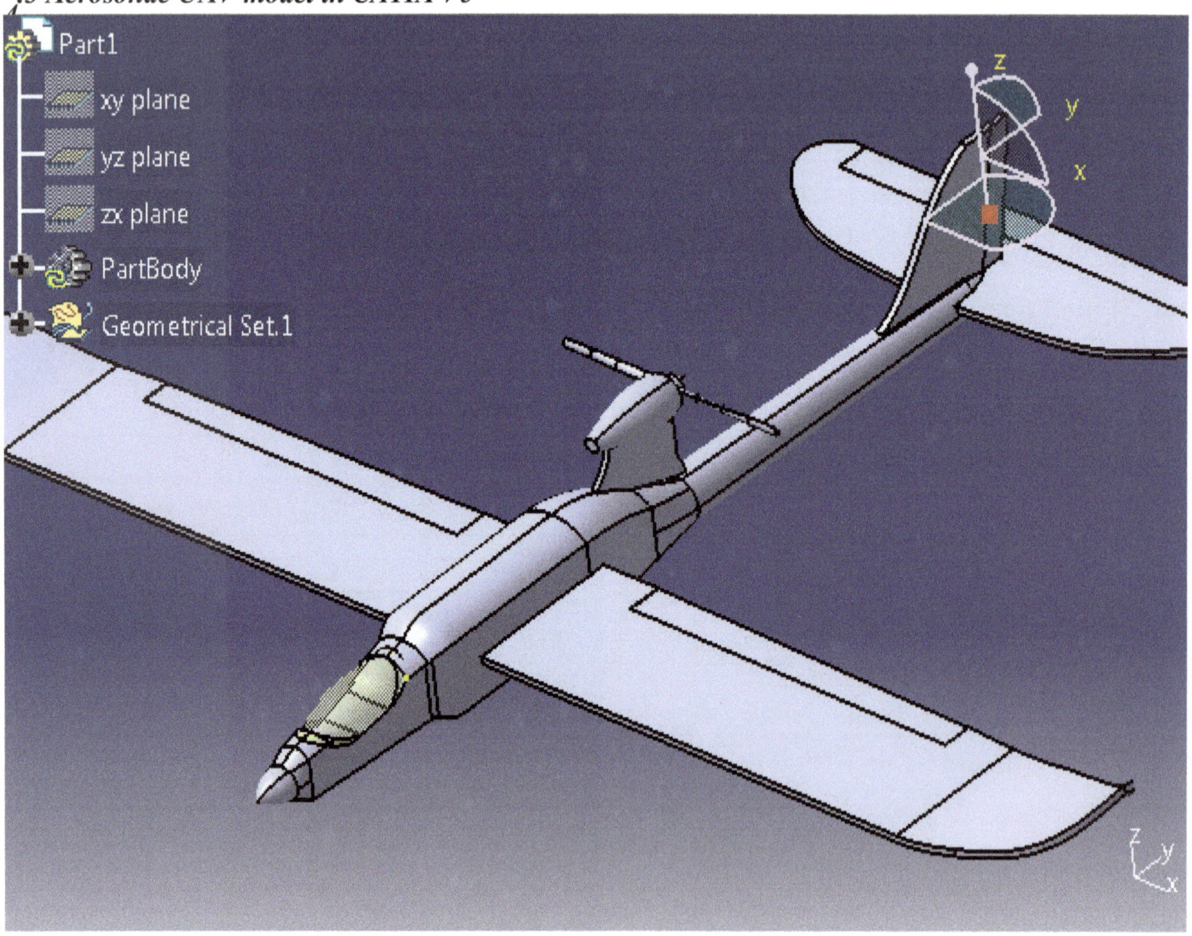

Figure 4: Design of Aerosonde UAV

4.4 Three View Diagram of Aerosonde UAV model

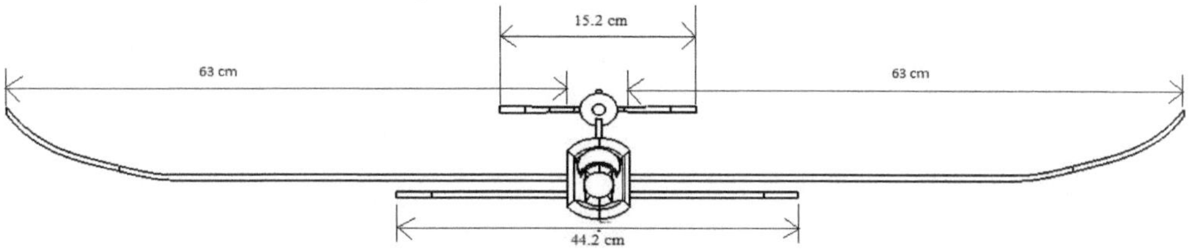

Figure 5: Front View of Modified UAV model

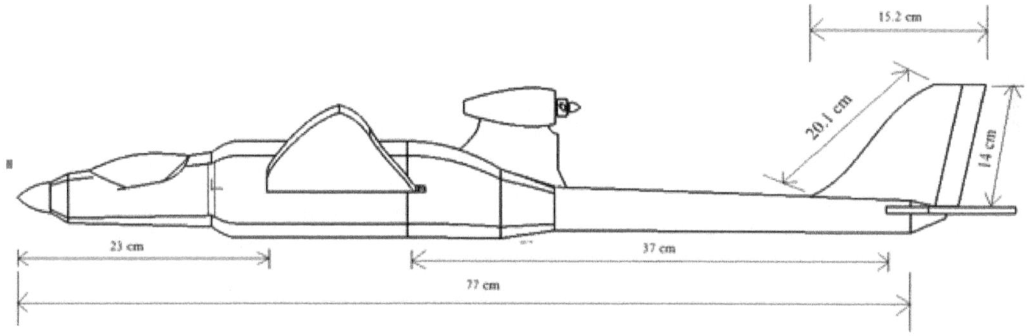

Figure 6: Side View of Modified UAV model

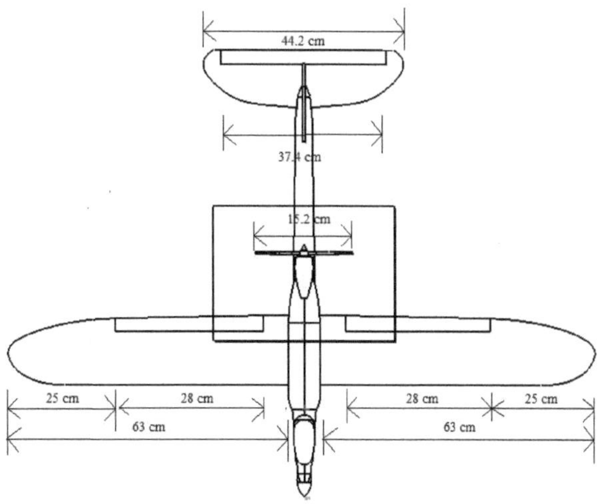

Figure 7: Top View of Modified UAV model

4.5 Fabricating Diagram of Aerosonde UAV

Figure 8: Modified Aerosonde UAV model

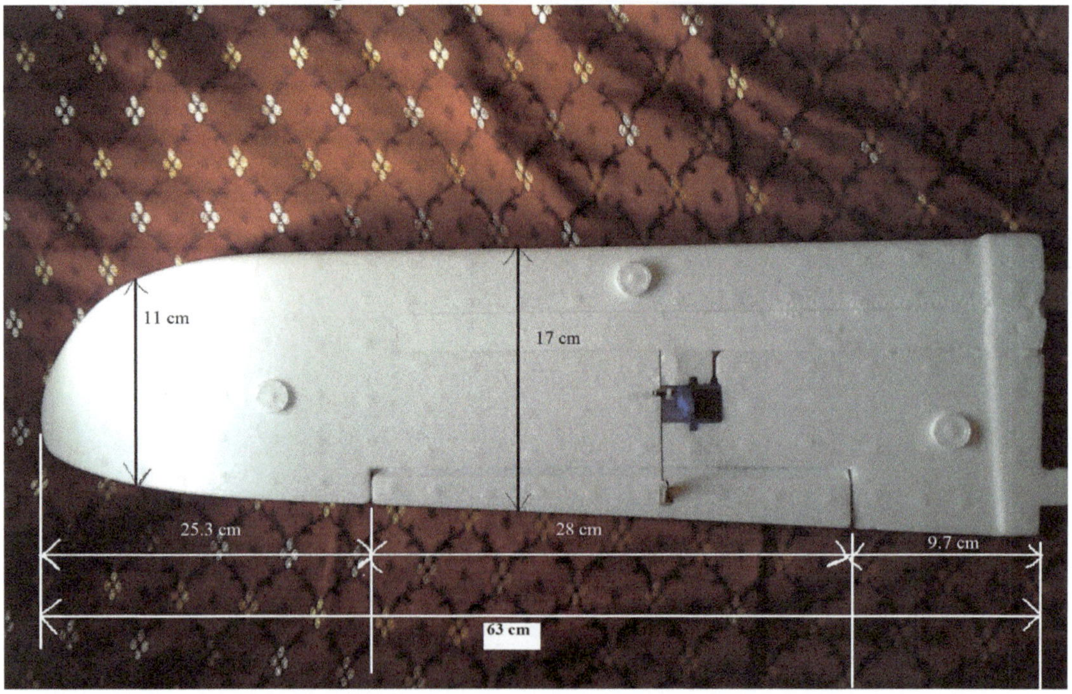

Figure 9: Modified Aerosonde UAV Wing Design

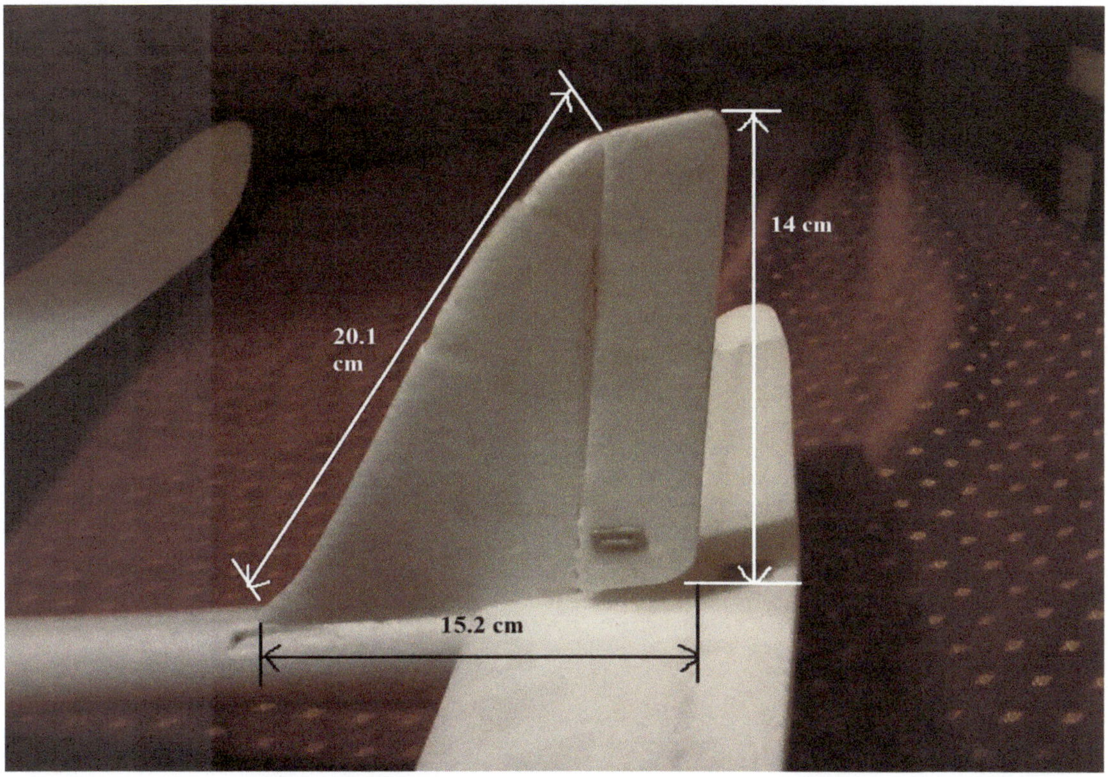

Figure 10: Modified Aerosonde UAV Empennage Design

5. Result and Discussion

TEST LOCATION 1

Test location	: Annanji, Theni 625531, India.
Latitude	: N 10^0 2' 33.2038"
Longitude	: E $77^0$30 12.2223
Time	: 9.30 – 11.00 AM

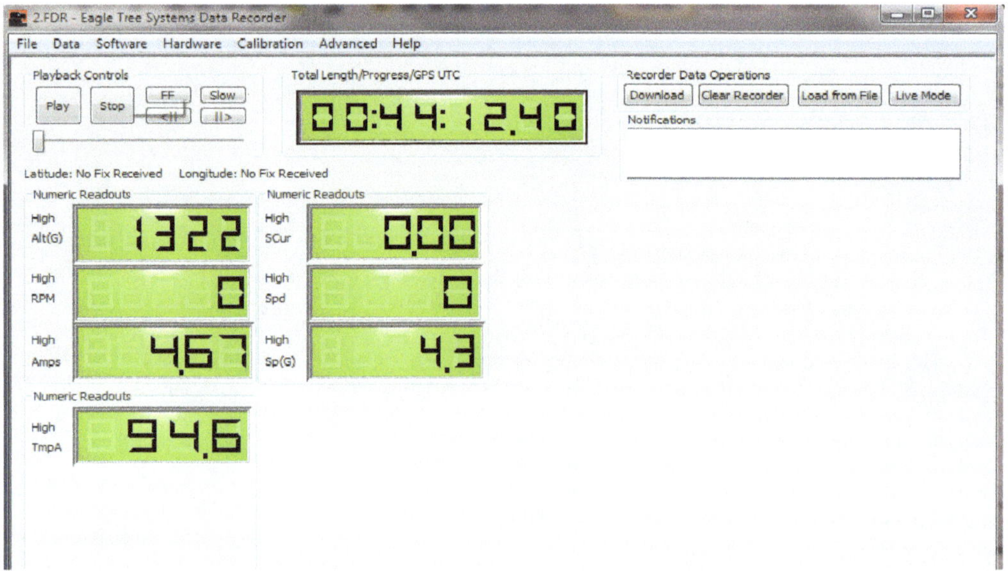

Figure 11: Recorded Parameters for 1st test location.

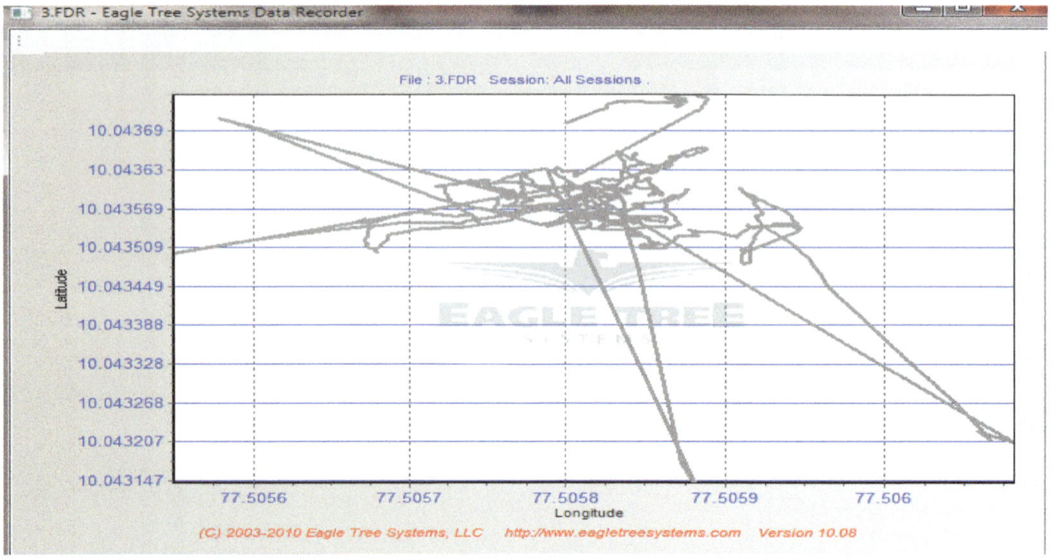

Figure 12: 2D GPS Data Chart.

This graph describes the relationship between the longitude and latitude. The X axis contains the longitude in east direction and the Y axis contains the latitude in north direction. The point at which the overlapping occurs shows the location of testing, the individual line shows the point of intersections.

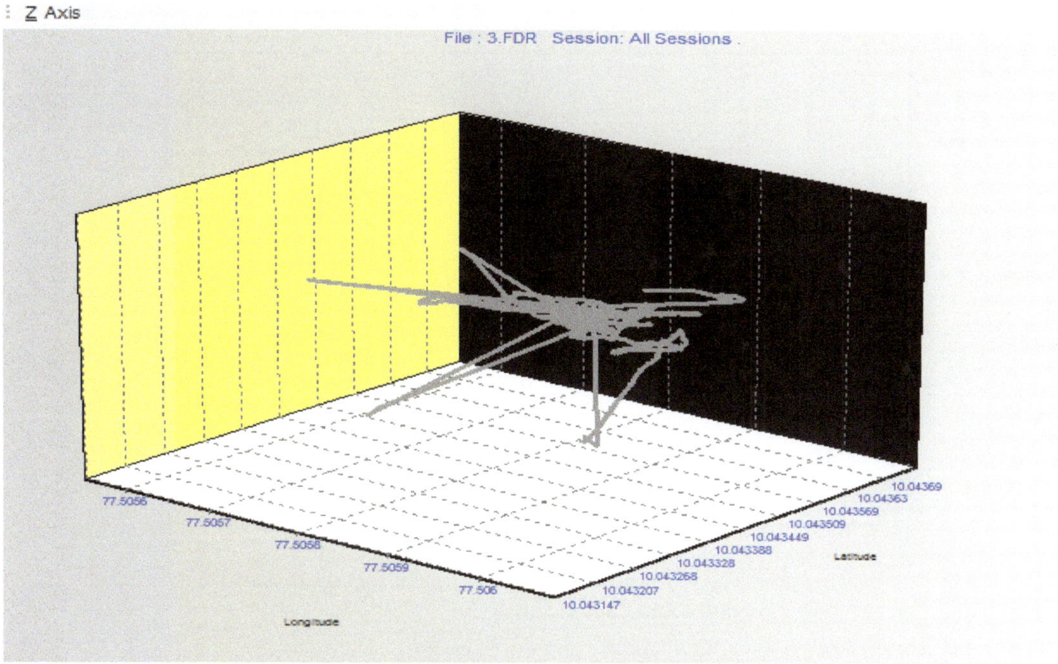

Figure 13: 3D GPS Data Chart.

The above figure depicts the 3d view of the location of the testing place plotted using its longitude and latitude. The black colour surface indicates the testing place whereas the yellow and white colour indicates the extensions

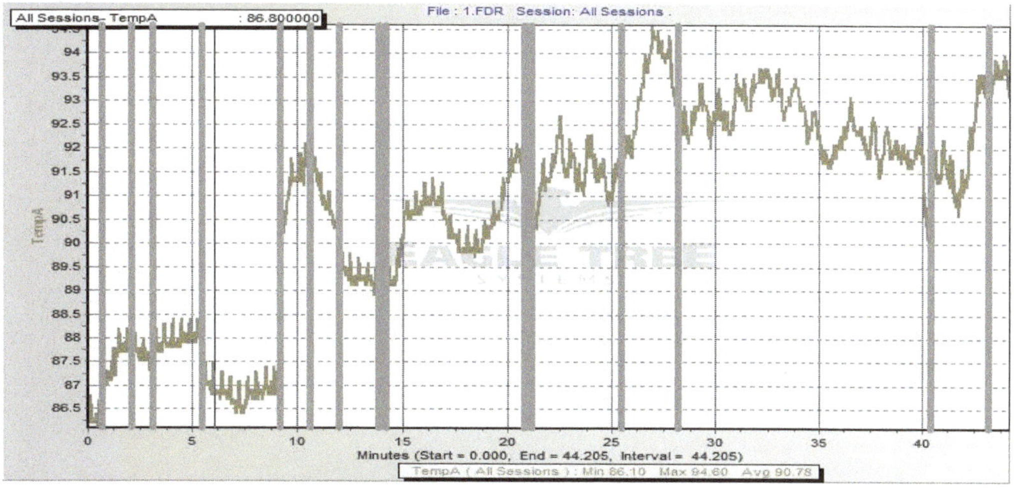

Figure 14: Time (Min) Vs Temperature.

The variation of temperature at various times is plotted in this graph. Since temperature depend on the time. The temperature increases with increase in time.

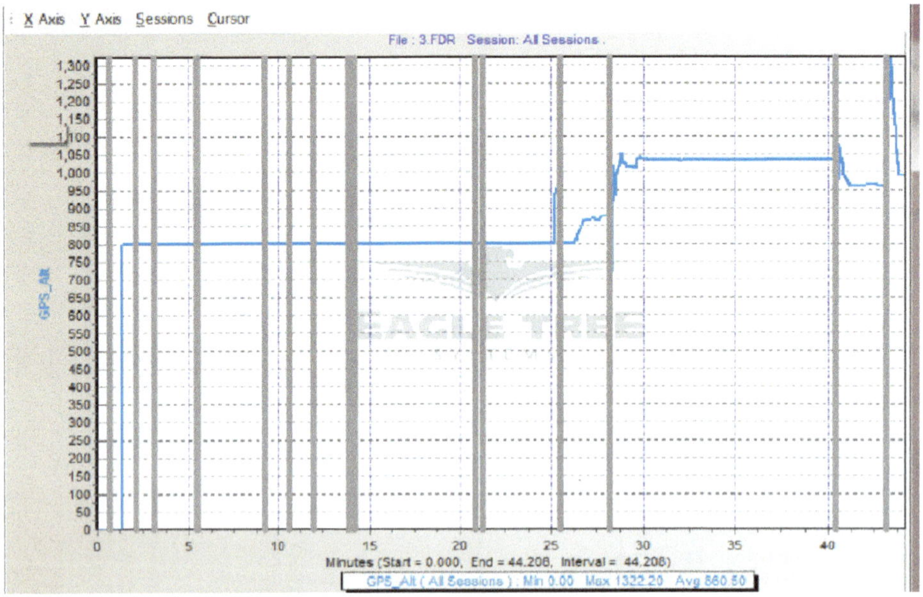

Figure 15: Time (Min) Vs GPS Altitude.

The Global positioning system is used to measure the latitude,longitude of the UAV .this graph shows the gps altitude of 800m for the first 25 minutes.then it gains an altitude of 1000m in the next 10 minutes of the UAV in the cruising condition.

TEST LOCATION 2

Test location : Karur - 639001, Tamil Nadu.

Latitude : N 10^0 57' 33.4664"

Longitude : E 78 4'47.8008"

Time : 4.00-5.15 PM

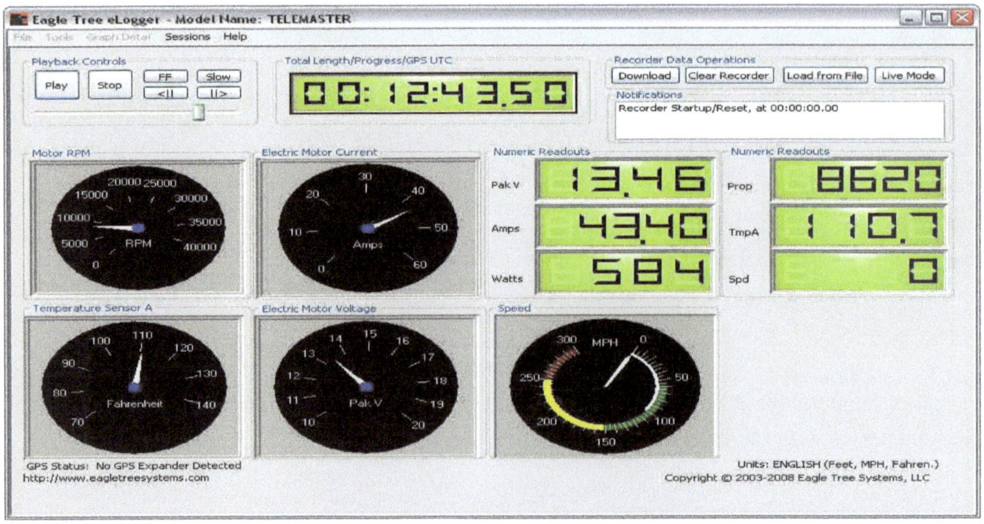

Figure 16: Recorded Parameters for 2nd test location.

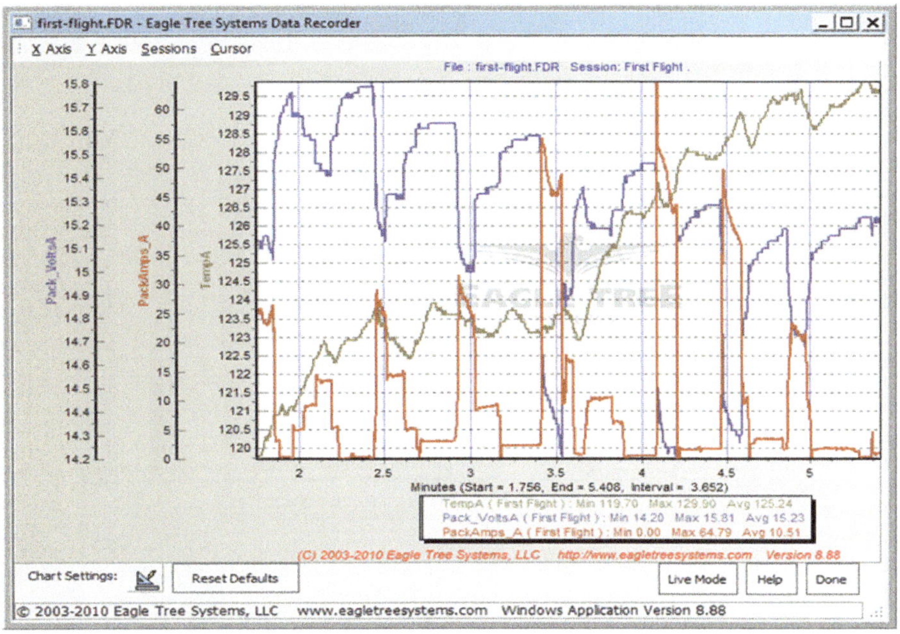

Figure 17: Temperature, Current, Voltage Vs Time (Min).

The variation of current used in the UAV with respect to time is shown in the graph in red colour. It is a continuous varying curve similar to the cos curve. The blue colour describes the plot between the voltage and time. Compare to current the voltage will be used at steady level in the range of 14 to 15.whereas the current in the range of 5 to 50.since the temperature is noted in the noon time .from the graph the brown colour indicates the temperature is directly proportional to the time.

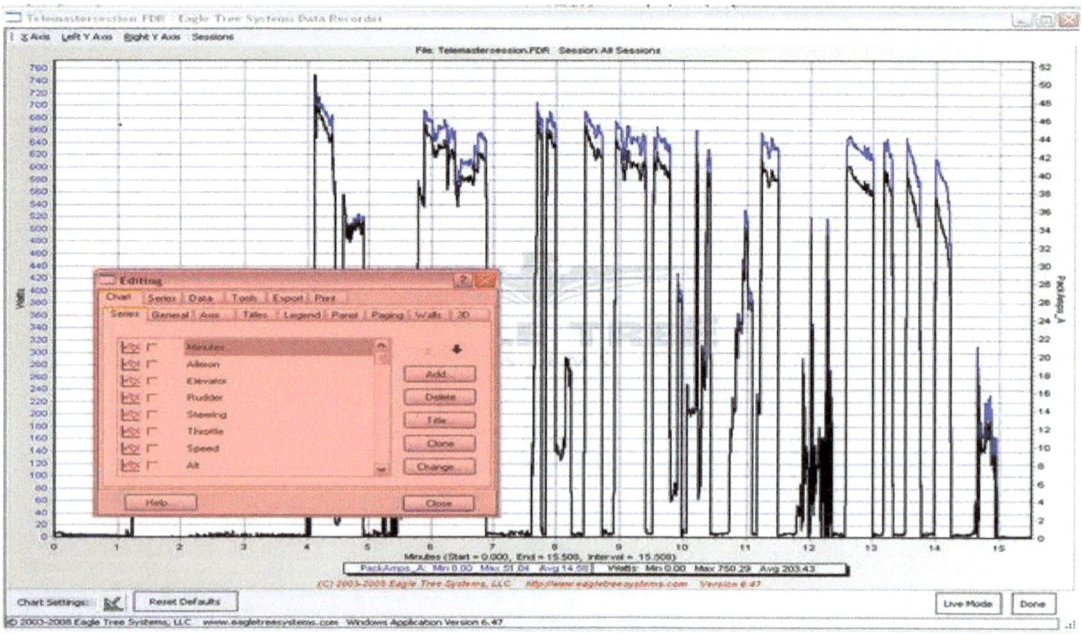

Figure 18: Current Vs Time (Min).

The usage of current will vary in a fraction of seconds. Hence the graph shows the sudden variation in the current level with respect to time. During the landing condition it requires only less amount of current so the graph descends at the final stage.

6. Conclusion

The recent meteoric rise of UAV development highlights an issue of the growing importance of UAVs in the future and leads to the corollary issue of whether UAVs will replace manned aircraft's roles and missions. The bottom line is that the Services still need an organic capability for a continuous, on-demand, all-weather platform to provide intelligence on weather forecasting. In addition, the Services must execute these operations quickly, safely, and cheaply. Due to the frequent change in weather conditions it is quite necessary to survey the environment conditions for the survival of human and other living beings. In remote areas such as deep forest regions and coastal areas predicting of weather conditions is possible only by UAVs. The UAV such as Aerosonde, plays a Vitol role in weather forecasting. Although the recovery of the mission's during system failure is risk in those model. Our proposed model can provide both recovery of the mission in emergency conditions and better in finding the environmental situations in reasonable time. With the aid of GPS system tracing of mission is also possible. There is no doubt that the unmanned aerial vehicle can provide a great contribution in Climatology in future.

References

1. Holland, G. J., and Coauthors, (2001) "The Aerosonde Robotic Aircraft: A New Paradigm for Environmental Observations", published in *Bull. Amer. Meteor. Soc.*, Vol.82, pp.889–901.
2. J.A.Curry.., et al., (2004) "Applications of Aerosondes in the Arctic", *Bulletin of the American Meteorological Society*, Vol.85, No.12, pp.1855-1861.

3. Tad Mcgeer.., (9 October 1998) "Regulatory Issues Involving Long-Range Weather Observation by Aerosonde Autonomous Aircraft".
4. Edward J patten.., et al., (12 June 1963) "Study for the Advancement of Unmanned Aircraft Systems in Southern", *Committee on Science and Astronautics*.
5. Po-Hsiung Lin and Cheng-Shang Lee.., et al, (2005) "The Eyewall-Penetration Reconnaissance Observation of Typhoon Longwang with Unmanned Aerial Vehicle, Aerosonde", *J. Atmos. Oceanic Technol.*, Vol.25, pp.15–25.
6. Jade R. Soddell.., et al., (2002) "An In-Flight Airframe Icing Nowcast for Use with the Aerosonde", Conference on Aviation, Range, and Aerospace Meteorology.
7. John Andrews, (2010) "System Reliability Assessment for Real-time UAV Mission Planning", conference in University of Nottingham.
8. Edward H. Teets.., et al., (1998) "Atmospheric Considerations for Uninhabited Aerial Vehicle (UAV) Flight Test Planning", NASA Dryden Flight Research Center.
9. Gloria Kulesa.., (2003) "Weather and Aviation: How Does Weather Affect the Safety and Operations of Airports and Aviation, and How Does FAA Work to Manage Weather-related Effects?", The Potential Impacts of Climate Change on Transportation.
10. Randy J. Lefevre.., et al., (January 2004) " Weather Forecast Uncertainty Management And Display", Proceedings of the 21st International Conference on Interactive Information Processing Systems (IIPS) for Meteorology, Oceanography, and Hydrology, San Diego, CA.
11. Stuart M. Adams.., et al., (2010) "Unmanned Aerial Vehicle Data Acquisition for Damage Assessment in Hurricane Events", presented in international workshop conducted by Department of Built Environment in Tokyo Institute of Technology.
12. Maryam kamgarpour.., et al., (15-17 December 2010) "Trajectory Generation for Aircraft Subject to Dynamic Weather Uncertainty", In Proceedings of the 49th IEEE Conference on Decision and Control, CDC, Atlanta, Georgia, USA, , IEEE, pp.2063-2068.
13. Anthony P.Tvaryanas.., (2004) "Visual Scan Patterns During Simulated Control of an Uninhabited Aerial Vehicle (UAV)", Aviation Space Environ Med, Research Article, Vol.75, pp.531-8.
14. Sooho Park.., et al., (2009-08) "Learning covariance dynamics for path planning of UAV sensors in a large-scale dynamic environment", American Institute of Aeronautics and Astronautics.
15. Dah-Jye Lee.., et al, (29 December 2004) "See and avoidance behaviors for autonomous navigation", this paper was published in SPIE Proceedings, Vol.5609, pp. 12.

15

Reduction of Jet Noise in the Aircraft Nozzle

Saurabh Shriwas[1], Sachin Shah[2] Prashant Singh[3], Kalpit P. Kaurase[4]

[1,2]Student, B.E. (Aeronautical), MATS University, Raipur saurabhshriwas1994@gmail.com
[3]Student, M. Tech. (Turbo machinery), MATS University, Raipur
[4]Assistant Professor, Dept. of Aeronautical Engg. MATS University, Raipur

Abstract

In this paper, the reduction of jet noise in exhaust nozzle of aircraft engines is discussed. This is due to the high turbulence structure created in the jet shear. These turbulence structures are generally caused by the flow instability due to pressure fluctuation created in nozzle. Here the effort is to study the way of noise reduction methods in the exhaust nozzle, and this can be done by using various methods such as the chevron nozzle, nano materials, extending the engine nacelle, increasing the bypass ratio in turbofans and acoustic lines of the main engine. Chevron nozzle are provided with the sharp saw tooth structure at its aft, which is responsible for the smooth mixing of hot and bypass stream hence the energy of the turbulence structure is reduced. Nano materials with very high density are able to absorb the noise within the nozzle. Increasing the bypass ratio is also a good option in turbofan engines; it reduces the energy of main stream by mixing cold stream into hot stream. Acoustic lines can also be the solution for the noise reduction; it converts the energy of turbulence structure into the form of heat. By increasing the size of the engine housing, noise can also be reduced. At this condition the hot jet will not directly expose to the atmosphere but passes through the nacelle first which can absorb the noise by some amount.

Keywords: Jet shear, Exhaust nozzle, Turbulence, Jet noise.

1. INTRODUCTION

As the cruise speed increases the noise of the jet also increases and due to the flow instability the turbulence is created which lead to the noise in the exhaust nozzle of the aircraft, but the problem is not associated with the time of cruising, the main problem occurs at the time of ground run, climbing and descent of the aircraft when it is more close to the population. Now the suppression has become the one of the most challenging research topic in the aviation industry in order to minimize its effects on people around the airport and the important breakthrough is the use of modern aircraft with various experimental designs like saw tooth pattern and other method described below. In this review paper our work is to study and analyze the method of noise reduction in aircraft nozzle and we will be dealing with the chevron nozzle and other available ideas. Here the problem is associated with the nozzle so it is mandatory to know about nozzle [1]. The exhaust nozzle is one of most essential part of an aircraft engine .basically the work of the nozzle is to collect the hot flow coming from the turbine and straightening or it increases the kinetic energy of the flow. For the higher value of the thrust the kinetic energy of the flow must be very high which can be done by very high exhaust velocity. The expansion of flow occur in the nozzle is due to the pressure ratio [2].

Some important functions of nozzle are described below:-

1. With minimum total pressure loss it accelerates the flow with very high velocity.
2. It can be used in thrust vectoring, by changing the direction of exhaust.
3. Use to suppress infrared radiation.
4. It allows the application of the thrust reversal; generally clamshell and cascade types are used. This arrangement allows the flow in reverse direction.
5. In turbofan engine it mixes the main stream and the bypass air.
6. One of the most important function of the nozzle is to match exit and the atmosphere pressure that is denoted by $P_e=P_a$, which is an optimum condition for the flow expansion from the nozzle [3].
7. It use also used for wall cooling.
8. Thrust augmentation can be performed by the use of nozzle, especially in afterburner when additional fuel is burn in the tail pipe then the mass flow rate of the hot jet increases and for the expansion of this addition hot jet variable area nozzle is required.

Generally for subsonic flow, subsonic nozzles (convergent type) and for supersonic flow, De Laval nozzle are used this selection is generally based upon the Area-Mach number relationship.

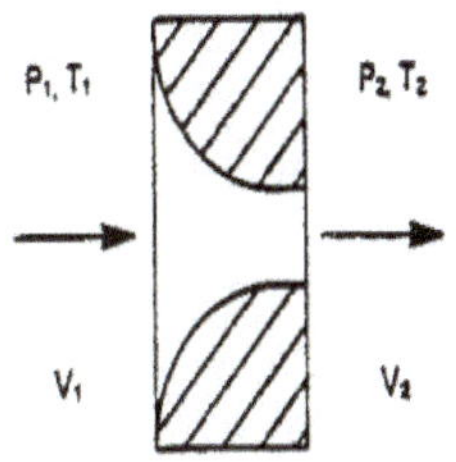

Figure 1: Flow through convergent nozzle

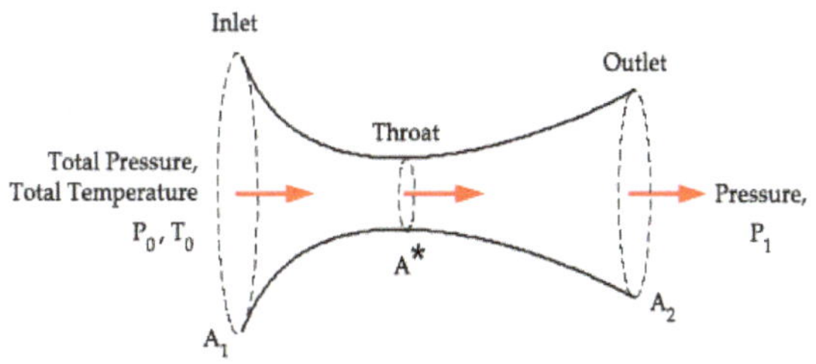

Figure 2: Flow through C-D nozzle

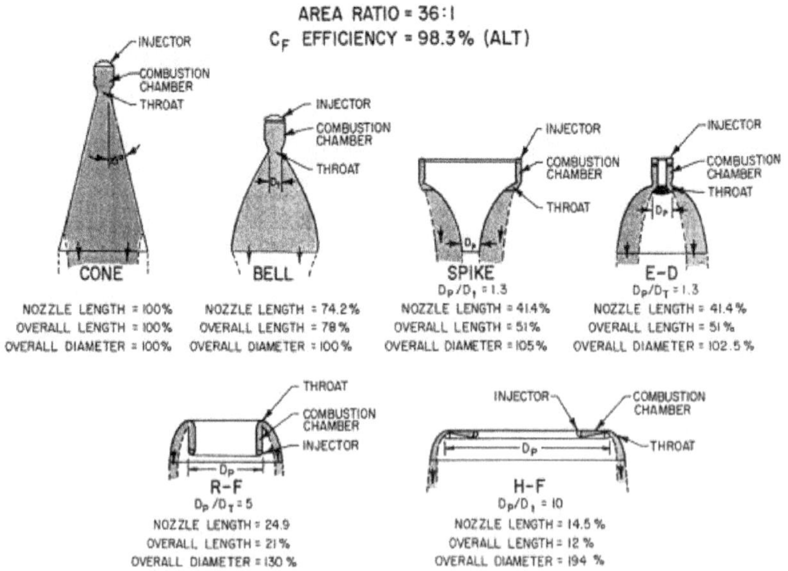

Figure 3: Different shapes of nozzle

2. Cause of jet noise in exhaust nozzle

In the field of aeronautics and aerospace the jet noise is a particular branch which deals with the generation of noise. In early aircrafts piston engine were used. These aircrafts usually produces very high noises which further reduced by the use of silencers. After that during Second World War jet engines used frequently which is also a noise producing engine. The initiative to decrease the noise in the jet engine was taken by the NASA, and they developed chevron nozzles but now here are several possible methods which can be used to reduce noise of the jet aircraft.

The generation of noise is caused by the velocity of jets and the turbulent eddies, developed from the shearing flow. This noise is also known as broadband noise and its frequency is around 100 kHz so it is well beyond the hearing of human. Jet noise sometimes produces the loudest sound ever produced by the humans [3].

In an exhaust nozzle when jet velocity reaches around 100 m/s it becomes the source for the jet noise and in supersonic flow the noise is associated with shocks along with this at the low speed some acoustic source in the tail pipe is responsible for the generation of the jet noise including combustion noise, turbine-compressor system and mixing of free stream with the fans.

When the ambient fluid is mixed with jet by turbulent mixing then the jet noise is created. The mixing initially occurs in the annular shear layer with enhances along with the length of the nozzle. The region of mixing usually fills the complete jet at four or five diameter from the nozzle. Most of the high frequency noises are close to the nozzle because here the dimension of turbulent eddies is very small. When we analyze further down the jet lower frequency starts because eddy size is similar to the diameter [3].

In supersonic jets the flow expands and contracts through the several cells these cells are generally known as the "shock cells". These cells can be seen extending up to 10 times of the jet diameter from the nozzle and it is responsible for the two types of jet noises "screech tones" and other is broadband "shock associated noise". Screech is produced by a feedback mechanism in which a disturbance convicting in the shear layer generates sound as it traverses the standing system of shock waves in the jet. Even though screech is a side effect of the jet's flight, it can be suppressed by an appropriate design for a nozzle [4].

Aircraft noise is also sometimes known as the "jet noise" when generates from jet powered aircraft regardless of the mechanism of noise production.

Possible methods to reduce jet noise

1. Chevron nozzles

Chevron nozzles are those nozzles having the saw tooth pattern in the trailing edge of the aircraft exhaust nozzle. Working of chevron nozzle is quite simple, the hot jet is mixed with the main cold air stream coming directly from the engine fan. The chevrons are generally triangular shaped structure having a base ,apex and a trailing edges in such a way that they converges there between and radially opposite first and second surface bounded there by. Each of the chevrons is triangular in configuration, with a base, apex, side trailing edges converging there between, and radially opposite first and second surfaces bounded thereby. The trailing edges of adjacent chevrons are spaced laterally apart to define respective diverging slots disposed in flow communication with the duct. The chevrons have a concave contour axially between the bases and apexes which promotes jet mixing through the slots. Shaped edges of the nozzle play an important part in smooth mixing of the flow which significantly reduces turbulence (pressure fluctuation), a cause of noise creation. Chevron nozzle was developed by the NASA. Some examples for the application of Chevron nozzles are GEnx, ROLLS-ROYCE Trent 1000 [5].

2. Nano-Technology

It is a one of the strongest innovation and it has made dramatic change in all kind of industry. Nano material is also used in different sectors of aviation industry because of its multi-functioning such as low weight, availability, eco-friendly and consumption of less fuel. Nano material also has various properties like highly responsive communication system, minimum requirement of repair.

Despite all above properties Nano material are quite useful in reduction of noise in exhaust nozzle of an aircraft. Nano materials has a property of high density (high strength) along with the very less weight, since if the exhaust nozzle of an aircraft is made of with the Nano material, so there is a chance of the absorption of maximum noise within the aircraft nozzle.

3. Aircraft Engine Configuration

In the engine of an aircraft after the nozzle portion if we further increase the length of the tail pipe longitudinally then the exhaust jet will not directly exposed to the atmosphere but it pass through the extended portion. While passing through this portion the most of the noise will be suppressed by the density of the extended part.

4. Acoustic Liner

In order to reduce the jet noise in exhaust nozzle the acoustic lines are used in various modern aircraft. Acoustic lines convert the energy of the jet particle into the form of the heat so that the reduction in heat of the particle leads to the suppression of noise in the exhaust nozzle.

5. Configuration of Nozzle

If the housing of the nozzle (nacelle) is further increased in the axis of the engine then the et will not directly exposed to the atmosphere and at the same instant the atmospheric air is allowed to mixed with the main jet stream with the use of the ports provided on the extended part of nozzle. In this way we will able to reduce the energy of the exhaust jet in the nozzle.

Figure 4: Saw tooth pattern improves noise profile of the Lufthansa 747-8

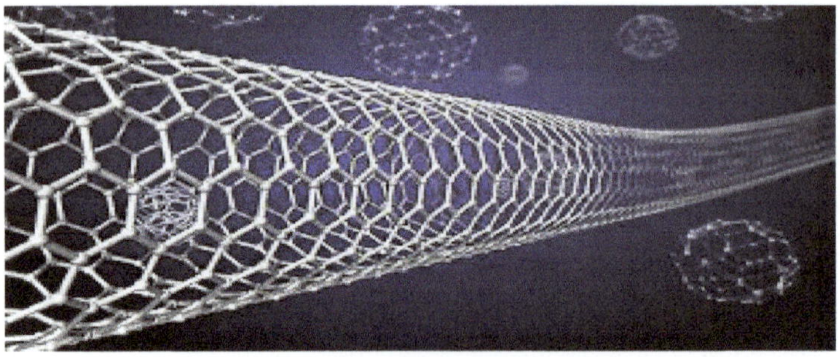

Figure 5: Structure of Nano material

Figure 6: Engine nacelle configuration

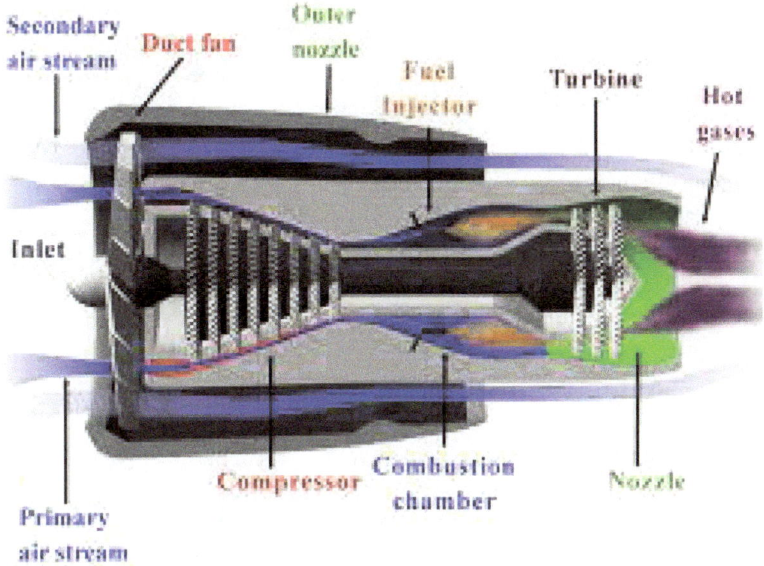

Figure 7: Schematic of turbofan engine

5. Increasing the bypass ratio in turbo fans

In turbofans the thrust is obtain by the hot jets as well as bypass air. Hot jet passes through the conventional jet engine and the cold stream through annular nozzle. If the bypass ratio ($\beta=\dot{m}_c/\dot{m}_h$) is increased with a certain amount and mixed with the hot stream that is obtain from the turbine, then the overall energy of exhaust gas can be reduced by some amount so that the jet noise of the aircraft can be reduce.

3. Conclusion

It is concluded that the paper describes the methods of prevention of noise in the exhaust nozzle of the aircraft, which is one of most interesting topic in aviation field. It is observed that noise arising from the exhaust jet can be reduced by using various techniques such as Chevron nozzles, nano materials, making changes in engine configurations, using acoustic liners, modifying nozzle configuration and in case of turbofan engines by increasing bypass ratio. These methods are easy to implement and can significantly reduce the noise in exhaust nozzle of the aircraft engine.

4. References

1. Thirumurthy, D., 2010, "Design and analysis of noise suppression exhaust nozzle system", M. S. Dissertation, Purdue University.
2. Ganeshan, V., 2010, "Gas Turbine", Tata McGraw Hill, Third Edition, ISBN: 0-07-068192-9.
3. Gp. Capt. Salam, M.A., Gp.Capt.Chattopadhyay, N. C., Jalal, M., Rumi, U., Eunus, I., 2013, "A Review on Jet noise reduction", Proceedings of 4th Global Engineering, Science and Technology Conference, BIAM Foundation, 63 New Eskaton, Dhaka, Bangladesh, 27 December, 2013, ISBN : 978-1-922069-43-6.
4. Schlinker, R. H., Simonich, J. C., Shannon, D.W., Reba, R.A., Colonius, T., Gudmundsson, K., Ladeinde, F., 2009, "Supersonic Jet Noise from Round and Chevron Nozzles: Experimental Studies", 15th AIAA/CEAS Aeroacoustics Conference (30th AIAA Aeroacoustics Conference), 11 - 13 May 2009, Miami, Florida.
5. Braush, J. F., Janardan, A., Barter, J. W., Hoff, G. E., 2002, Chevron Exhaust Nozzle for a Gas Turbine Engine, United States Patent, Date of Patent: March, 26, 2002, Patent No. US 6,360,528 B1.

16

STRESS MAGNIFICATION ON THE CONTROL SURFACES DUE TO MANEUVER LOADS BY FINITE ELEMENT FORMULATION

[1]G. Elavarasi, [2]J. Bruce Ralphin Rose

[1] *PG Scholar, Department of Aeronautical Engineering, Regional Centre of Anna University, Tirunelveli, 627007.*

[2]*J. Bruce Ralphin Rose, Assistant Professor, Department of Aeronautical Engineering, Regional Centre of Anna University, Tirunelveli, 627007*

[1]*aeromanikarnika0108@gmail.com*, [2]*bruceralphin@gmail.com*

Abstract

The interaction between a flexible structure and the surrounding fluid is a critical flow field to assess the structural and aerodynamic properties. In the present article, the influence of maneuver loads on the aeroelastic properties of the control surfaces are investigated by Finite Element formulation. The interaction among the structural and aerodynamic forces at various Angle of Attack will produce severe hinge moments. To study the fluid structure interaction phenomena, it requires the modeling of both fluid medium and structure coupled together. An efficient moving grid technique can be used to account for structural deformation in the robust coupled aero elastic model. Fluid-Structure Interaction governing equations are either loosely coupled or strongly coupled models. Here, the structural response lags behind the flow field solution because of the implementation of partly coupled technique. A doublet hybrid method is used to apply the aerodynamic forces in terms of the parameters involved in the Navier stokes equations. The wing model preferred is a flexible high aspect ratio wing that is applicable to a high altitude long endurance flight vehicle. For high Aspect Ratio wings, the geometrical deformation experienced by the structure is high and the computation process is time consuming. Therefore, a methodology to compute the maximum stress exerted by the maneuver loads is implemented through the Finite Element Approach. The control surfaces enhance the flutter suppression modes because of their different locations along the span and by producing required moments.

Keywords: Fluid –Structure Interaction, Aeroelastic stress, Control design, Maneuver loads.

1 .Introduction

Aeroelasticity is a multidisciplinary science dealing with the interactions of aerodynamic forces and structural deformations. As a structure moves through the stationary air, the motion will induce the aerodynamic loads that leading to deformations of the structure. The deformation in turn has a reactionary force on the airflow, thus changing the aerodynamic loading. Aerodynamic and structural interactions occur apparently depending on the physical properties of the structure. Aeroelastic analysis can be divided in two fields of study such as Static aeroelasticity and Dynamic aeroelasticity. Dynamic aeroelastic effects result from the interaction of the aerodynamic, elastic, and inertial forces. When an aircraft is subjected to dynamic instabilities, the interaction of the control system with aeroelasticity must also be considered. Accurate prediction of this phenomenon is complex, but the undesirable effects of aeroelasticity include airframe fatigue, loss of control, unacceptable vibration can be minimized up to certain extend. Therefore, it is most important to predict the aeroelastic characteristics accurately to prevent the aeroelastic instabilities such as control reversal and fatigue.

Several methods have been implemented in the past to predict the aeroelastic instabilities occurring in aircraft and few problems are sorted out by some unique techniques. In the previous studies, the harmonic balance method (exact formulation) and numerical simulation method are adopted with different fractional stiffness on the aeroelastic system. In the prediction of stable boundary, the results of these of methods are in good concord with the aeroelastic system. The change of friction stiffness has little influence on the stable boundary of the system. [1] The non-planar Vortex Lattice Method (VLM) could be used to solve the aeroelastic deformation with great accuracy. [2] Higher order panel method provides a more accurate static aeroelastic response analysis for the initial and detailed design stage of an aircraft. [3]The current work presents a framework to implement the unsteady VLM and how it can be improved for control system optimization. [4] Yang Ning has presented an iterative method for wave investigation with the authority of dynamic stiffness in incidence domain. [5] Nonlinear aeroelastic analysis for such a wing is carried out by using a loosely-coupled method. A non uniform Euler solver and a nonlinear CSD solver are fixed collectively by the 3-D integral constant level tetrahedron interfacing technique. [6] In this article, an attempt has been made to build up a method for high fidelity simulation of aircraft wing with flap combination. [8] The influence of control surface loads on the hinge moment parameters are quantified for a static aeroelastic phenomenon. The airfoil with control surface is prepared using CATIA design tool and the airfoil is prepared with the aluminum material the control surface is made of composite material (Graphite epoxy). To calculate the stresses produced on the Graphite epoxy material, the theoretical strength of Graphite epoxy is computed by MATLAB program. The wing with control surface is subjected to a fluid flow analysis to acquire the varying stress of the wing.

2. Static aeroelastic behavior of a two dimensional flexible airfoil

The aeroelastic problems arise mainly because of the insufficient torsional stiffness and non-uniform load distributions on the wings. Torsional divergence is a critical static aeroelastic phenomenon that is encountered at cruising flight. In this article, partially coupled method is utilized to attain the characteristics of aeroelastic wing divergence. The prediction is made by modelling a high aspect ratio aircraft wing, installed with a control surface (flap). Control surface is fixed with respect to the aerodynamic centre and it will be deflected downwards. It is designed to have up to 3^0 upward deflections to quantify the possibility of control operation at extreme conditions. The flap is designed along with an arbitrary wing using ANSYS and the material properties are assigned as graphite epoxy composite. Aerodynamic lift force distribution acting over a flexible wing that is fixed at the root will be considered for the stress analysis. The control surfaces are deflected against the rigid wing assumption to get the stress distribution at various initial conditions. It is then compared to a flexible wing with fixed control angles to identify the possibility of divergence speed enhancement. The static aeroelastic behavior is considered initially using an iterative approach and then a direct numerical approach.

Iterative Analysis

A 2-D rigid airfoil with unit span and chord 'c' is considered as symmetric with torsional spring of stiffness K_θ at a distance '*ec*' aft of the aerodynamic center on the quarter chord. The slope of lift curve is a_1 and the airfoil has an initial incidence of θ_0 , an elastic twist through angle θ due to the aerodynamic loading.

The lift force acting on the airfoil at the air speed 'V' and initial angle of incidence θ_0 causes a pitching moment of about the flexural axis.

$$M = qec^2 a_1 \theta_o \tag{1}$$

Where, '*q*' is the dynamic pressure. The equation of motion of 2-D airfoil is obtained using Lagrange's equations. Since, only static aeroelastic influences are being considered, the kinetic energy term is ignored. The potential energy '*U*' is found from the angle of twist of the torsional spring, namely

$$U = \frac{1}{2} K_\theta \theta^2 \tag{2}$$

The generalized moment is obtained from the incremental work done by the pitching moment acting through the incremental angle of incidence by Eq. (3),

$$Q_\theta = qec^2 a_1 \theta_0 \tag{3}$$

Then, the application of Lagrange's equation for coordinates θ gives,

$$\theta = \frac{qec^2 a_1 \theta_0}{K_\theta} \tag{4}$$

Thus, the initial aerodynamic loading is applied on the airfoil and it is twisted by angle θ. While doing this calculation it is assumed that the pitching moment is not being modified because of the wing twist. However, as a consequence of twist, the aerodynamic moment changes to allow the new angle of incidence. This new loading in turn causes the change in airfoil twist and again leading to further modification in the aerodynamic loading and so on. The stepping between applications of the aerodynamic load on the airfoil and changing the angle of twist illustrates the fundamental interaction between a flexible structure and aerodynamic forces that gives rise to aeroelastic phenomena.

First iteration

By repeating the above procedure, the updated elastic twist value in the pitching moment and work expression, leading to an infinite series expansion for the elastic twist in the form of Eq. (5).

$$\theta = qR\left[1 + qR + (qR)^2 + (qR)^3 + (qR)^4 + \dots\dots\dots\right]\theta_0 \tag{5}$$

The Lagrange's equation gives a revised elastic angle of twist

$$\theta = qR(1 + qR)\theta_0 \tag{6}$$

If advanced static aeroelastic calculation involving the coupling of Computational Fluid Dynamics (CFD) with Finite Element Methods (FEM) is applied for an entire aircraft then it requires a partially coupled approach similar to the iterative process. If the mathematical model for aileron reversal problem is represented in terms of iterative process then it will be strictly static case. To obtain a convenient index for the aeroelastic characteristics of the wing, consider the lift coefficient and coefficient of moment about the aerodynamic centre is

$$C_l = a\alpha + \beta \frac{\partial C_l}{\partial \beta} \tag{7}$$

$$C_m = \beta \frac{\partial C_m}{\partial \beta} + C_{m_0} \tag{8}$$

The elastic moment induced by a rotation of the airfoil through an angle 'α' is α*k. Hence,

$$\alpha k = qc^2(eC_l + \frac{\partial C_m}{\partial \beta} + C_{m_0}) = qc^2(ea\alpha + e\beta \frac{\partial C_l}{\partial \beta} + \beta \frac{\partial C_m}{\partial \beta} + C_{m_0}) \tag{9}$$

Hence the reversal speed is

$$U_{rev} = (-\frac{1}{a}\frac{\partial C_l}{\partial \beta})^{1/2} (\frac{2k}{\rho c^2})^{1/2} \tag{10}$$

Hence, the rate of change of lift due to aileron deflection is given by

$$\frac{dL^{'}}{d\beta} = \frac{qc[aqc^2(e\frac{\partial C_l}{\partial \beta} + \frac{\partial C_m}{\partial \beta})]}{(k - qc^2 ea) + \frac{\partial C_l}{\partial \beta}} \tag{11}$$

This shows the mathematical model of the aileron reversal in two dimensional as similar to divergence problem.

3. Wing Model Design

A high aspect ratio wing model with control surfaces such as flap is designed using CATIA V5 with required dimensions as shown in Fig (1). The control surface is fixed at the hinge point at appropriate distances from the root and leading edge. Firstly the NACA airfoil (64- 009) coordinates are plotted and extruded to the specified span by the design tool. Cut sections are prepared about 25% of the chord to locate the flap or aileron control based on the iterative process involved.

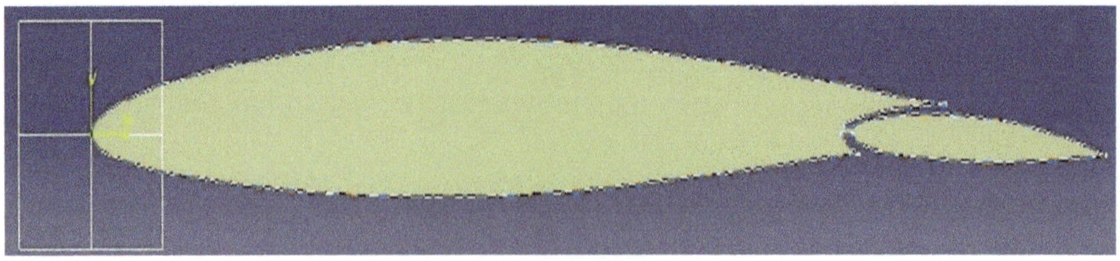

Fig (1). Airfoil with control surface at 3 degree deflection

Flap sections are designed separately using the NACA 0012 coordinate and then assembled with the wing occupying the space provided. To deflect the flap control, suitable hinge with appropriate length is fixed at about 25% of the cut section. The wing model is then imported to the analysis tool to carry out mesh and for applying the boundary conditions. Fig (1) is a sample image and all degree of control deflection is being taken into account in the current analysis.

4. Results and Discussion

The FEM analysis is done for various flap deflections and the corresponding stresses, strains and deformations are evaluated. The results show the variation of these physical properties against each deflection angle of the control surface. A sample normal stress output for each control deflection angle is illustrated below. Fig (2) presents the normal stress distribution at 0^0 flap deflection case. Here, the maximum stress is produced about 2.5 MPa near the mid-chord location of the flap. The angle of incidence of the oncoming wing is the critical reason behind this finite stress build-up.

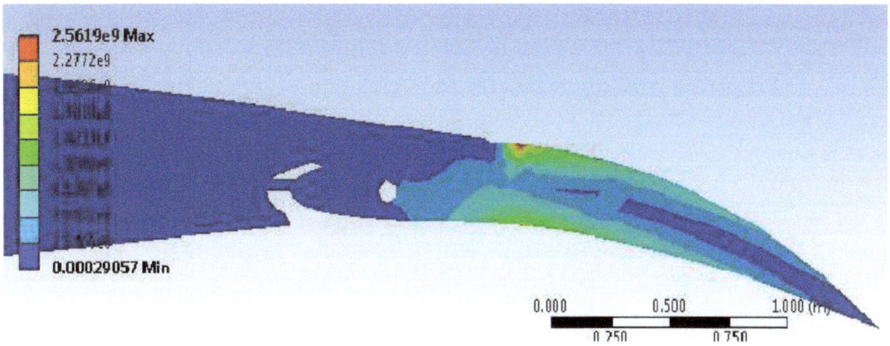

Fig (2). Stress at 0 deg flap deflection with 5^0 angle of incidence

The main wing airfoil is assigned with aluminium 2024-T3 alloy material properties. The main wing airfoil is meshed with coarse mesh and the control suface meshing is fine one. The stress regions of interest are meshed with adaptive mesh generation technique to quantify the spatial variation of stress at each angle of incidence.

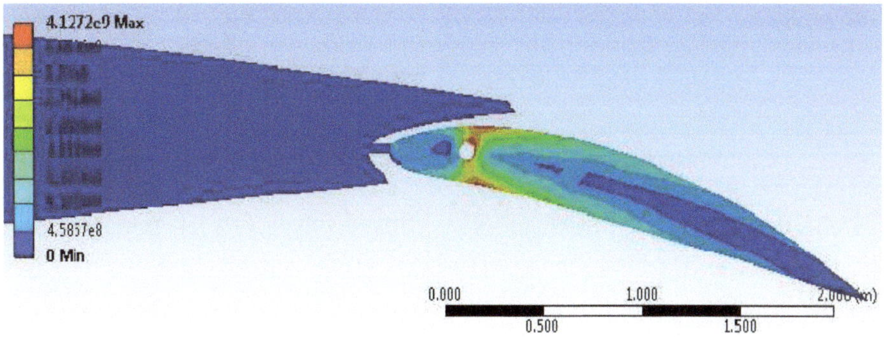

Fig (3). Stress at 3 deg flap deflection with 5^0 angle of incidence

The deformation profiles of the controls about its hinge also computed using ANSYS to ensure its rigidity at various aerodynamic pressures. All the stress levels are computed at 5^0 angle of incidence to give an insight about the influence of wind direction on the control stress. When the flap is deflected about 3 degrees the stress magnitude increases to 41.2 MPa. This is an expected result because the control is deflected downwards. A linear stress increment is observed from 1^0 and 2^0 control deflections which conveys the influence of control rotation. The applied pressure on the control surface is the ¾ th flap deflection load usually computed for its model selection. The corresponding strains are significant because of the varying atmospheric characteristics.

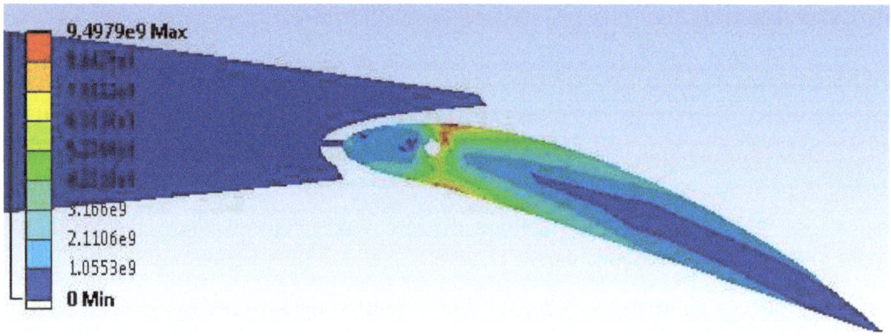

Fig (4). Normal stress at 6 deg flap deflection with 5^0 angle of incidence

At high angle of deflections the mesh is further refined since the load magnitude is very severe. For this iterative process, the model design and meshing is performed separately for each angle of deflection. It is a tedious process but the less computing power is sufficient to attain optimum solutions. At 6 deg flap deflection the stress level reaches 94.9 MPa. It is almost two times of the previous case and the stress concentration is very high near the hinge point. This result focuses the importance of aerodynamic balancing in the stability characteristics of any airplane at high cruising speeds.

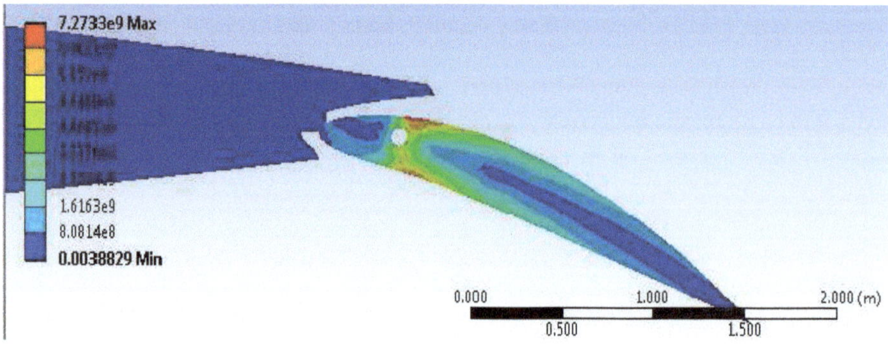

Fig (5). Normal stress at 9 deg flap deflection with 5^0 angle of incidence

S.No	Deflection angles	Stress values	Strain values ($\mu\varepsilon$)	Deformations (cm)
1	0 degree	25.619e9	0.016652	0.061675
2	3 degree	41.272e9	0.0268277	0.1969
3	6 degree	94.979e9	0.061732	0.43683
4	9 degree	72.733e9	0.33018	0.33018

Table (1). Results for stresses, strains and deformations

As the control surface is deflected about 9 degree and the main airfoil is fixed to the aerodynamic centre the stress level decreases (Fig 5). Because of the initialization of elastic rotation produced on the wing structure the resulting normal stress decreases. It reveals an important fact that the maximum control deflection can induce the elastic rotation in opposition the wind induced twist. However, for high aspect ratio wings it is difficult to incorporate this methodology because of its flexibility nature. However, for better maneuvering capabilities it is vital to assess the control loads to avoid the over balancing and loss of control.

5. CONCLUSIONS

In the present article, the influence of control surface loads on the static aeroelastic phenomena is discussed. Here, initially the aircraft wing with flap as control surface is modeled in CATIA using specific coordinates. Then the model is meshed with tetrahedral elements using ANSYS work bench. It is assumed that the arbitrary wing is said to be fixed and the flap is movable. The structural analysis is carried out at fixed Mach number and at fixed angle of incidence to quantify the control stresses. The airfoil designed using aluminum material to provide the practical insights about the implementation. Flap control is assigned with graphite epoxy material properties because of the high strength to weight ratio requirements. The aerodynamic pressure load is applied on the flap and corresponding stresses are evaluated at various chord points. Corresponding strains and deformations are computed for different angle of attacks to ensure the reliability of this approach. The stress value is kept in minimum while the flap is deflected about 0 degree compared to the other angles. The increased angle of incidence enhances the stress value and has significant influence on the aeroelastic wing divergence.

References

[1] Tan Tiancai, Li Min, Liu Baihui, "Stability analysis of an aeroelastic system with friction", Chinese Journal of Aeronautics, 2013, 26(3): 624–6302013.

[2] Xie Changchuan, Wang Libo, YangChao, LiuYi, "Static aeroelastic analysis of very flexible wings based on non-planner vortex lattice method", Chinese Journal of Aeronautics, 2013, 26(3): 514–521, 2013.

[3] YaokumWang, Zhiqiang Wan, Chao Yang, YunzhenLiu, "Static Aeroelastic Response Analysis of Aircrafts Based on CFD Pressure Distribution", AASRI conference on modeling, identification and control, 2012.

[4] Joseba Murua, Rafael Palacios, J. Michael R. Graham, "Applications of the unsteady vortex-lattice method in aircraft aeroelasticity and flight dynamics", Progress in aerospace sciences, 2012.

[5] YANG Ning, WU zhigang, YANG Chao, "Structural Nonlinear Flutter Characteristics Analysis for an Actuator-fin System with Dynamic Stiffness", Chinese journal of aeronautics 24, 2011.

[6] Dang Huixue , Yang Zhichun Li Yi, "Accelerated loosely-coupled CFD/CSD method for nonlinear static aeroelasticity analysis", Aerospace science and technology 14, 2010.

[7] Benjamin P. Hallissy, Carlos E.S. Cesnik, "High-fidelity Aeroelastic Analysis of Very Flexible Aircraft", University of Michigan, Ann Arbor, MI 48109, 2011.

[8] Xiangying Chen, Ge-Cheng Zha, Ming-Ta Yang, "Numerical simulation of 3-D wing flutter with fully coupled fluid-structural interaction", computers and fluids 36, 2007.

[9] LIU Xiang ning, XIANG Jirrwu, "Stall Flutter Analysis of High Aspect Ratio Composite Wing", Chinese journal of aeronautics, 2006.

[10] Davide Fransos, Luca Bruno, "Determination of the aeroelastic transfer functions for streamlined bodies by means of a Navier-Stokes solver", mathematical and computer modeling, 43, 2006.

[11] Luciano Demasi and Eli Livne, "Aeroelasticity of Structurally Nonlinear Lifting Surfaces Using Linear Modally Reduced Aerodynamic Generalized Forces", AIAA, 2006.

[12] Manoranjan Majji , Bong Su Koh, John L.Junkins , Othon K. Rediniotis, "Morphing Wing A Determination of Aero Servo Elastic Distributed Sensing and control", S.C Patel et al.TiiMS, 2005.

[13] ZHAO Yong-hui, HU Haiyan, "Structural Modelling and Aeroelastic Analysis of High-Aspect Ratio Composite Wings", Vol 18, Chinese journal of aeronautics, 2005.

[14] M. Abdo, D. Mateescu., "Low-Reynolds Number Aerodynamics of Airfoils at Incidence", AIAA, 2005.

[15] N. Siepenkotter, W.Alles, "Stability analysis of the nonlinear dynamics of flexible aircraft", Aerospace science and technology 9, 2005.

[16] J.S. Bae, D.J. Inman, I. Lee, "Effects of structural nonlinearity on subsonic aeroelastic characteristics of an aircraft wing with control surface", Journal of fluids and structures 19, 2004.

[17] M. Kampchen, A. Dafins, H.G. Reimerdes, G. Britten, J. Ballmann, "Dynamic aero-structural response of an elastic wing model", Journal of fluids and structures 18, 2003.

[18] E J Berger, "Friction modeling for dynamic system simulation", CAE Laboratory, Department of Mechanical, Industrial, and Nuclear Engineering, University of Cincinnati, 2002.

[19] M.J Smith , M.J. Patti, D.H. Hodges, "CFD-Based Analysis of Nonlinear Aeroelastic Behavior of High-Aspect Ratio Wings", AIAA 2001.

[20] F.Liu J. Cai and Y.Zhu H.M.Tsai and A.S.F.Wong, "Calculation of wing Flutter by a Coupled Fluid-Structure Method", Journal of aircraft, Vol 38, 2001.

[21] Mayuresh J. Patil, Dewey H. Hodges, "Limit Cycle Oscillations of a Complete Aircraft", AIAA2000.

[22] Mayuresh J. Patil and Dewey H. Hodges, "On the Importance of Aerodynamic and Structural Geometrical Nonlinearities in Aeroelastic Behavior of High-Aspect-Ratio Wings", AIAA, 2000.

[23] Mayuresh J. Patil, Dewey H. Hodges, Carlos E.S. Cesnik, "Limit Cycle Oscillations in High-Aspect –Ratio Wings", AIAA 1999.

[24] Mayuresh J. Patil, Deway H. Hodges, Carlos E.S. Cesnik, "Nonlinear Aeroelasticity and Flight Dynamics of High-Altitude Long-Endurance Aircraft", AIAA, 1999.

[25] Mayuresh J. Patil, Deway H. Hodges , Carlos E.S.Cesnik, "Characterizing the Effects of Geometrical Nonlinearities on Aeroelastic Behavior of High-Aspect Ratio Wings" 1999.

[26] Todd O' Neil and Thomas W.Strganac, "Aeroelastic Response of a Rigid Wing Supported by Nonlinear Springs", Journal of aircraft, Vol. 35, No 4, July-august, 1998.

17

Application of hybrid technique in medical image compression based on SVM and clustering method

Varsha.C.K[1], R.Manoharan[2]

[1]PG Student,varshacknambiar@gmail.com
[2] Associate Profesor, ECE Department
Sri Lakshmi Ammaal Engineering College,Chennai-126

Abstract

The Project proposes the region based Image compression technique based on clustering model and hybrid compression technique. The primary and secondary region of interest will be selected automatically by clustering algorithm. The lifting based discrete wavelet and NSCT is used to decorrelate the pixels into fine as well as redundant or noisy data and edge details. The regions are encoded by lossless and lossy technique to increase the compression ratio and preserve the image quality. These methods are useful to compress data for transmission and telemedicine application.

Key Terms— Medical Image, Image compression technique, clustering model, hybrid compression technique, clustering algorithm, discrete wavelet, data transmission

1. Introduction

A large amount of image data is produced in the field of medical imaging, in the form of computed tomography (CT), magnetic resonance imaging (MRI), positron emission tomography (PET) and ultrasound images, which can be stored in picture archiving and communication system (PACS) or hospital information system (HIS). A medium scale hospital with the above facilities produces on an average 5–15 GB of data per day. So, it is very difficult for hospitals to manage the storing facilities for the same. Moreover such high data demands a high-end network, especially for transmitting the mages over the network such as in telemedicine. This is a very significant to the limitations of the transmission media in information and communication technology (ICT), especially for rural areas. Image compression is useful in reducing the storage and transmission and width requirements of medical images. If the image is compressed by 8:1 compression without any perceptual distortion, the capacity of storage increases eight times. Compression methods are classified into lossless and lossy methods. In the medical imaging scenario, lossy compression schemes, even though it gives up to 10% compression ratio, are not generally used. This is due to the possible loss of useful clinical information that may influence diagnosis. In addition to these reasons, there can be legal issues. Storage of medical images is generally problematic because of the requirement to preserve the best possible . Image quality which is usually interpreted as a need for lossless compression. Three-dimensional magnetic resonance imaging (MRI) contains multiple slices representing a part of a body, requires all information of that part.The storage size for such 3D images is huge.

2. Review of Related Research

A handful of researches are available in the literature for encoding an image based on ROI. Recently, utilization of wavelets and techniques like SPIHT and AT-SPIHT for encoding of ROI has received a great deal of attention among the researchers. A brief review of some of the recent research works is presented here.

V.K. Bairagi and A.M. Sapkal[1] *et* al. This paper proposes an automated, efficient and low complexity, lossless, scalable RBC for Digital Imaging and Communications in Medicine (DICOM) images using SPIHT coding. Region Based Coding (RBC) technique is significant for medical image compression and transmission. Lossless compression in these 'regions' and lossy compression for rest of image can helps to achieve high efficiency and performance in telemedicine applications. For the medical images, only a small portion of the image might be diagnostically useful, but the cost of a wrong interpretation is high.

Abdul Sattar[2] *et* al. This paper proposes the multi-iteration algorithm and particularly the two iteration EZW for a given image quality produce lower bit rate. It is applied for medical images and here, the thorax radiology is chosen as a sample image and the good performance is codified. The embedded zerotree wavelet algorithm (EZW) is used for image coding. It is designed to optimize the combination of zerotree coding and Huffman coding.

Calderbank, A.R., Daubechies, I., Sweldens, W., Yeo, B.L[8] *et* al. This paper describes the Use of lattice vector quantizers in conjunction with a quadtree-based sorting algorithm for the compression of multidimensional data sets, as encountered, for example, when dealing with hyperspectral imagery.The SPECK algorithm is presented that deals with vector samples and i used to encode a group of successive spectral bands extracted from the hyperspectral image original block. Finally provide comparison against state-of-the-art encoders, both 2D and 3D ones, showing the proposed encoding method is very competitive, especially at small bit rates.

Eva Cavero, A´ lvaro Alesanco, Lena Castro, Jose´ Montoya, Isaac Lacambra and Jose´ Garc *et* al[3]. The paper deals with an echocardiogram coding method that takes into account the visualization modes in order to compress efficiently the echocardiogram, a methodology to evaluate compressed echocardiograms, and the evaluation of the compression method using the proposed evaluation methodology. The evaluation of the echo cardiograms compressed with the proposed method has been carried out. The recommended transmission rates have been established as follows: 200 kb/s for the 2-D and the color Doppler modes, and 40 kb/s for theMand the pulsed/continuous Doppler modes. These rates, especially the latter, are very low to previous results. These recommendations are valid for all devices and images compressed with the proposed method. The evaluation process can be applied to any compression method.

Jamuna.M, A.M.Vijaya Prakash, J.Pushpanjali[13] *et* al. This paper presents implementation of 2 methods of DWT, one is conventional method and the other one is lifting scheme. Since conventional method requires more memory, area and power, lifting scheme is used as an enhanced method. DWT is the most widely used image compression technique and it is the most efficient algorithm used in JPEG image compression. The Inverse Discrete Wavelet Transform (IDWT) is also obtained in a similar way to get back the image matrix. This architecture enjoys reduced memory referencing.

3. Proposed Method

Automated Region based Image compression technique based on clustering model, lossless and lossy coding algorithms for efficient medical image transmission in telemedicine applications. Region based hybrid image compression through, Clustering model for region selection, Non-Sub sampled Contour let Transformation (NSCT), Entropy coder-Adaptive arithmetic coder (AAC), Lifting wavelet with SPECK Coding for Lossy compression and Parameter Analysis(MSE, PSNR, CR, Correlation and Elapsed Time).

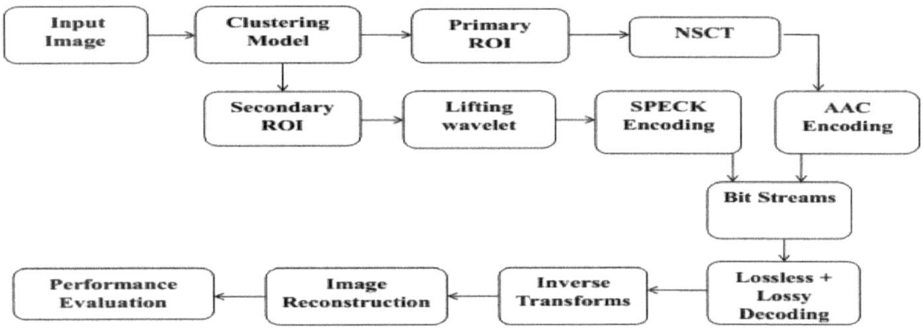

Figure.1: Block diagram of proposed method

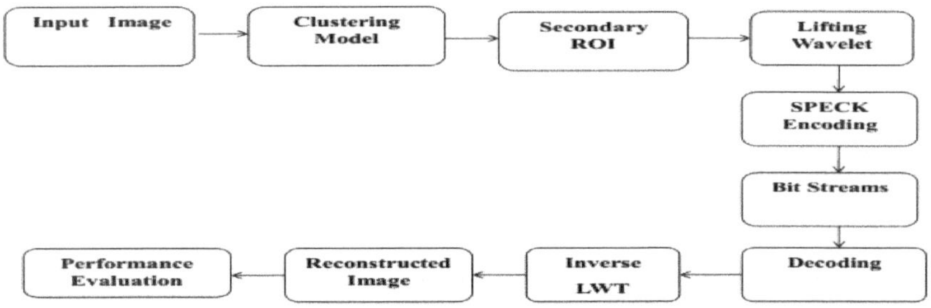

Figure2:flow of lossy compression

3.1 Region Selection

The first step is to select the desired regions from input automatically based on identifying an intensity variations. Here, grouping of similar region will be performed by Kmeans clustering algorithm. Here grouping of similar region will be performed by Kmeans clustering algorithm. It is an unsupervised clustering algorithm that classifies the input data points into multiple classes based on their inherent distance from each other. The algorithm assumes that the data features form a vector space and tries to find natural clustering in them. It is a well suited method for this process on MRI brain samples and it consumes less time and provides better result for better contrast image.

3. 2 Clustering algorithm

The K-means algorithm is an iterative technique that is used to partition an image into K clusters. The basic algorithm is Pick K cluster centres, either randomly or based on some heuristic. Assign each pixel in the image to the cluster that minimizes the distance between the pixel and the cluster centre. Re-compute the cluster centres by averaging all of the pixels in the cluster. Repeat steps 2 and 3 until convergence is attained (e.g. no pixels change clusters).

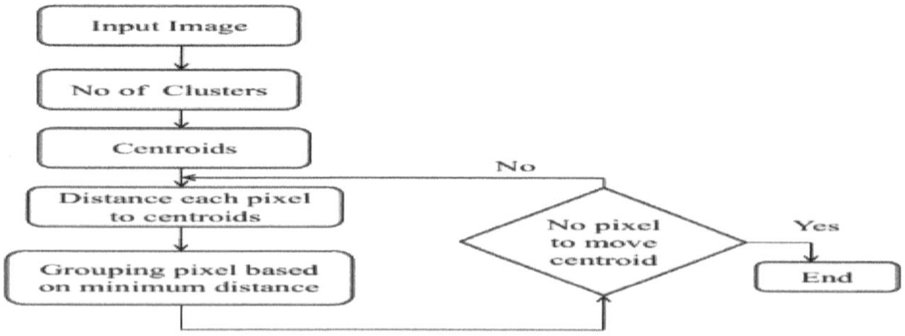

Figure3: Process Flow –kmeans algorithm

3.3. Orginal And Region Splitted Images

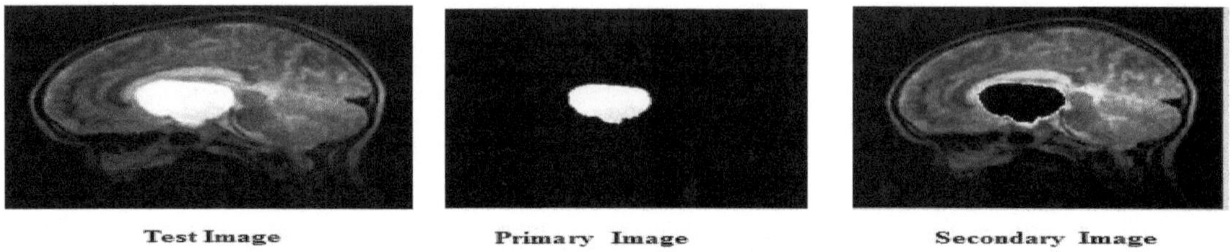

Figure4: Process Flow

4. Lifting Wavelet Transform

LWT decomposes the image into different subband images,namely, LL, LH, HL, and HH for getting pixel coefficients of images. LL subbands contains the significant part of the spatial domain image. High-frequency subband contains the edge information of input image. Integer wavelet transform can be obtained through lifting scheme. Liftin scheme is a technique to convert DWT coefficient to Integer coefficients without losing information.

This will be used to provide unique values for quantization and encoding proces

4.1 Wavelet Computation

Forward Lifting in IWT

Step1: Column wise processing to get H and L

H = (Co-Ce) and L = (Ce-~~[H/2]~~) (1)

Where Co and Ce is the odd column and even column wise pixel values,

Step 2: Row wise processing to get LL,LH,HL and HH,

Separate odd and even rows of H and L,Namely, Hodd – odd row of H, Lodd- odd row of L Heven- even row of H, Leven- even row of L,

LH = Lodd-Leven ,LL = Leven – [LH / 2] (2)

HL = Hod –Heven ,HH = Heven–[HL / 2] (3)

4.2 Reverse Lifting Scheme In Iwt

Set partitioning in embedded coder is based on multiscale 2D DWT and exploits the self-similarity across scales by using set partitioning. After transformation, the coefficients are ordered into a tree structure, called spatial orientation tree (SOT). The SOT is defined by each wavelet coefficient (parent) in a certain decomposition scale has either no child (i.e., tree leave) or four children in the next finer scale and the coefficients in the low-frequency subband are the tree roots. The coefficients are quantized by partitioning them into different set likes significant information and insignificant informations and performs the priority based transmission.

5. SPECK Algorithm

SPECK algorithm is based on 3 concepts

- Ordered bit plane progressive transmission.
- Set partitioning sorting algorithm.
- Spatial orientation trees

In a practical implementation the significance information is stored in three ordered lists

- LIS-list of insignificant sets
- LIP-list of insignificant pixels
- LSP-list of significant pixels

5.1 spatial orientation trees

The following sets of coordinates are used to present the new coding method

◦ O (i, j) : Set of coordinates of all offspring of node (i, j);

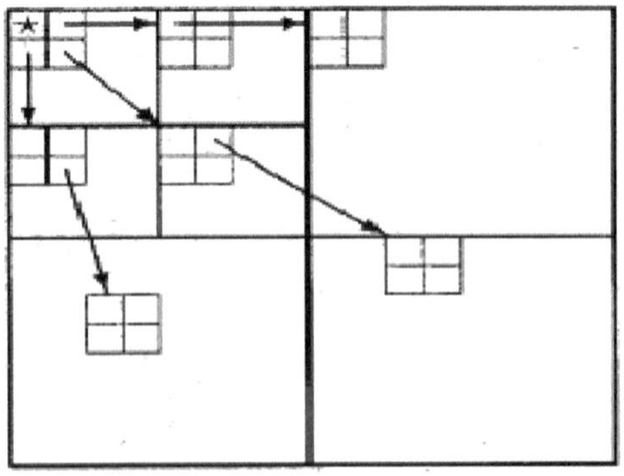

Figure4: splitting into child co-ordinates

5.2. Set partioning algorithm

The sorting algorithm divides the set of pixels into partitioning subsets T_m and performs the significance test by using the function

$$S_n(T) = \begin{cases} 1, \max(i.j) \in T|Ci,j| > 2 \\ 0, \text{otherwise}, \quad n \text{ is the \# of pass} \end{cases}$$

6. Experimental Analysis

The test image of 69kb is used as input image here in the experiment. This image will be converted into frequency domain and then it will be compressed by using SPECK Algorithm. The input test image is of the size 822*446 dimension. after applying lifting wavelength algorithm the image split into different frequency range. This will help to identify where the effected part is and where the normal part in the image is. Then we can locate the exact tumour region(effected region)by using k-means algorithm.it will split into clusters.

SPECK Algorithm help us to perform lossy compression which have a good compression region. When i take the image of 69kb ,got a compression ratio of 1.7360.image size reduced to 24kb.And got 47.6602 as PSNR value.

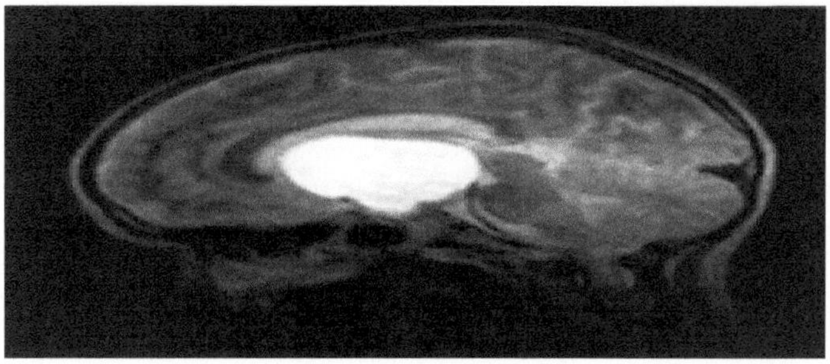

Figure.5: input image

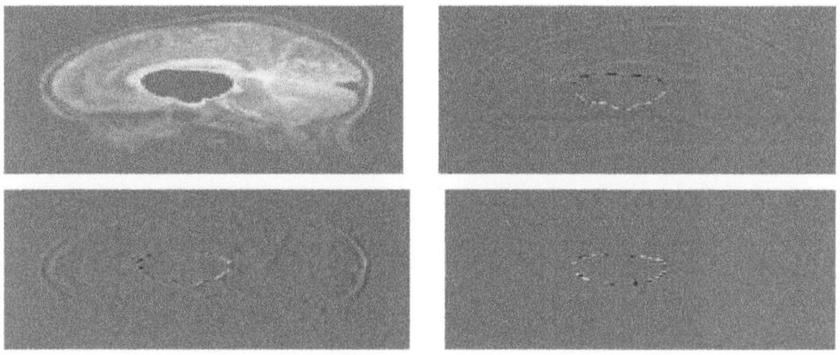

Figure 6:after applying lwt algorithm

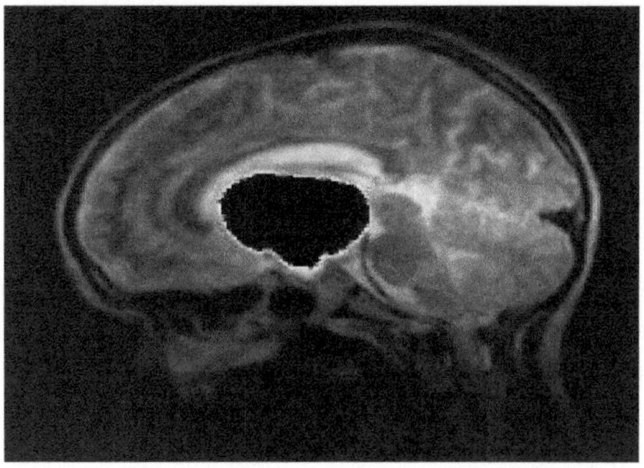

Figure7: Reconstructed image

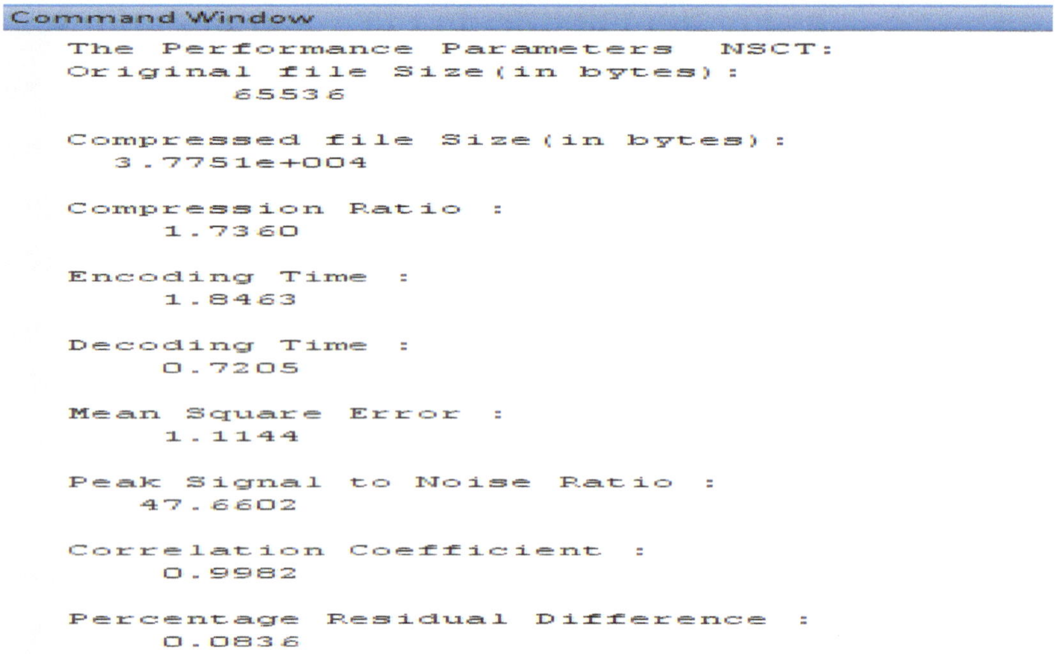

```
Command Window
The Performance Parameters NSCT:
Original file Size(in bytes):
        65536

Compressed file Size(in bytes):
   3.7751e+004

Compression Ratio :
      1.7360

Encoding Time :
      1.8463

Decoding Time :
      0.7205

Mean Square Error :
      1.1144

Peak Signal to Noise Ratio :
     47.6602

Correlation Coefficient :
      0.9982

Percentage Residual Difference :
      0.0836
```

Figure.8: parameter analysis

After compression we got a good compression ratio and MSE value. A good PSNR value, correlation coefficient value could achieve. After combining with lossless compression for tumour affected area by using AAC Algorithm with NSCT technique we can make a detailed comparison with existing methods.

7. Conclusion

This paper presented to provide solutions for efficient region based image compression for increasing the compression ratio with less mean square error at minimum processing time based on NSCT transform with adaptive arithmetic coding, Lifting wavelet transform with set partitioning embedded block coding. This paper heavily utilized for compressing medical images to transmit for telemedicine application. To minimize the information loss, arithmetic entropy coding was used effectively. It will be enhanced by combining speck coding for compressing the secondary region and this hybrid approach was increased the CR and reduce the information loss. Here performance will be analysed through determining the image quality after decompression, compression ratio, correlation and execution time.

8. Acknowledgment

I would like to thank my project guide as well as my project co-coordinator for his valuable help and support. And my head of the department, all the faculties in my department for the support. And I would like to thank all authors of the papers which have been used.

9. References

[1]V.K.Bairagi A.M.Sapkal ,Automated region based hybrid for imaging and communications in medicine magnetic resonance imaging images for telemedicine applications,2012 january.

[2]Abdul Sattar, The medical image compression with embedded zerotree wavelet,2013 February
[3] Calderbank, A.R., Daubechies, I., Sweldens, W., Yeo, B.L.: 'Wavelet transforms that map integers to integers', Appl. Comput. Harmon. Anal., 2006, 5, pp. 332–369
[4] Eva Cavero, A´ lvaro Alesanco, Lena Castro, Jose´ Montoya, Isaac Lacambra, and Jose´ Garc´ıa SPIHT-Based Echocardiogram Compression: Clinical Evaluation and reccomendations of Use, ieee journal of biomedical and health informatics, vol. 17, no. 1, january 2013

[5]Jamuna.M, A.M.Vijaya Prakash, J.Pushpanjali, Low Power VLSI Architecture for Image Compression System Using Discrete Wavelet Transform,2012 june
[6]Mrs. S. Sujatha and Mrs. A. Shanthi Sona, New Fast K-Means Clustering Algorithm using Modified Centroid Selection Method,2013 February

[7] Sayood, K.: 'Introduction to data compression' (Elsevier Publication, 2010, 3rd edn.), pp. 473–512

[8] Zhang, Q., Xiaio, X.: 'Extracting regions of interest in biomedical images'. Int. Seminar on Future BioMedical Information Engineering, 2011, pp. 3–6

[9] Yu, Y., Wang, B.: 'Saliency based compressive sampling for image signals', IEEE Signal Process. Lett., 2010, 17, (11), pp. 973–976 [10] Doukas, C., Maglogiannis, I.: 'Region of interest coding techniques for medical image compression', IEEE Eng. Med. Biol. Mag., 2007, 26, (5), pp. 29–35

[11]Deepak.S.Thomas, M.Moorthi and R.Muthalagu, medical image compression based on automated roi selection for telemedicine application,2014 january.

[12] t. celine therese jenny and g. muthulakshm,

a modified embedded zero-tree wavelet method for medical image compression,2010 november

[14] Zdenek Prusa and Pavel Rajmic, Real-Time lifting wavelet transform algorithm,2011 september.

[15] Zdenek Prusa and Pavel Rajmic, Real-Time lifting wavelet transform algorithm, 2011 september.

[16] Jignesh Sarvaiya, Suprava Patnaik & Hemant Goklani, Image Registration using NSCT and Invariant Moment

[17] Arthur L. da Cunha, Jianping Zhou, Minh N. Do

The Non sub sampled Contourlet Transform: Theory, Design, and Applications, 2006 October

ABOUT AUTHOR

Varsha.C.K [1]was born in Kannur, Kerala in 1991.she received bachelor of engineering degree from GKM College of engineering and technology, Chennai in 2013. Currently doing her master of engineering in applied electronics in Sri Lakshmi Ammaal Engineering college, Chennai.

R.Manoharan[2] received bachelor of engineering degree in electronics and communication engineering from Regional engineering college,Trichy. And completed master of technology degree in VLSI Design technique from Sathyabhama university. He is doing his PhD in image processing. He has been working since 1988.

.

18

THEORETICAL AND EXPERIMENTAL VALIDATION OF PERFORMANCE AND EMISSION CHARACTERISTICS OF NANOADDITIVE BLENDED DIESEL ENGINE

K.Ramesh Babu [1*], R.Bharathi Raja [2]

1 - Associate Professor, SVS College of Engineering - Coimbatore-09, Email: rabakrishnan@gmail.com*
2 - Assistant Professor, Mechanical Engineering, SVS College of Engineering-Coimbatore-09 Email: rbraja23@gmail.com

Abstract

Worldwide energy demand has been growing steadily during the past five decades and experts believe that this trend will continue to rise. An experimental investigation is carried out to establish the performance, emission and combustion characteristics of a diesel engine using diesel fuel and alumina-nanoparticle blended diesel fuel. The nanoparticles of alumina are mixed with the diesel fuel in the mass fractions of 25 ppm to 75 ppm systematically. The investigation is carried out using an experimental set-up consisting of a single-cylinder diesel engine coupled with an eddy current loading device, an AVL DiGas 4000 Light analyzer, an AVL smoke meter, and a data-acquisition system. Theoretical Investigation was carried out by creating a 2D IC Engine model using the ICEM CFD software and solved using the Fluent Software. The blending of nanoparticle reduces viscosity, shortens the ignition delay and increase the atomization which in turn will reduce NOx and increase the efficiency. The experimental results were compared with that of simulated values and a reasonable agreement between them was noticed.

Keywords: Diesel Engine; CFD; Nano; AVL.

1. Introduction

The compression ignition engines are widely used due to its reliable operation and economy. As the petroleum reserves are depleting at a faster rate due to the growth of population and the subsequent energy utilization, an urgent need for search for a renewable alternative fuel arise. Also the threat of global warming and the stringent government regulation made the engine manufacturers and the consumers to follow the emission norms to save the environment from pollution. It is necessary to improve the combustion process to reduce exhaust emissions.

Among the various techniques available to reduce exhaust emissions, the use of fuel-borne catalyst is currently focused due to the advantage of increase in fuel efficiency while reducing harmful greenhouse gas emissions and the health-threatening chemicals such as NOx and particulate matter. The influence of aluminum oxide additive on ultrafine diesel particle emissions and kinetics of oxidation are studied.

Computational Fluid Dynamics is popularly used in different stages of engine design and optimization. Combustion research is more extensive, diverse and interdisciplinary due to powerful modelling tool like CFD. The challenge in using CFD is the complexity of interaction of flow, turbulence, spray and combustion inside the IC engine cylinder. High pressure Injection and modification of combustion chamber geometry were tried to reduce particulate emissions.

In the present work, the stable diesel fuel blends are prepared and the emission reduction potential are investigated using alumina (aluminum oxide) nanoparticles as fuel borne additive with neat diesel on the compression ignition engine. The CFD simulation of a diesel engine combustion process to predict the temperature distribution inside the cylinder and NOx emission with different properties of the pure and blended diesel fuel with Nano alumina. A 2D IC Engine model is created using the ANSA CFD software and solved using the Fluent Software for different propositions of Nano alumina in 25ppm to 75ppm with diesel fuel. The observations of the experimental analysis have been presented.

2. ENGINE PARAMETERS:

- TABLE I
- ENGINE SPECIFICATIONS

Type	Single Cylinder, Vertical, Four Stroke Cycles, Water Cooled Diesel Engine
Bore (mm)	87.5
Stroke (mm)	110
Displacement (cm3)	379
Compression ratio	17.5:1
RPM	1800
HP	7.5
Fuel nozzle injection pressure(bar)	180

3. METHODOLOGY:

1. Experimental Setup and Procedure

The experimental investigations were carried out in two phases. In the first phase, the various physicochemical properties of modified diesel were determined and compared with those of the base fuels. The properties flash, fire points, and viscosity. Standard ASTM test procedures were used in the experiments. In the second phase, extensive performance tests were conducted on a single cylinder compression ignition engine using the modified and base fuels, in order to evaluate the engine performance as well as the emission characteristics using an exhaust gas emission analyzer.

A) Preparation of Modified Fuels (Ultrasonification)

The fuel used for the current investigation is diesel. The viscosity, density, and Calorific value of the diesel are 3.2 cSt at 40°C, 883 kg/m^3 and 42.5 MJ/Kg, respectively. The fuel additive used is Aluminum oxide, in the form of commercially available nanoparticles of size > 50 nanometers and density of 3.97 g/mL. The dosing level of the Aluminum oxide nanoparticle samples (by weight) in the base fuel was varied from 25 to 75 ppm. The required quantity of the nanoparticle sample required for each dosing level was measured using a precision electronic balance and mixed with the fuel by means of an Ultrasonicator (**UP 400S – ULTRASONIC PROCESSOR)** , applying a constant agitation time of 30 minutes to produce a uniform suspension in 2L beaker. Normally Ultrasonification carried out ordinarily with liquid but in this case we are using diesel fuel. Since the beaker gets heated during the process, ice bath was used to cool the beaker. The modified fuel should be used within 8 days, in order to avoid any settling or for sedimentation to occur.

TABLE II.
SPECIFICATION OF ULTRASONICATOR

Rated Voltage	200-400 V
Rated Current	4A
Rated Frequency	50/60 Hz
Amplitude	20% to 100 %

B) Determination of Fuel Properties.

The viscosity, flash and fire points were measured using standard test methods. The viscosity was measured using the Redwood viscometer. A Cleveland open cup flash and fire point apparatus was used for measuring the flash point, and fire point.

Fig .1 Photographic View of Ultrasonicator during Fuel Preparation

C) Instrumentation System

The instrumentation system consists of AVL DiGas 4000 Light analyzer, AVL 437Smokemeter (0-100%), data-acquisition system comprising Kistler piezoelectric pressure transducer and a crank-angle encoder. The DiGas 4000 Light analyzer is used to measure the exhaust emissions such as NO*x*, CO, Unburnt HC and CO2, and the smoke opacity of the exhaust gases is measured by the Smoke meter.

TABLE III.
SPECIFICATIONS OF EXHAUST GAS ANALYZER

Measuring item	Measuring method	Measuring range (vol)	Resolution (vol)
CO	NDIR (Non Dispersive Infra Red)	0 -10 %	0.01%
HC	NDIR	0 - 20,000 ppm	1 ppm
CO_2	NDIR	0-20 %	0.1%
O_2	Electrochemical	0-25 %	0.01%
NOx	Electrochemical	0-5000 ppm	1 ppm

D) Test Procedure

All the tests are conducted by starting and warming up the engine with neat diesel and then switching to modified fuel (nanoparticles-blended diesel fuels). At the end of the test, the engine is made to run with neat diesel to blush out the nanoparticles-blended diesel fuels from the fuel line systems. The experiments are conducted for the tested fuels at a speed range of 1500 – 1900 rpm by varying the loads at a constant injection timing of 26° bTDC.

2. Model Generation and Post processing

A) **Solid Model Generation using Gambit software.**

The analysis has to be started by creating the geometry of the model. There is lot of options through which the geometry of the model is created. This generally involves modeling the geometry with a design software package and approximations of the geometry and simplifications may be required to allow an FLUENT with reasonable effort. In this case the cross section of IC engine is designed 2 dimensionally using ANSA software. The geometry and flow domain are modeled in such a manner as to provide input for the grid generation.

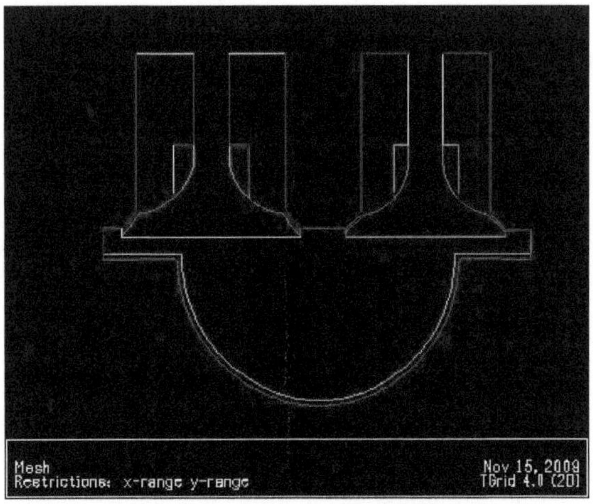

Fig .2. Two-Dimensional model of IC Engine

B) **Meshing of 2D Model**

An ANSA model includes a mesh of nodes and elements. The best way of creating mesh is to create the part's geometry, and then generate a mesh on the geometry. Since the finite element model is associated with the part, any change to the part is automatically reflected in the nodes and elements of the mesh. In mapped mesh the points of the mesh are arranged in a regular way all through the continuum and can be stretched to fit a given geometry where as in free mesh the points fill the space to be considered but is not connected with the regular topology.

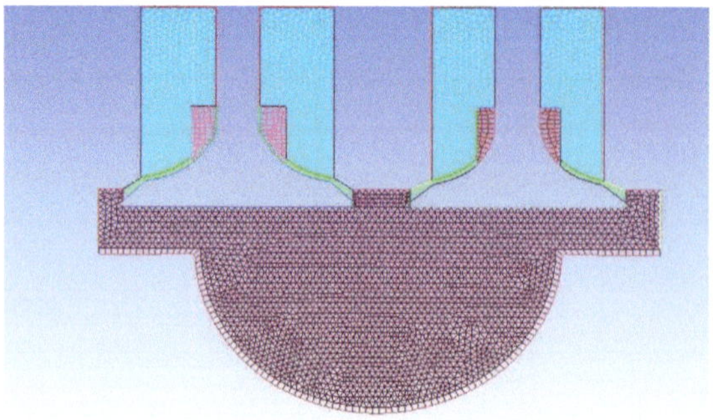

Fig .3. Meshed IC Engine model

C) **Importing the Model to Analyzing Software**

The meshed 2-D IC engine model is than imported to analyzing software such as ANSYS, FLUENT, CFX etc. In this case the FLUENT software is used to analyze the model and hence these designed 2-D IC engine model is imported in FLUENT and further analysis is performed.

D) **Establishing the Intial and Boundary conditions**

Since a finite flow domain is specified, physical conditions are required on the boundaries of the flow domain. The simulation generally starts from an initial solution and uses an iterative method to reach a final flow field solution. The valve timings and temperature of the wall are given as initial and boundary conditions.

E) Performing the Simulation & Monitoring It

The simulation is performed with various possible options for interactive or batch processing and distributed processing. As the simulation proceeds, the solution is monitored until a "converged" solution has been obtained, which is iterative convergence.

F) Post-Processing the Simulation for Results

Post-Processing involves extracting the desired flow properties (temperature, velocity, turbulence, etc...) from the computed flow field.

G) Computational Domain

The 2-Dimensional model shown in figure 2 is used for the analysis. During the solution, as the piston moves, the internal mesh structure deforms automatically to minimize the distortion of each individual cell. Figure 2 and 3 shows the geometry and mesh distribution when piston at a bottom dead center (BDC) and the injection parameters are given in Table III. The starting Crank angle when piston @ BDC is 180^0. For each stroke corresponding crank angle is 180^0c, total for 4 strokes 720^0 CA. Injection starts @ 343^0 CA means, when piston moves from BDC to TDC, the fuel injection starts 17^0 CA before TDC.

TABLE IV
INJECTION PARAMETERS

Type	Direct injection with single central injector
Number of nozzles	2
Total flow rate (g/cycle)	0.05
Start time (CA)	350
Stop time (CA)	370

4. RESULTS AND DISCUSSION:

The ASTM standard tests to determine various physicochemical properties of the base fuels (diesel) as well as the modified fuels were carried out under identical laboratory condition. The primary objectives of this investigation were to determine the variations in the properties of the fuels, due to the addition of the Aluminum oxide (Alumina) nanoparticles and to estimate the effect of the level of inclusion of the additives (dosing level). The combustion performance and the emission characteristics of the diesel engine are investigated. Performance tests were conducted on the diesel engine using the modified fuel samples and compared with those with the base fuels, to determine the engine performance enhancement and the reduction of emissions, due to the addition of the catalyst. The performance parameters, such as BTE (Brake Thermal Efficiency), Brake Specific Fuel Consumption, and the emission characteristics such as NO_x, CO, HC, CO_2 and smoke opacity are plotted against loading condition.

A) Fuel Properties

The flash point of the fuel gives an indication of the volatility of a fuel. The lower the volatility, the higher the flash and fire points. Table IV shows the variation of the flash point of the diesel with modified fuel as a function of the dosing level. As illustrated, the modified fuel shows an increasing trend for the flash point with the dosing level, which indicates a successive decrease in the volatility of the fuel with increases in the quantity of the fuel additive. Higher flash point temperatures are desirable for safe handling of the fuel. The influences of the dosing level of the additive and the temperature on the kinematic viscosity of modified fuel are illustrated in Table IV, which indicates that the viscosity of the fuel decreases with an increase in the temperature for all dosing levels. The change in the viscosity of the fuel affects the engine performance as well as the hydrocarbon emissions. Lower fuel viscosities may not provide sufficient lubrication of fuel injection pumps or injector plungers resulting in leakage or increased wear thus reducing the maximum fuel delivery. This imposes a limitation on the quantity of the fuel additive that can be used to enhance the combustion performance of the fuel. The fuel atomization is affected by the fuel viscosity, and the fuel with higher viscosity tends to form larger droplets on injection, which can cause poor combustion and increased exhaust smoke and emissions.

TABLE V
FUEL PROPERTIES

PROPERTIES	UNITS	DIESEL	DIESEL + 25 PPM AL	DIESEL + 50 PPM AL	DIESEL + 75 PPM AL
Kinematic Viscosity at 40°C,cSt	mm^2/s	896	895	896	896
Density at 15°C	kg/m^3	3.20	3.05	2.98	2.80
Flash Point,	°C	55	57	59.5	62

B) Variation of Cylinder Peak Pressure

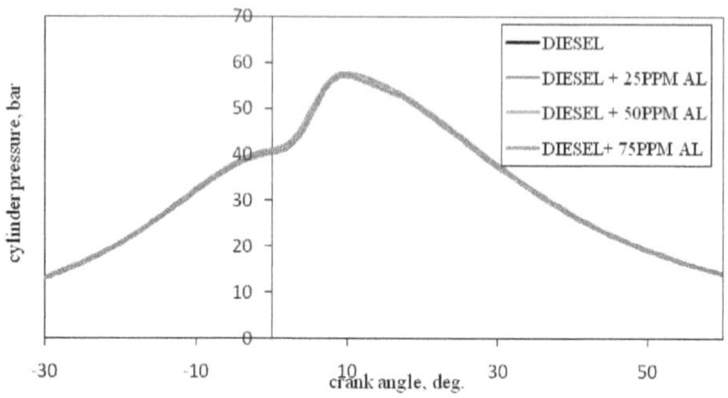

Fig 4: Variation of Cylinder Peak Pressure

The variation of cylinder gas pressure with crank angle is shown in figure 4. It is observed that the peak pressure increases with the addition of aluminum oxide in diesel. The addition of aluminum oxide with neat diesel accelerates early initiation of combustion and the ignition delay decreases, hence more fuel is accumulated in the premixed combustion phase is the cause for faster combustion which results in higher peak pressure. The highest peak pressure is observed as 10.2MPa for the D+75ppm Al blend, whereas is 8.4MPa for neat diesel.

C) Variation of Brake Thermal Efficiency

The variation of Brake Thermal Efficiency with load is shown in figure 5. The results show that the brake thermal efficiency of the diesel engine is improved by the addition of aluminum oxide in the fuel. The Aluminum oxide nanoparticles present in the fuel promote longer and more complete combustion, compared to the base fuel as aluminum oxide acts as an oxygen buffer and thus increases the efficiency.

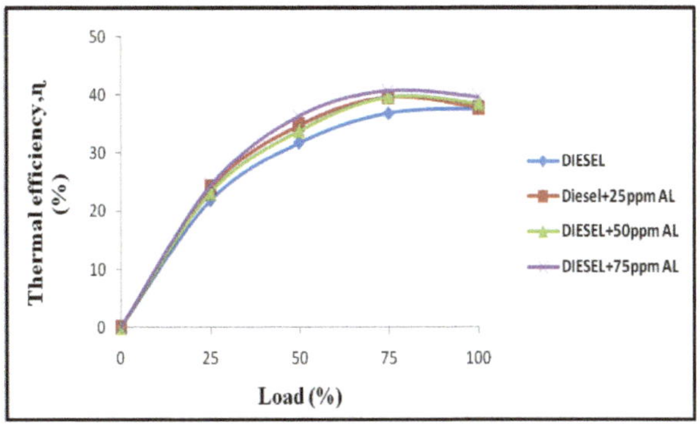

Fig 5: Variation of Brake Thermal Efficiency

It has also been observed that the improvement in the efficiency generally increases with the dosing level of nanoparticles. A maximum brake thermal efficiency was obtained when the dosing level was varied from 25 to 75 ppm, with a maximum improvement observed at a dosing level of 75 ppm. The maximum BTE at the full load is observed as 39.6%.

D) Variation of Brake Specific Fuel Consumption

The SFC is higher for the diesel - alumina nanoparticle blends than neat diesel at all loading condition. The lowest SFC is observed as 0.2240 kg/kW.hr for the D+75ppm Al blend whereas it is 0.23 kg/kW.hr for neat diesel at the load of 75 %. This phenomenon is due to the aluminum oxide addition which promotes combustion.

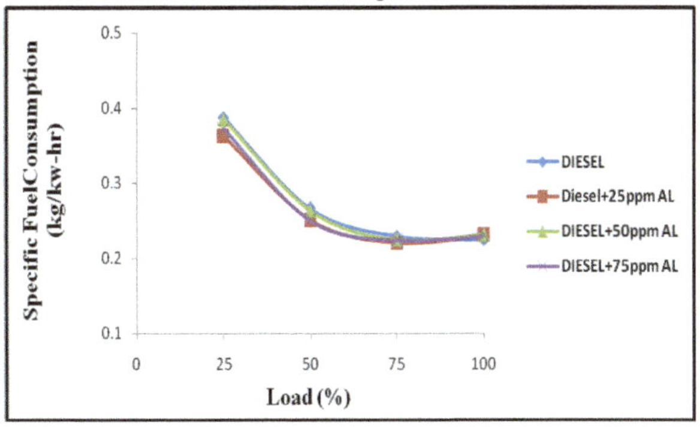

Fig 6: Variation of Brake Specific Fuel Consumption

E) Variation of Nitrogen Oxide (No_x) Emissions

The variation of nitrogen oxide with load is shown in figure 7. The NOx emission is higher for neat diesel when compared to all the fuel blends. The effect of oxygenated additives enhances combustion & The magnitude of NO*x* emission observed is 983 ppm for D+75ppm Al, The reductions in NOx emission is due to complete combustion of modified fuels with the help of catalyst effect of nanoparticle additions which promotes heat transfer in the combustion chamber

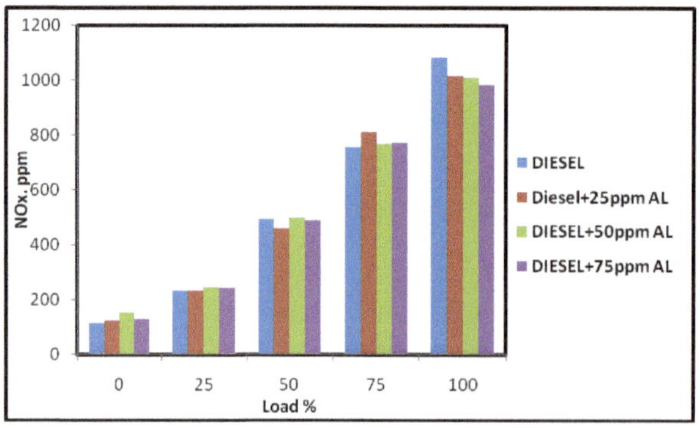

Fig 7: Variation of Nitrogen Oxide (Nox) Emissions

F) Variation of Smoke Opacity

The variation of smoke opacity with engine load is shown in the figure 8. The smoke decreases with diesel – alumina blends when compared with neat diesel. The use of oxygenated fuel improves better combustion is the cause for the smoke reduction. The least smoke is observed as 80 ppm for the D+75ppm Al blend at higher load.

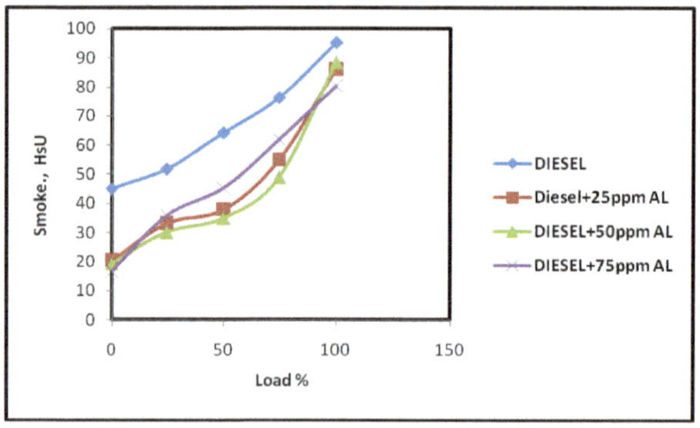

Fig 8: Variation of Smoke Opacity with various blend ratios.

G) Variation of Carbon Monoxide (Co) Emissions

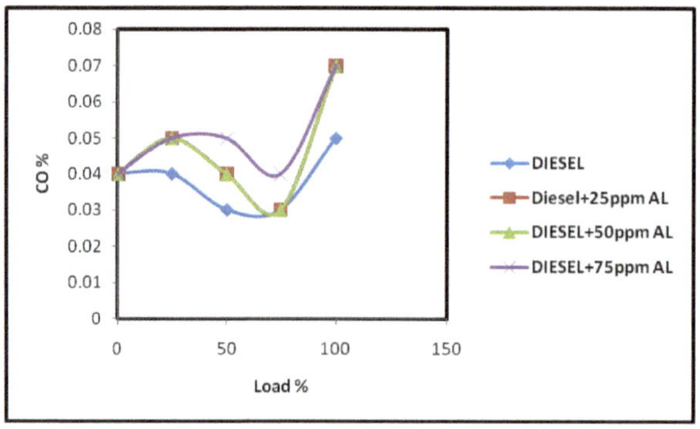

Fig 9: Variation of Carbon monoxide (CO) with various blend ratios

Carbon monoxide (CO) emission of the diesel fuel, which is formed due to incomplete combustion of fuel air-mixture in the combustion chamber was found to decrease with the inclusion of the alumina nanoparticle additives. Figure 9, demonstrates the carbon monoxide (CO) emissions versus engine load with the variation of additive dosage. The carbon monoxide emission increases with the use of diesel – alumina blends than neat diesel. The CO emission is marginal up and then increases rapidly with higher load.

H) Variation of Hydrocarbon (Hc) Emissions

The variation of hydrocarbon emission with engine load is shown in figure 10. It is observed that the level of Unburned Hydrocarbon (HC) emission is higher for diesel – alumina blended fuel compared to base engine, this is because the faster evaporation of the blended fuel hence temperature inside the combustion chamber is lower and it leads to high percentage of Hydrocarbon (HC) emission

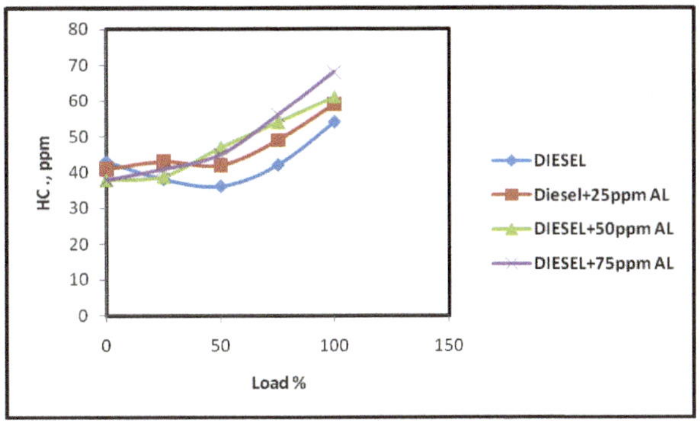

Fig 10: Variation of Hydrocarbon (HC) emission with various blend ratios

I) Variation of Carbondioxide (Co$_2$) Emissions

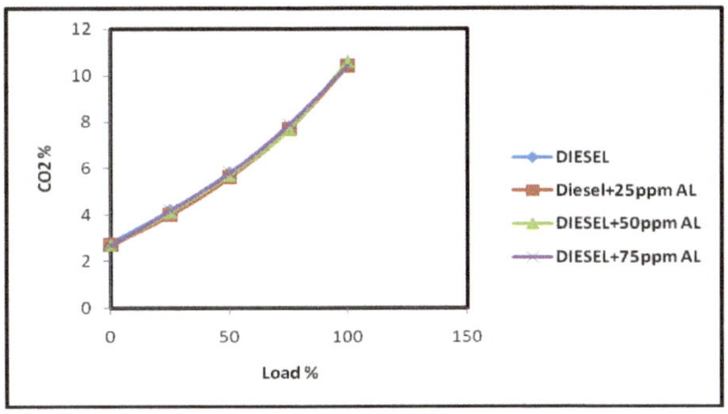

Fig 11: Variation of Carbon dioxide (CO_2) with various blend ratios

The variation of $CO_2$with engine load is shown in the figure 11. It is observed that at all operating conditions the emissions of carbon dioxide is slightly increased while using the diesel – alumina blended fuel compared to conventional diesel fuel. The CO_2 emissions are considered as GHGs (Green House Gases), their higher magnitude indicates the complete combustion in the engine cylinder on par with the reduction of other hazardous emissions such as NO_X, CO, Unburnt HC and Smoke.

J) COMPARISON OF RESULTS

Nox Variations

The NO_X from emission test is 983 ppm at 1800 rpm for 75 ppm Alumina blended diesel. The detailed time history of spray, fuel mass fraction and temperature distributions provided by the CFD simulation are valuable towards gaining a better understanding of the features of combustion for given engine configurations.

The temperature distribution inside the cylinder for different crank angles, and NOx emissions are given above. NOx formation is highly sensitive to temperature and also effected by species concentration. Table VI presents the comparison of simulated and measured emission NOx for various fuel cases. The deviations from experiment and simulation results of NOx emission are around 4.5 to 13.2 %. It was found that the general agreement between prediction and engine test was reasonable for the base and modified fuels.

DIESEL

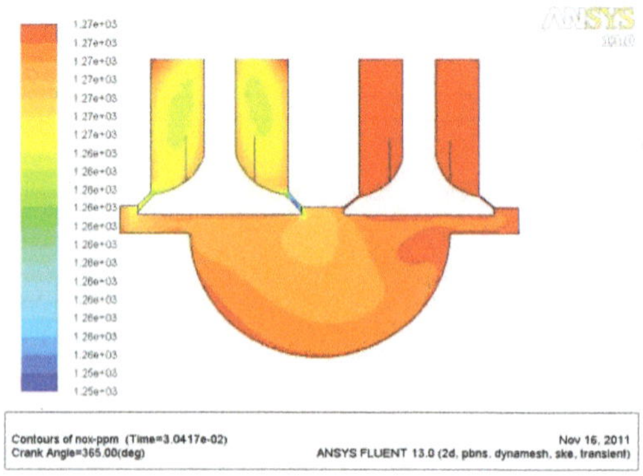

25 PPM ALUMINA BLENDED DIESEL

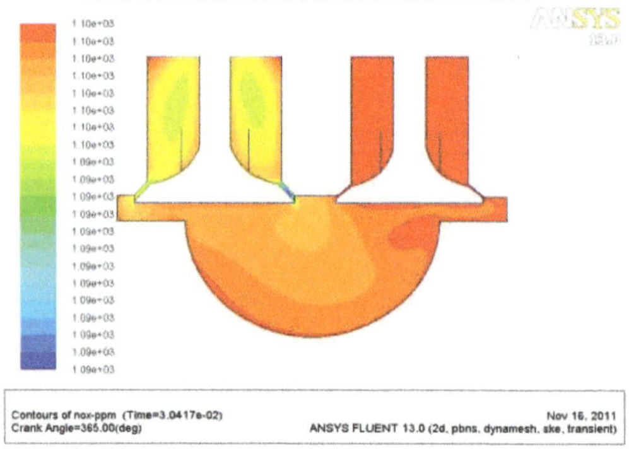

50 PPM ALUMINA BLENDED DIESEL

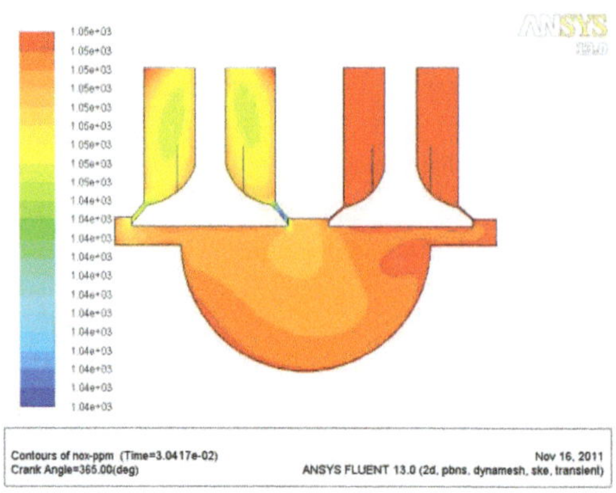

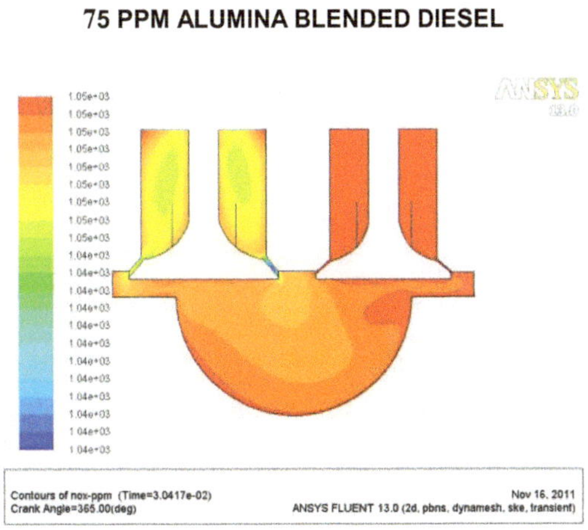

Fig 12 : NO_X Variations at different crank angles for various blend ratios

TABLE VI Comparison of Results

Fuel Used	Experiment	CFD Simulation	% Variations
Diesel	1084	1250 -1270	13.2
25 ppm Alumina Blended Diesel	1016	1090 - 1110	6.7
50 ppm Alumina Blended Diesel	1010	1040 - 1050	4.0
75 ppm Alumina Blended Diesel	983	1030 - 1040	4.5

5. CONCLUSION:

One of the methods to vary the physicochemical properties and combustion characteristics of a hydrocarbon fuel is the use of additives, which are found to be especially effective in nanoparticle form, due to the enhancement of the surface area to volume ratio. Engine performance tests were conducted for diesel and modified fuel by the addition of Aluminum oxide (Alumina) nanoparticles. CFD analyses along with experimental investigations were carried out to compare emissions of diesel and modified fuel. The observations are as follows. The thermal efficiency of modified diesel is higher than pure diesel at all the loads. The NO_X emission is lower for diesel- alumina blends than neat diesel. The smoke opacity decreases with the diesel - alumina blends. The addition of aluminum oxide (alumina) nanoparticles in neat diesel proportionately decreases the smoke further. Temperature profiles as well as NO_X emissions contour from CFD have good agreement with experimentally measured data.The combustion process can be better understood by precise estimation of flame temperature and studying reaction chemistry for using the blended fuel. This is suggested for future work.

REFERENCE

[1].Sadhik Basha, J. and Anand, R.B. (2010a) "Applications of nanoparticle/nanofluid in compression ignition engines – a case study", International Journal of Applied Engineering Research, Vol. 5, pp.697–708.
[2]. Sajith, V., Sobhan, C.B. and Peterson, G.P. (2010) "Experimental investigations on the effects of Cerium oxide nanoparticle fuel additives on biodiesel", Advances in Mechanical Engineering, ID 581407.

[3 Tayfun Ozgur, Mustafa Ozcanli, Kadir Aydin., 2011, "Investigation of nanoparticle additives to biodiesel for improvement of the performance and exhaust emissions in a compression ignition engine", IGEC-VI – 027.

[4] Mohamed Yusuf Ali, M.Masood, S.N.Mehdi., 2010, "A CFD Combustion analysis of Hydrogen – Biodiesel dual fuel system" , World Applied Science Journal 9 (2): 144-150.
[5]. K. Kannan and M. Udayakumar., 2009, "NOx and HC Emission Control Using Water Emulsified Diesel in Single Cylinder Diesel Engine", ARPN Journal of Engineering and Applied Sciences, Vol. 4, No. 8, ISSN 1819-6608.
[6]. R. Bhoobathi , Shaik Amjad , and Dr. R. Rudramoorthy., 2010, "Diesel Engine Combustion Simulation using Computational Fluid Dynamics", Process of International Conference on Advances in Mechanical Engineering : 18-22.

[7]. T.S. Strauss, G.W. Schweimer and U. Ritscher, 1995,
"Combustion in a Swirl chamber Diesel engine Simulation by Computation of Fluid Dynamics," SAE Trans. 104, Section: Journal of Engines, pp.519-530.

[8].C.C.Enweremadu et al, 2011, "Investigation of the relationship between some basic flow properties of shea butter biodiesel and their blends with diesel fuel", International Journal of the Physical Sciences Vol. 6(4), pp. 758-767, 18 February,

[9]. M. T. Naik ,G.Ranga Janardhana, K.Vijaya Kumar Reddy., 2010, "Experimental Investigation into Rheological Property of Copper Oxide Nanoparticles suspended in Propylene Glycol-Water based fluids", ARPN Journal of Engineering and Applied Sciences, Vol. 5, No. 6 : ISSN 1819-6608.
[10] Y. Kidoguchi, M. Sanda and K. Miwa, "Experimental and theoretical optimization of combustion chamber and fuel distribution for the Low Emission Direct – Injection Diesel Engines," Engg for Gas turbines & Power, Vol.125/351, 2003.
[11]. Barman S.C., N. Kumar, R. Singh, G.C. Kisku, A.H. Khan, M.M. Kidwai, R.C. Murthy, M.P.S. Negi, P. Pandey, A.K. Verma, G. Jain, and S.K. Bhargava, 2010, "Assessment of urban air pollution and its probable health impact". J of Envr Biology 31:913-920.
[12]. Selvan A.M.V., Anand, R.B. and Udayakumar, M. (2009) "Effects of cerium oxide nanoparticle addition in diesel and diesel-biodiesel-ethanol blends on the performance and emission characteristics of a CI engine", Journal of Engg and Applied Sciences, Vol. 4, pp.1–6.

[13]. Reitz R.D. and Rootland C.J., 1995, "Development and testing of diesel engine CFD model". Progess. Combustion. Science. vol.21, pp 173196.

[14]. J. Sadhik Basha and R.B. Anand., 2011, "An experimental investigation in a diesel engine using carbon nanotubes blended water-diesel emulsion fuel", Journal of Power and Energy : 225 - 279.

[15]. Kao, M. J., Lin, B. F., and Tsung, T. T., "The Study of High-Temperature Reaction Responding to Diesel Engine Performance and Exhaust Emission by Mixing Aluminum Nanofluid in Diesel Fuel," 18th Internal Combustion Engine Symposium Jeju, Korea, 2005.

[16]. V.J.Mohanraj and Y Chen, 2006, "Nanoparticles – A Review", Tropical Journal of Pharmaceutical Research, 5 (1): 561-573.

[17]. Usman Asad, Raj Kumar, Xiaoye Han, Ming Zheng., 2011, "Precise instrumentation of a diesel single-cylinder research engine", Measurement
44: 1261–1278.

[18]. G.D'Errico, D. Ettorre, T. Lucchini., 2007, "Comparison of Combustion and Pollutant Emission Models for DI Diesel Engines", SAE Technical Paper 07NAPLES - 90.

[19]. K.M.Pandey, D.H.Das and B. Acharya., 2010, "Effects of Variation of Specific Heat on Temperature in Gaseous Combustion with Fluent Software", International Journal of Environmental Science and Development 5: 419-422.

[20]. Yanan Gan and Li Qiao., 2011, "Combustion characteristics of fuel droplets with addition of nano and micron-sized aluminum particles", Combustion and Flame 158

Structural & Mode Shape Analysis Of Roller Conveyor Using FEA

Rajratna A. Bhalerao [1], Dr. R.J. Patil [2]

PG student, Dr. D.Y.Patil Inst. of Engineering & Tech, Ambi,Pune, India. [1] rajbhalerao1988@gmail.com[1]

Principal, Dr. D.Y.Patil Inst. of Engineering & Tech, Ambi,Pune, India. [2]

Abstract

Over the years a lot of work has done and is still continuing with great effort to save weight and cost of applications. The current trend is to provide weight/cost effective products which meet the stringent requirements. The aim of this paper is to study existing conveyor system and optimize the critical parts like roller, shafts, C-channels for chassis and support, to minimize the overall weight of assembly and material saving.

Keywords: Optimized design, APDL programming, material handling systems.

1. Introduction

1.1 Conveyors

A conveyor system is a common piece of mechanical handling equipment that moves materials from one location to another. Conveyors are especially useful in applications involving the transportation of heavy or bulky materials. Conveyor systems allow quick and efficient transportation for a wide variety of materials, which make them very popular in the material handling and packaging industries. Many kinds of conveying systems are available, and are used according to the various needs of different industries. There are chain conveyors (floor and overhead) as well. Chain conveyors consist of enclosed tracks, I-Beam, towline, power & free, and hand pushed trolleys. Conveyor systems are used widespread across a range of industries due to the numerous benefits they provide.

- Conveyors are able to safely transport materials from one level to another, which when done by human labor would be strenuous and expensive.
- They can be installed almost anywhere, and are much safer than using a forklift or other machine to move materials.
- They can move loads of all shapes, sizes and weights. Also, many have advanced safety features that help prevent accidents.

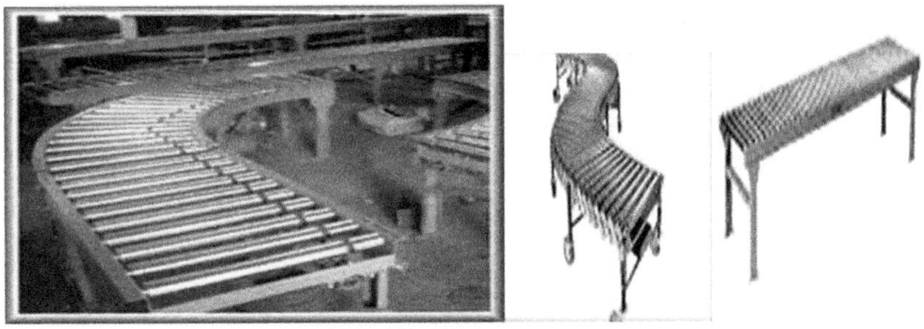

Fig. 1.1 Conveyor System

There are a variety of options available for running conveying systems, including the hydraulic, mechanical and fully automated systems, which are equipped to fit individual needs.

Conveyor systems are commonly used in many industries, including the automotive, agricultural, computer, electronic, food processing, aerospace, pharmaceutical, chemical, bottling and canning, print finishing and packaging. Although a wide variety of materials can be conveyed, some of the most common include food items such as beans and nuts, bottles and cans, automotive components, scrap metal, pills and powders, wood and furniture and grain and animal feed. Many factors are important in the accurate selection of a conveyor system. It is important to know how the conveyor system will be used beforehand. Some individual areas that are helpful to consider are the required conveyor operations, such as transportation, accumulation and sorting, the material sizes, weights and shapes and where the loading and pickup points need to be.

1.2 Types of Conveyor Systems

- Gravity Conveyor systems
- Powered Belt Conveyor systems
- Pneumatic conveyor systems
- Vibrating conveyor systems
- Flexible conveyor systems
- Vertical conveyor systems and spiral conveyors
- Live Roller Conveyor systems

2.0 Problem Definition

The aim of this project is to redesign existing gravity roller conveyor system by designing the critical parts (Roller, Shaft, Bearing & Frame), to minimize the overall weight of the assembly and to save considerable amount of material.

Gravity roller Conveyor has to convey 350 kg load, 30 inch above ground and inclined at 4 degree. Fig. 2.1 shows roller conveyor assembly.

Components of conveyor are as follows,

Sr. No.	Component	Material	Qty.
1	C-Channels for Chassis	ISMC 100	2
2	Rollers	Mild Steel	15
3	Bearing	Std.	30
4	C-Channels for Stand	ISMC 100	4
5	Shaft	Mild Steel	15

Design roller conveyor to reduce weight.

3.0 Objective of the Study

The following are the objectives of the study:

1. Study existing roller conveyor system and its design.
2. Geometric modeling of existing roller conveyor.
3. To generate parametric model using ANSYS Parametric Design Language (APDL) program.
4. To carry out linear static, modal, transient and optimization analysis of existing roller conveyor.
5. Modification of critical conveyor parts for weight optimization.
6. To carry out Analysis of Modified design for same loading condition.
7. Recommendation of new solution for weight optimization.

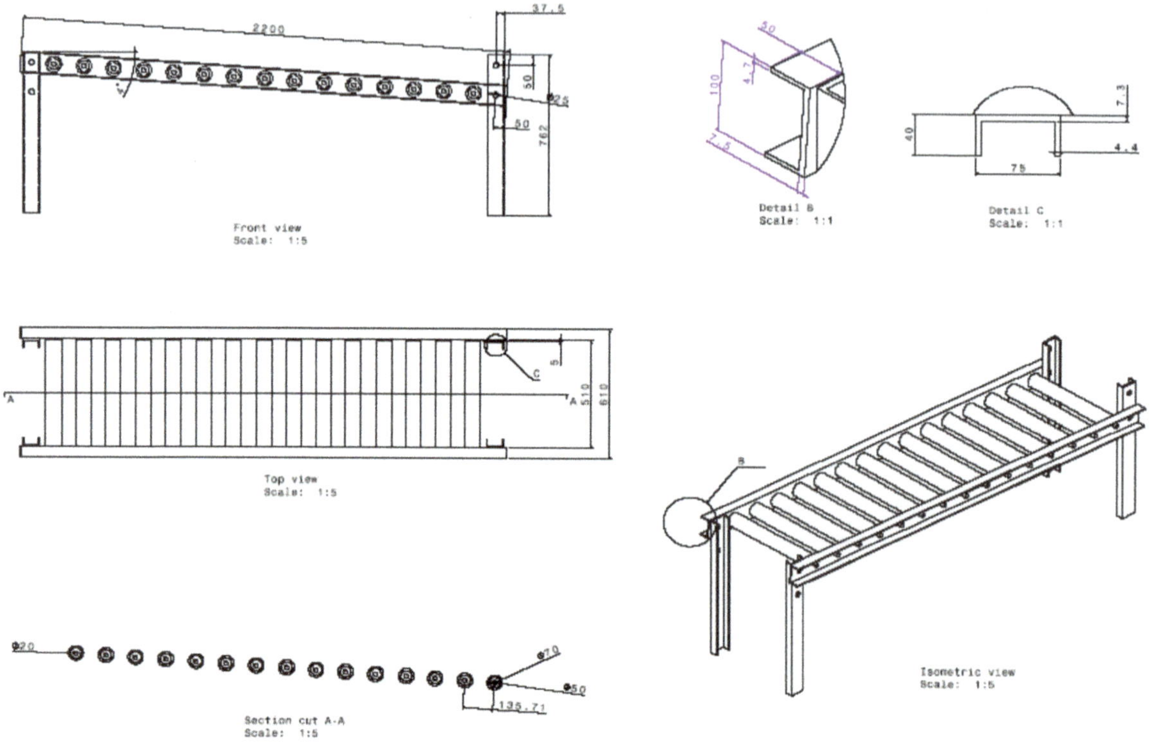

Fig. 2.1 Gravity Roller Conveyor Assembly

4.0 Scope of Present Study

1. Check design of existing conveyor system.
2. Simulation method applied to optimize parameters web thickness, flange thickness, web height, roller thickness and shaft diameter.
3. ANSYS APDL codes applied for linear static, modal, transient and optimization analysis.
4. 150 simulations for linear static Analysis.
5. 150 simulations for Modal Analysis.
6. Optimization of conveyor assembly for weight reduction.
7. Comparison between existing and optimized design.

5.0 Study of the Existing Assembly of Conveyor System

5.1 Total Weight of Existing Conveyor Assembly

Sr.No.	Name of Component	Weight (Kg)
1	C- Channel for Chassis	36.7548
2	Rollers	111.1181
3	Shafts	20.7421
4	Bearing	2.994
5	C- Channel for Supports	20.81
	Total weight of assembly	**192.419 kg**

5.2 Geometric Modeling

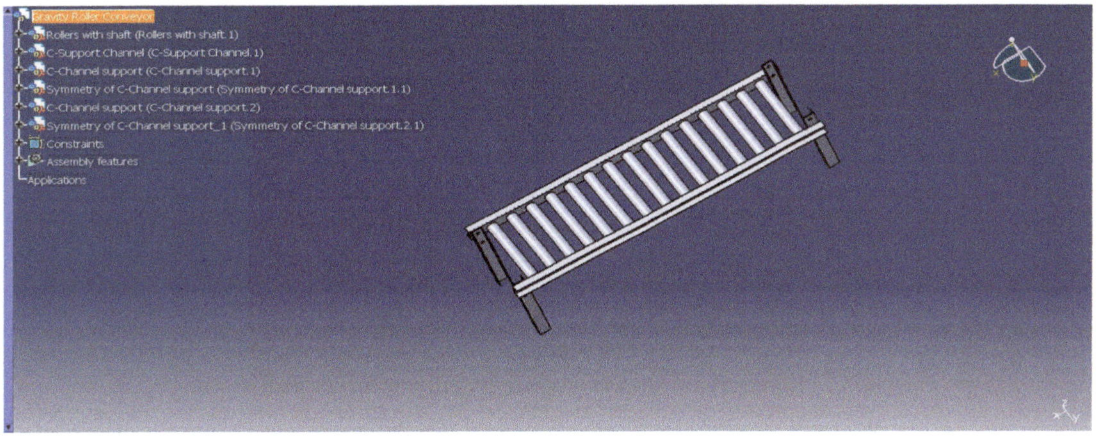

Fig. 5.1 Geometrical modeling using Catia

5.3 Finite Element Modeling

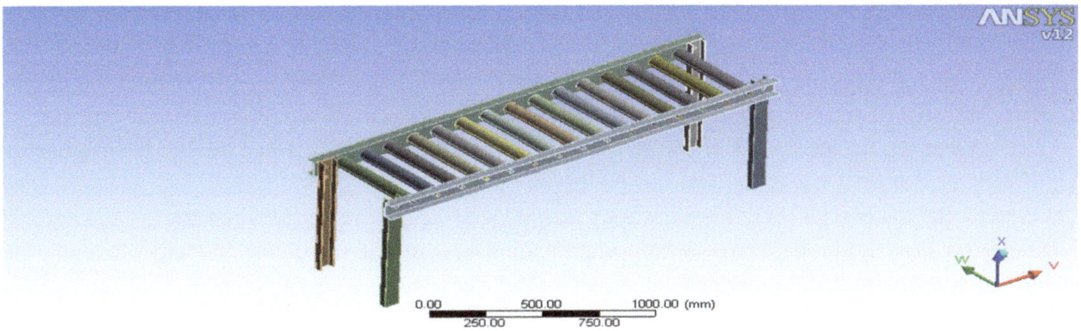

Fig. 5.2 Geometrical modeling using ANSYS 14.5

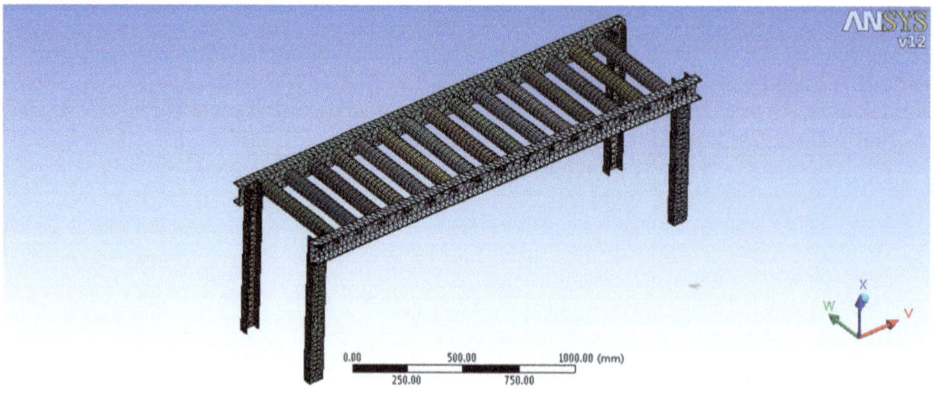

Fig. 5.3 Finite element mesh of the model

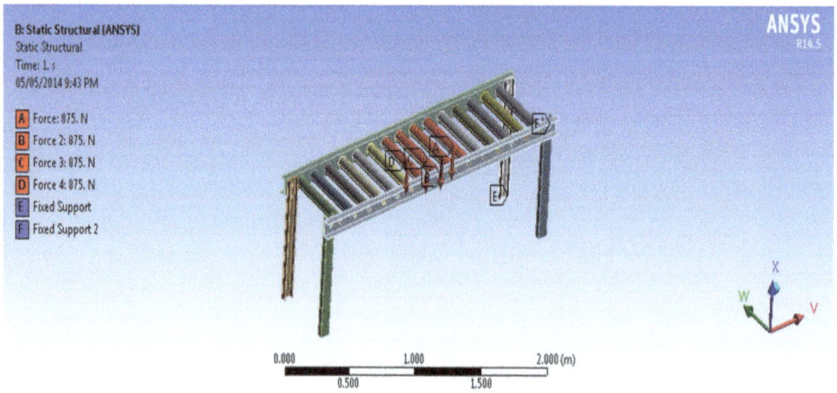

Fig. 5.4 A static load of 3500 N (approx 350 kg) is applied on the 4 rollers at the centre, as the deflection will be maximum, when the load is applied at the centre.

5.4 Static Structural Analysis

A static analysis calculates the effect of steady loading condition on a structure, while ignoring inertia and damping effects, such as those caused by time varying loads.

A static analysis can, however, include steady inertia load (such as gravity and rotational).

Design and analysis of roller conveyor for weight optimization & material saving (velocity), and time varying load that can be approximated as static equivalent loads (such as static equivalent wind and seismic loads commonly defined in many building codes). Select element and apply material properties. Static analysis determines the displacements, stresses, strains, and forces in structures or components caused by lodes that do not induce significant inertia and damping effects. Steady loading and response conditions are assumed; that is, the loads and the structure's response are assumed to vary slowly with respect to time.

Critical load condition-

Load act on any four rollers hence by considering 350 kg load act on four rollers maximum deflection, maximum stress values are checked for existing design.

5.5 Procedure of Static analysis consists of these tasks

1. Build the Model
2. Set Solution Controls
3. Set Additional Solutions Options
4. Apply the loads
5. Solve the Analysis
6. Review the Results

Results for static analysis,

-Weight = the weight of the model is 203.1 kg

- Maximum deflection plot shown in fig 5.5
- Maximum stress plot

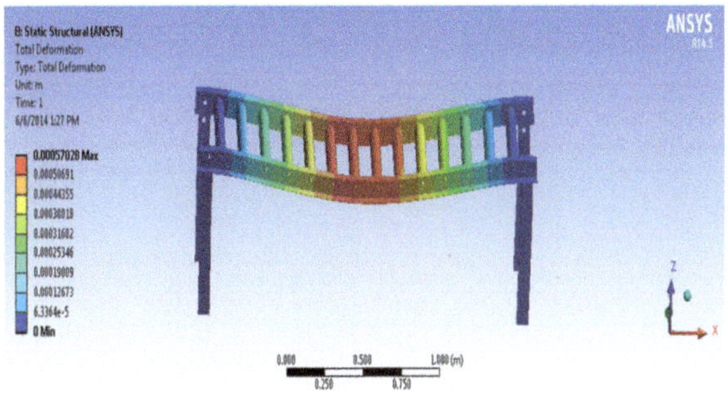

Fig. 5.5 Deflection plot

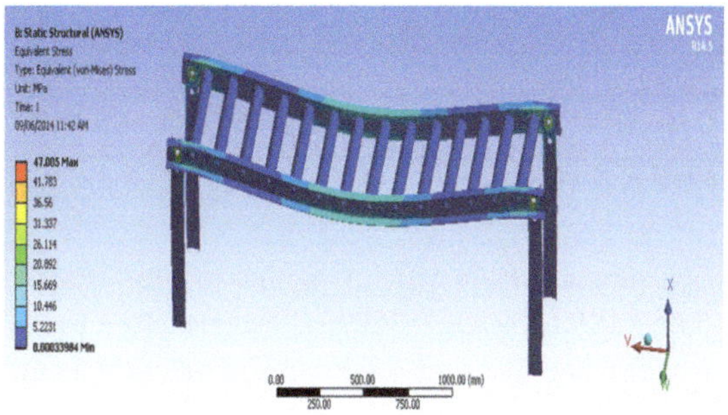

Fig. 5.6 Stress Plot

5.5 Modal analysis

- Modal analysis is carried out to find natural frequency and mode shapes.
- As the loading will be in vertical direction (gravity) the mode shape which will show movement in vertical direction is important.

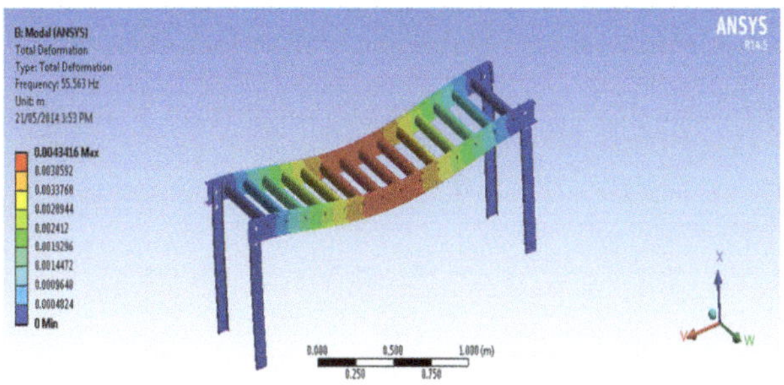

Fig. 5.6 Critical Mode Shape

Result from Modal analysis-

- From the results it is clear that the third mode shape will have maximum motion in vertical direction. So third natural frequency should be greater than the excitation frequency.
- Third natural frequency is 54.34 Hz.

6.0 Need of Optimization

As factor of safety of C-Channels and Rollers is very high there is scope of weight reduction in this component.

6.1 Selection of Critical Parameter

- Flange Width
- Flange thickness
- Web height
- Web thickness
- Roller Outer diameter
- Roller thickness

7.0 Optimized design

- Selecting available which is similar to optimized design.
- Select ISJC 100 and ISJC 75 C-channels for chassis and supports respectively
- Roller Outer diameter is 60 mm and roller thickness 5 mm

Sr No	Name of Component	Max. Bending Moment (Nm)	Max. Bending Stress N/mm^2	Deflec-tion (mm)	Actual Factor of Safety
1	C- Channel for Chassis	858.375	34.139	1.108	14.1389
2	Rollers	26.8242	9.95	1.605	59.2955
3	C- Channel for Supports	224.256	8.1567	0.0501	31.8353

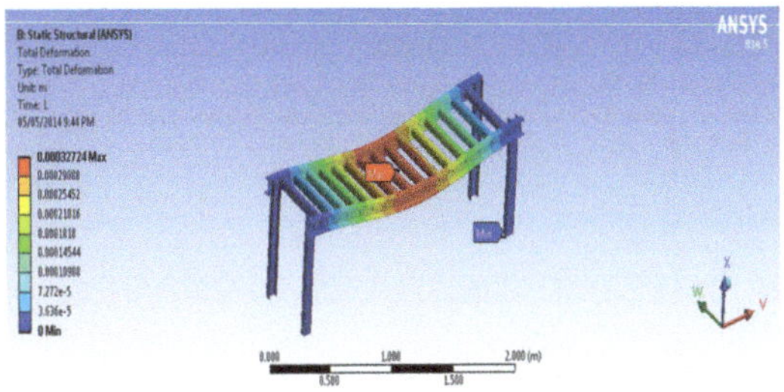

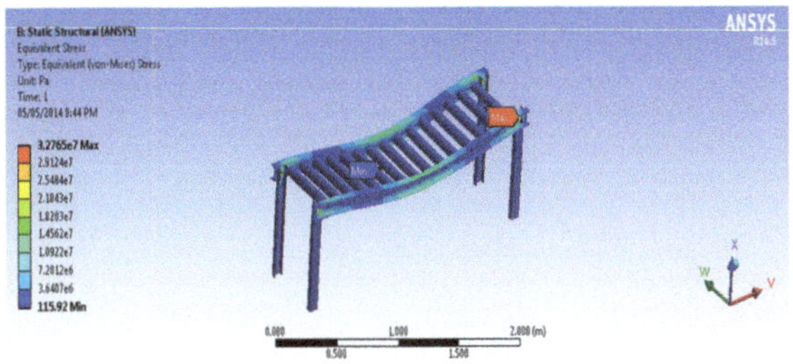

Fig 7.1 Linear Static Analysis of Optimized design

Deflection plot Fig 7.2 Stress Plot

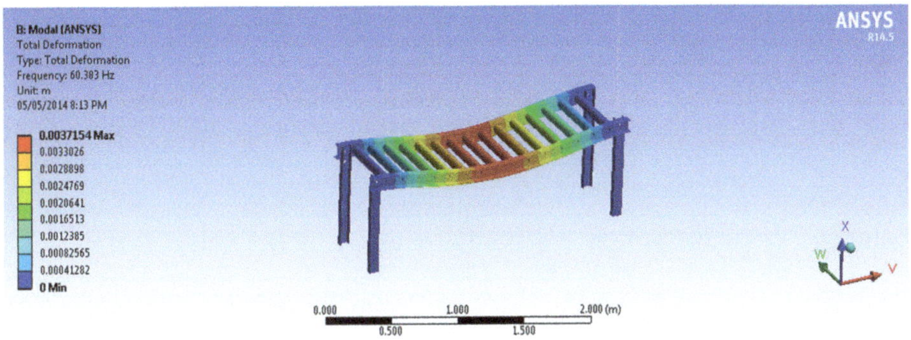

Fig. 7.3 Critical Mode Shape of Optimized design

8.0 Results- Effect of Optimized Design:

Sr. No.	Name of Component	Weight (Kg)
1	C-Channels for Chassis	25.020
2	Rollers	71.96
3	Shafts	20.7421
4	Bearings	2.994
5	C-channels for Supports	12.187
	Total Weight of Conveyor	**132.9031**

Design	Max. Def (mm)	Natural Freq. (Hz)	Max. Stress (N/mm^2)
Existing	0.3272	60.383	32.762
Optimized	0.87	55.563	47.005

8.1 Weight reduction due to Optimization

Design	Weight (Kg)	% Material required compared To Existing design	% Material save compared To Existing design
Existing	192.419	100	--
Optimized	132.9031	69.069	30.931

9.0 Conclusions

- Existing design calculation shows the factor of safety is very greater than requirement and there is a scope for weight reduction.
- Critical parameter which reduces the weight is C-channels, roller outer diameter and roller thickness.
- Though value of deflection, stress is more in case of Optimized design, but it is allowable.
- 30.931 % of weight reduction due to Optimized design.

10.0 References

[1]. M. A. Alspaugh, "Latest Developments in Belt Conveyor Technology" MINExpo 2004, Las Vegas, NV, USA. September 27, 2004

[2]. S.H. Masood · B. Abbas · E. Shayan · A. Kara "An investigation into design and manufacturing of mechanical conveyors Systems for food processing", Springer-Verlag London Limited 2004

[3]. Dima Nazzal , Ahmed El-Nashar "Survey Of Research In Modeling Conveyor-Based Automated Material Handling Systems In wafer fabs" Proceedings of the 2007 Winter Simulation Conference.

[4]. Chun-Hsiung Lan, "The design of a multi-conveyor system for profit maximization" International Journal Adv Manuf Technol (2003) 22: 510–521.

[5]. John Usher, John R , G. Don Taylor "Availability modeling of powered roller conveyors".

[6]. Espelage W, Wanke E."Movement minimization for unit distances in conveyor flow shop processing",

[7]. C.Sekimoto "Development of Concept Design CAD System", Energy and Mechanical Research Laboratories, Research and Development Center, Toshiba Corporation.

[8]. Ying WANG, Chen ZHOU "A Model and an analytical method for conveyor system in distribution centers", J Syst Sci Syst Eng (Dec 2010) 19(4): 408-429.

[9]. R. Long, T. Rom, W. H¨ansel, T.W. H¨ansch, J. Reichel "Long distance magnetic conveyor for precise positioning of ultra cold atoms" Eur. Phys. J. D 35, 125–133 (2005).

A Brief Author Biography

Author 1 : M.E. Mechanical Design.

Author 1 : PHD

20

Numerical Scheme for 2-Dimension Optimum Length Nozzle Contour design imposing Viscous Correction to Method of Characteristics

Premchand C P[1], Ramanathan V[2]

[1] *Research Associate, DRDO-GTRE, Bangalore, Karnataka -560093.*
[2] *Post Graduate, Indian Institute of Science, Bangalore, Karnataka – 560012.*

[Premchand C P - Research Associate, DRDO-GTRE, Bangalore, Karnataka -560093, prem.tech123@gmail.com]

Abstract

A new numerical method is proposed to design optimum length nozzle contours imposing viscous correction to Method of Characteristics. Proposed method utilizes the information from the characteristic lines to design the contour which has wide applications in the field of Aerospace and Aeronautics. Nozzle contours for both circular and rectangular cross sections are developed and various numerical experimentations were performed to analyze the performance of the scheme proposed. A comparative study on optimum number of characteristics required for optimum design is presented. 3D contour design will be carried out as future research.

***Keywords**:* Optimum length nozzle, Method of characteristics, Contour design, 2D nozzles, Characteristic lines.

1. Introduction

Method of characteristics forms the major branch in the numerical techniques used to analyze steady supersonic flows. The keystone of theoretical fluid dynamics is a set of conservation equations that describe the physics of fluid motion; These set of conservation equation states that Mass is conserved (Continuity Equation), Momentum is conserved (Navier stokes Equation), Energy is conserved (Energy Equation). They describe the variation in the fluid pressure, temperature, density, velocity throughout the entire flow field both in space and time. Since the time of formulation of these general conservation equations, attempt to find an analytical solution were begun. Though analytical solutions were formulated for special cases of flows like flow over thin airfoil, flow over a flat plate, flow over round corner etc., a general solution to the entire equation was never formulated.

The method of characteristics is an older and more developed, and is limited to inviscid flows, whereas finite-difference techniques are still evolving as computational fluid dynamics grows and matures, and have much more general application to inviscid and viscous flows. So, a new numerical scheme was developed for Supersonic nozzle contour design using method of characteristics was developed.

2. Formulation of method of characteristics

Though the method of characteristics is limited to Inviscid flows, Continuity and Momentum equations are alone required to predict the parameters in the grid points. By combining both these equations, the velocity potential equation is given as,

$$\left(1-\frac{\Phi_x^2}{a^2}\right)\Phi_{xx}+\left(1-\frac{\Phi_y^2}{a^2}\right)\Phi_{yy}+\left(1-\frac{\Phi_z^2}{a^2}\right)\Phi_{zz}$$

$$-\frac{2\Phi_x\Phi_y}{a^2}\Phi_{xy}-\frac{2\Phi_x\Phi_z}{a^2}\Phi_{xz}-\frac{2\Phi_y\Phi_z}{a^2}\Phi_{yz}=0$$

Flow field properties for any grid shown below are determined based on their ability to be expanded using Taylor series as follows,

$$u_{i+1,j}=u_{i,j}+\left(\frac{\partial u}{\partial x}\right)_{i,j}\Delta x+\left(\frac{\partial^2 u}{\partial x^2}\right)_{i,j}\frac{(\Delta x)^2}{2}+\cdots$$

Neglecting the higher order terms, the expansion series can be written as,

$$u_{i+1,j}=u_{i,j}+\left(\frac{\partial u}{\partial x}\right)_{i,j}\Delta x$$

The value of the derivative can be obtained from the general conservation equations. For example, consider a two-dimensional Irrotational flow, so that the velocity potential equation yields, in terms of velocities,

$$\left(1-\frac{u^2}{a^2}\right)\frac{\partial u}{\partial x}+\left(1-\frac{v^2}{a^2}\right)\frac{\partial v}{\partial y}-\frac{2uv}{a^2}\left(\frac{\partial u}{\partial y}\right)=0$$

Solving for $\partial u/\partial x$ gives:

$$\frac{\partial u}{\partial x}=\frac{\left[\frac{2uv}{a^2}\left(\frac{\partial u}{\partial y}\right)-\left(1-\frac{v^2}{a^2}\right)\frac{\partial v}{\partial y}\right]}{\left(1-\frac{u^2}{a^2}\right)}$$

Now, assume that we know the velocity V and its components (u,v) along a vertical line (say inlet). Thus we know the value of $u_{i,j}$ By finding the gradient in the x-direction from the above equation the value of $u_{i+1,j}$ can be found using Taylor series expansion about the point (i,j). However there is one notable exception, if the denominator in the above equations tends to zero, the derivative does not exist. The determinant is zero when u=a, i.e., when the component of flow velocity perpendicular to inlet is sonic. The line that makes a Mach angle with respect to the streamline direction at a point is also a line along which the derivative of u is indeterminate, and across which it may be discontinuous. These lines exist, and they are called the CHARACTERISTIC LINES or Mach Lines.

Let us move towards the supersonic 2d inviscid Irrotational flow in the divergent section of the nozzle. The problem can be solved in 3 steps.

2.1 Algorithm:

1. Find some particular lines (directions) in the x y space where flow variables (p, T, u, v, ρ) are continuous, but along which the derivatives ($\partial u/\partial x$, $\partial p/\partial x$ *etc.*) are indeterminate, and in fact across which the derivatives may even sometimes be discontinuous. As already defined, such lines in the xy space are called characteristic lines.

2. Combine the partial differential conservation equations in such a fashion that ordinary differential equations are obtained that hold only along the characteristic lines. Such ordinary differential equations are called the compatibility equations.
3. Solve the compatibility equations step by step along the characteristic lines, starting from the given initial conditions at some point or region in the flow. In this manner, the complete Flow field can be mapped out along the characteristics. In general, the characteristic lines depend on the Flow field, and the compatibility equations are a function of geometric location along the characteristic lines' hence, the characteristics and the compatibility equations become algebraic equations explicitly independent of geometric location.

2.2 Determination of characteristic lines

Consider steady, adiabatic, two-dimensional, Irrotational supersonic flow. The governing non-linear equations are as follows,

$$\left(1-\frac{\Phi_x^2}{a^2}\right)\Phi_{xx}+\left(1-\frac{\Phi_y^2}{a^2}\right)\Phi_{yy}-\frac{2\Phi_x\Phi_y}{a^2}\Phi_{xy}=0$$

$$d\Phi_x=\frac{\partial\Phi_x}{\partial x}\,dx+\frac{\partial\Phi_y}{\partial y}\,dy=\Phi_{xx}\,dx+\Phi_{xy}\,dy$$

$$d\Phi_y=\frac{\partial\Phi_y}{\partial x}\,dx+\frac{\partial\Phi_y}{\partial y}\,dy=\Phi_{xy}\,dx+\Phi_{yy}\,dy$$

Where, $\Phi_x = u$; $\Phi_y = v$

This provides us a means to calculate the equations of the characteristic lines.

$$\begin{vmatrix} 1-\frac{u^2}{a^2} & -\frac{2uv}{a^2} & 1-\frac{v^2}{a^2} \\ dx & dy & 0 \\ 0 & dx & dy \end{vmatrix}=0$$

The above system yields the slope of the characteristic line as follows since it is a quadratic equation in slope of the characteristic line.

$$\left(\frac{dy}{dx}\right)_{char}=\frac{-\frac{uv}{a^2}\pm\sqrt{M^2-1}}{\left(1-\frac{u^2}{a^2}\right)}$$

Tthe supersonic flows are hyperbolic in nature and two characteristics runs at every point in the flow, this makes the method of characteristics a practical technique for solving supersonic flows. In contrast, since the characteristics are imaginary the method of characteristics is not used for subsonic solutions.

Since the contour design problem for a supersonic nozzle is a 2 dimensional, steady, supersonic flow problem. Method of characteristics is the tranquil way of determining the divergent portion of supersonic nozzle. At any point u = V cos(θ), v = V sin(θ). Hence the slope of the characteristic line can be said to be equal to

$$\left(\frac{dy}{dx}\right)_{char} = \tan(\theta \mp \mu)$$

2.3 Compatibility equation

$$d\theta = \mp\sqrt{M^2 - 1}\frac{dV}{V}$$

This equation is called as the compatibility equation. Two values of the $d\theta$ are obtained among which the negative one corresponds to the negative characteristic line or left running characteristic and so on. The above equation is similar to Prandtl Meyer expansion flow equation, so they can be integrated to give the Prandtl Meyer function $\nu(M)$. Therefore they can be replaced by algebraic compatibility equation.

$\theta + \nu(M) = const = K_-$ (along the C_- characteristic)

$\theta - \nu(M) = const = K_+$ (along the C_+ characteristic)

The above equations relate the velocity magnitude and the direction along the characteristic lines. For this reason, they are sometimes identified in the literature as 'hodograph characteristics'. Since the compatibility equations are independent of the spatial coordinates x and y. Hence, they can be solved without requiring knowledge of the geometric location of the characteristic lines. This is peculiar case applicable only to two_ dimensional Irrotational flow.

3 Viscous correction to inviscid method of characteristics

Since the Method of Characteristics is only applicable for Inviscid flows, a viscous correction is required to account for Viscous induced Effects.

Due to the viscous effects, the effective turn angle at the wall is reduced, thereby reducing the expansion angle, where a viscous correction angle is introduced as follows,

$$\theta_{viscous} \pm \nu(M) = const$$

$$\theta_{viscous} = \theta - \delta\left(\left(\frac{\mu}{\rho VL}\right), x\right)$$

Where,

δ – Correction function for viscous effects

μ – Dynamic Viscosity

Now assuming Laminar flow inside nozzle, and boundary layer growth and other viscous related effects occur only in the wall of nozzle, correction function can be approximated to displacement boundary layer thickness. Thus the Viscous correction is made only in the wall grid points. Final expression for viscous correction function can be stated as (from Blasius Solution for Flat Plates).

$$\delta(Re, x) = \frac{5x}{\sqrt{\frac{\mu}{\rho V L}}}$$

This assumption leads to an approximation, that consecutive grid points are assumed to be joined by linear lines (flat plates) and the distance x varies from 0 to Δx, where Δx is the grid size.

Finally the Viscous Corrected Method of Characteristics Scheme can be summarized as follows

On wall Grid points	$\theta_{viscous} \pm \nu(M) = const$
On internal Grid points	$\theta \pm \nu(M) = const$

4 Numerical experiments

Proposed Scheme is tested for various inlet conditions at throat of nozzle, and for various number of characteristic line designs. One of the results for a specific Test condition is presented in this paper. 10, 20 and 100 characteristic designs for the Throat Inlet condition given below are presented.

Throat Inlet Condition

Temperature	: 1200 K
Pressure	: 1.2E6 bar
Ambient Pressure	: 1.01325E5 bar
Ambient Temperature	: 288.16 K
Specific Ratio	: 1.4
Molecular Mass of gas	: 25.4
Number of Characteristics	: Chosen (10,20,100)
Initial Theta	: 0.03 radian
Area of the Throat	: 0.1 m^2

5 Results and discussions

Figure 1, shows the 2D contour design using viscous corrected scheme of Method of Characteristics. Approximately all the three designs produce contour with exit area 0.58 sq. m. Thus the scheme achieves grid independence for even less number of characteristic lines (i.e. very less number of characteristics lines are adequate to achieve high accurate contour design).

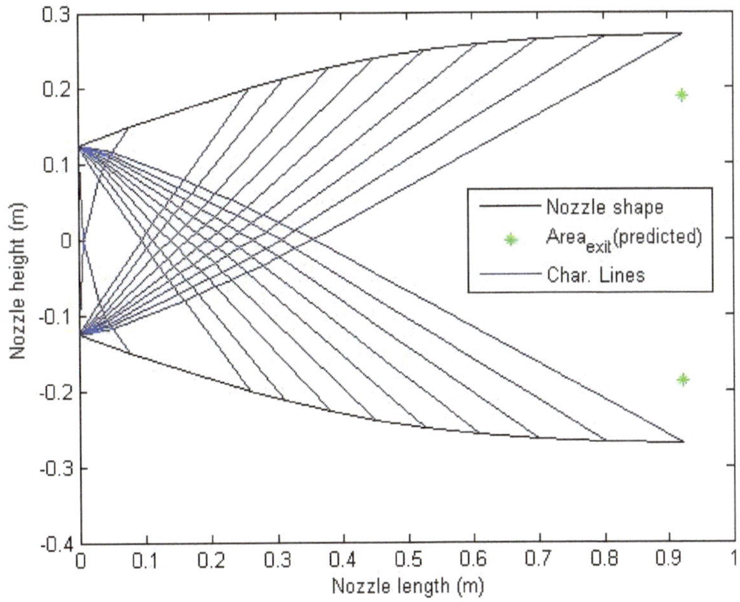

Figure 1: Contour design using 10 Characteristic lines

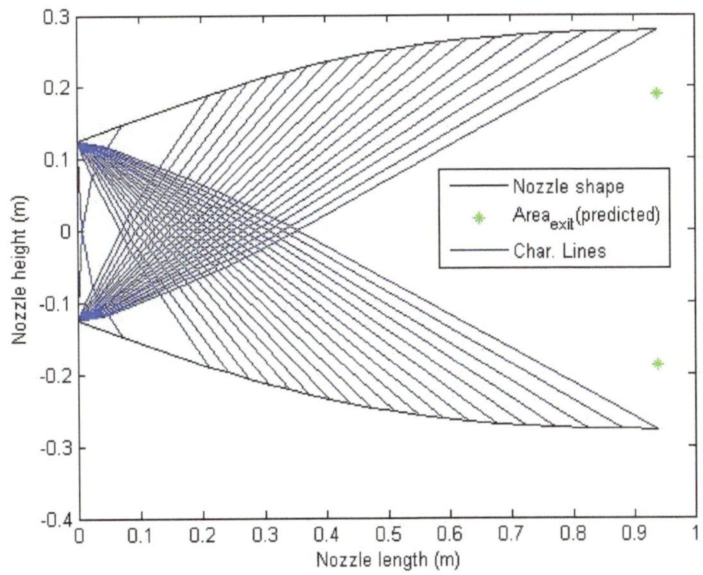

Figure 2: Contour design using 20 Characteristic lines

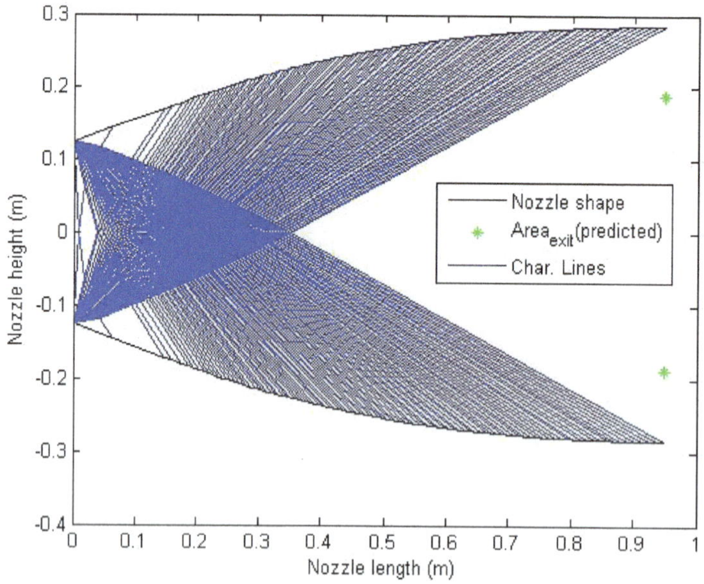

Figure 3: Contour design using 100 Characteristic lines

6 Conclusion

Viscous corrected method of characteristics scheme creates nozzle contour where there isn't much deviation for increased number of characteristic lines. Solutions using 10, 20 & 100 characteristic lines were presented shows not much deviation. Viscous corrected scheme accounts for the viscous boundary layer effects inside the diverging region of the nozzles. Proposed scheme performs better than most existing schemes used for 2D Nozzle contour design.

7 References

[1] M. Al-Ajlouni, "An Automaic Method for Creating the Profile of Supersonic Convergent-Divergent Nozzle", Department of Mechanical Engineering, Faculty of Engineering, Mu'tah University, Mu'tah, Al-Karak 61710 Jordan

[2] Fredrick L Shope, "Contour design Techniques for supersonic/hypersonic wind Tunnel", Aerospace Testing Alliance Arnold Engineering development center, Tennessee 37389

[3] Khin Oo Mon and Changjin Lee, "Optimal design of supersonic nozzle contour for altitude test facility", Department of Aerospace Engineering, Konkuk University, Seoul, 143-701, Kore (Manuscript Received September 15, 2011; Revised January 3, 2012; Accepted March 22, 2012)

[4] Jan Östlund, "Flow processes in rocket engine nozzle with focus on flow separation", Licentiate Thesis Stockholm, 2002, Royal Institute of Technology, Department of Mechanics, TRITA-MEK, Technical Report 2002:09, ISSN 0348-467X, ISRN KTH/MEK/TR--02/09-SE

[5] Dr. Stuart i. Brodsky, "Supersonic nozzle design", United Nations Navy Ordnance Laboratory, White oak, Maryland

[6] George E. Gomf, "Supersonic nozzle design for viscous fluids", Degree of Aeronautical Engineer, California Institute of Technology, Pasadena, California - 1949

[7] W.F. Louisos et al., "Design considerations for supersonic micro-nozzles", School of Engineering, Mechanical Engineering Program, University of Vermont, 301 Votey Hall, Burlington, Vermont 05405, USA.

[8] David J. Cerantola, "Rocket Nozzle Design with Ejector Effect Potential", Ottawa-Carleton Institute for Mechanical and Aerospace Engineering, Department of Mechanical and Aerospace Engineering, Carleton University, Ottawa, Ontario, Canada, November 2007

[9] M. Moin Hussaini and John J. Korte, "Investigation of Low-Reynolds – Number Rocket Nozzle Design Using PNS-Based Optimization Procedure", Langley Research Center, Hampton, Virginia, National Aeronautics and Space Administration, Langley Research Center, Hampton, Virginia 23681-0001

[10] T. Zebbiche and Z. Youbi, "Supersonic two-dimensional minimum length nozzle design at high temperature", Department of Aeronautics, Faculty of Sciences of Engineer, University SAAD Dahleb of Blida, B.P. 270 Ouled Yaich, 09470 Blida, Algeria.(Received November 2005 and accepted May-2006)

[11] Donald W. Hapke, "Jet engine two dimensional, asymmetric afterburner nozzle", Issaquah, Wash., The Boeing Company, Seattle, Wash. Appl. No.: 306,746. Filed: Sep. 29, 1981

[12] J. C. Crown and W. H. Heybey, "Supersonic Nozzle Design",U. S. Naval Ordnance Laboratory, White Oak, Silver Spring 19, Maryland

[13] Gerald Hagemann, "Advanced Rocket Nozzles", DLR, German Aerospace Research Center, Lampoldshausen 74239, Germany. Hans Immich, Daimler – Benz AG, Munich 81663, Germany. Thong Van Nguyen, GenCorp Aerojet, Sacramento, California 95813 and Gennady E. Dumnov, Keldysh Research Center, Moscow 125438, Russia.

Permissions

All chapters in this book were first published in IJRAME, by IJRAME; hereby published with permission under the Creative Commons Attribution License or equivalent. Every chapter published in this book has been scrutinized by our experts. Their significance has been extensively debated. The topics covered herein carry significant findings which will fuel the growth of the discipline. They may even be implemented as practical applications or may be referred to as a beginning point for another development.

The contributors of this book come from diverse backgrounds, making this book a truly international effort. This book will bring forth new frontiers with its revolutionizing research information and detailed analysis of the nascent developments around the world.

We would like to thank all the contributing authors for lending their expertise to make the book truly unique. They have played a crucial role in the development of this book. Without their invaluable contributions this book wouldn't have been possible. They have made vital efforts to compile up to date information on the varied aspects of this subject to make this book a valuable addition to the collection of many professionals and students.

This book was conceptualized with the vision of imparting up-to-date information and advanced data in this field. To ensure the same, a matchless editorial board was set up. Every individual on the board went through rigorous rounds of assessment to prove their worth. After which they invested a large part of their time researching and compiling the most relevant data for our readers.

The editorial board has been involved in producing this book since its inception. They have spent rigorous hours researching and exploring the diverse topics which have resulted in the successful publishing of this book. They have passed on their knowledge of decades through this book. To expedite this challenging task, the publisher supported the team at every step. A small team of assistant editors was also appointed to further simplify the editing procedure and attain best results for the readers.

Apart from the editorial board, the designing team has also invested a significant amount of their time in understanding the subject and creating the most relevant covers. They scrutinized every image to scout for the most suitable representation of the subject and create an appropriate cover for the book.

The publishing team has been an ardent support to the editorial, designing and production team. Their endless efforts to recruit the best for this project, has resulted in the accomplishment of this book. They are a veteran in the field of academics and their pool of knowledge is as vast as their experience in printing. Their expertise and guidance has proved useful at every step. Their uncompromising quality standards have made this book an exceptional effort. Their encouragement from time to time has been an inspiration for everyone.

The publisher and the editorial board hope that this book will prove to be a valuable piece of knowledge for researchers, students, practitioners and scholars across the globe.

List of Contributors

M. M Kaushik
M.Tech. Student (Machine Design), Department Of Mechanical Engineering, KSIT, Bangalore-62

M. M. M. Patnaik
Associate Professor, Department of Mechanical Engineering, KSIT, Bangalore-62

R. Venkateswaran
Head, OEND, Scientist/Engineer-SG, LEOS-Indian Space Research Organisation, Bengalore-58

Laurent Dala
CSIR, DPSS/Aeronautics Systems, Pretoria 0001, South Africa
Mechanical & Aeronautical Engr. Dept., University of Pretoria, Hatfield 0083, Pretoria, South Africa

Vatsal Sheth
Undergraduate student Mechanical Engineering Dept. Dwarkadas J. Sanghvi College of Engineering, Mumbai

Abhay Sharma
UG students, School of Mechanical Engineering, Chitkara University, Punjab

Manpreet Singh Sarwara
UG students, School of Mechanical Engineering, Chitkara University, Punjab

Harsimranjeet Singh
UG students, School of Mechanical Engineering, Chitkara University, Punjab

Lakshya Swarup
UG students, School of Mechanical Engineering, Chitkara University, Punjab

Rajeev Kamal Sharma
Associate Professor, School of Mechanical Engineering, Chitkara University, Punjab
UG students, School of Mechanical Engineering, Chitkara University, Punjab

Yassir A. Awad Elkarim
PhD student, Aeronautical Engineering Department, Engineering College, Cairo University Giza State, Egypt

Mohammed M. Abd Elrahman
Professor, Aeronautical Engineering Department, Engineering College, Cairo University Giza State, Egypt

L. M. Karthikeyan
Department of Aeronautical Engineering, Techno Global University, India

Abhishek Patil
Undergraduate Student, Department of Mechanical Engineering Nitte Meenakshi Institute of Technology, Karnataka, INDIA

Piloo Raghu Kumar
Undergraduate Student, Department of Mechanical Engineering Nitte Meenakshi Institute of Technology, Karnataka, INDIA

Aditya Kodkani
Undergraduate Student, Department of Mechanical Engineering Nitte Meenakshi Institute of Technology, Karnataka, INDIA

Sekhar Majumdar
Professor Department of Mechanical Engineering Nitte Meenakshi Institute of Technology, Karnataka, INDIA

Y B Sudhir Sastry
Dept. of Aeronautical Engg. Institute of Aeronautical Engineering, Hyderabad, India

Y krishna
scientist G, Head, Structural Test Facility, DRDL, Hyderabad, India

S S Prasad
Dept. of Aeronautical Engg. MLR Institute of Technology, Hyderabad, India

A Q Syed Mohsin
Dept. of Aeronautical Engg. MLR Institute of Technology, Hyderabad, India

Tukesh Thakur
PG Scholar (Department of Mechanical Engineering, Mats University, Raipur)

Husain Mehdi
Assitant Professor, MED, Meerut Institute of Technology, Meerut (U.P), 250002/ INDIA

Sayed Fahad Anwer
Department of Mechanical Engineering ZHCET, AMU Aligarh, 202002/ INDIA

Akhlaque Ahmad
Department of Mechanical Engineering ZHCET, AMU Aligarh, 202002/ INDIA

Aldin Justin
Assistant Professor, Karunya University, Karunya Nagar, Coimbatore

Robin James
Assistant Professor, Karunya University, Karunya Nagar, Coimbatore

Shruti Panicker
Student, Karunya University, Karunya Nagar, Coimbatore

A. N. Subash
Student, Karunya University, Karunya Nagar, Coimbatore

Saravana Kumar
Student, Karunya University, Karunya Nagar, Coimbatore

Husain Mehdi
Assitant Professor, MED, Meerut Institute of Technology, Meerut (U.P), 250002/ INDIA

Fahad Anwer
Department of Mechanical Engineering ZHCET, AMU Aligarh, 202002/ INDIA

Akhlaque Ahmad
Department of Mechanical Engineering ZHCET, AMU Aligarh, 202002/ INDIA

Dr Ian R. McAndrew
Embry Riddle Aeronautical University, College of Aeronautics Earnley Place, Clappers Lane, Chichester, Hants PO20 7JL UK

Elena Navarro
Embry Riddle Aeronautical University, College of Arts &Science, Earnley Place, Clappers Lane, Chichester, Hants PO20 7JL UK

Dr Kenneth Witcher
Embry Riddle Aeronautical University, College of Aeronautics, Earnley Place, Clappers Lane, Chichester, Hants PO20 7JL UK

R. Sriram
PG- SVCE, Sriperumbudur, 8/42 Salem mail road, Elavanasur Kottai, Ulundurpet TK, Villupuram

S. Nandhini
PG- SVCE, Sriperumbudur 8/42 Salem mail road, Elavanasur Kottai, Ulundurpet TK, Villupuram

Saurabh Shriwas
Student, B.E. (Aeronautical), MATS University, Raipur

Sachin Shah
Student, B.E. (Aeronautical), MATS University, Raipur

Prashant Singh
Student, M. Tech. (Turbo machinery), MATS University, Raipur

Kalpit P. Kaurase
Assistant Professor, Dept. of Aeronautical Engg. MATS University, Raipur

G. Elavarasi
PG Scholar, Department of Aeronautical Engineering, Regional Centre of Anna University, Tirunelveli, 627007

J. Bruce Ralphin Rose
J. Bruce Ralphin Rose, Assistant Professor, Department of Aeronautical Engineering, Regional Centre of Anna University, Tirunelveli, 627007

C. K Varsha
PG Student, Sri Lakshmi Ammaal Engineering College, Chennai-126

R. Manoharan
Associate Profesor, ECE Department Sri Lakshmi Ammaal Engineering College, Chennai-126

K. Ramesh Babu
Associate Professor, SVS College of Engineering - Coimbatore-09

R. Bharathi Raja
Assistant Professor, Mechanical Engineering, SVS College of Engineering-Coimbatore-09

Himanshu Dudeja
Dept. of Mechanical Engineering, Dr. D.Y. Patil School of Engineering & Technology Pune, India

Laxman Bagal
Dept. of Mechanical Engineering, Dr. D.Y. Patil School of Engineering & Technology Pune, India

Nityanand Zunjur
Dept. of Mechanical Engineering, Dr. D.Y. Patil School of Engineering & Technology Pune, India

Prof. S.S. Jagadale
Dept. of Automobile Engineering, Dr. D.Y. Patil School of Engineering & Technology Pune, India

C P Premchand
Research Associate, DRDO-GTRE, Bangalore, Karnataka -560093 [Premchand C P - Research Associate, DRDO-GTRE, Bangalore, Karnataka -560093

V Ramanathan
Post Graduate, Indian Institute of Science, Bangalore, Karnataka – 560012 [Premchand C P - Research Associate, DRDO-GTRE, Bangalore, Karnataka -560093